What people are saying about
Hey, Doc! Are you listening to YOUR heart?

I'm a changed person from reading *Hey, Doc!*

Roger Jellinek
Executive Director, Hawaii Book and Music Festival

Dr. Sheff's gripping stories about what happens inside physicians and between physicians and patients help us all create a new reality for health and healing. Through the voice of a mature physician reflecting on the formative experiences that mold every doctor, he provides a new way for patients to better understand physicians and physicians to better understand themselves. Science cures. Spirituality heals. Dr. Sheff has captured the powerful interplay of both in a book all of us, healers as well as those in need of healing, should read.

Leland R. Kaiser, Ph.D.
Author and Healthcare Futurist

Out of his pain, his wisdom, and his humor, Dr. Sheff gives readers an understanding of the deepest sources of great doctoring as well as the pitfalls when doctors receive, not a true medical education, but medical information, which drains the true capacity to heal out of today's doctors. In order to coach and guide every patient into becoming an exceptional patient who utilizes the self-healing abilities within each of us, every patient, doctor in training, and practicing physician should read this book.

Bernie Siegel, MD
Author of *Love, Medicine and Miracles*
and *Love, Magic and Mudpies*

Hey, Doc!

Are You Listening to YOUR Heart?

Richard Sheff, MD

www.listentoyourheartmedicine.com

Listen to Your Heart Press
35 South County Way, Suite D12
South Kingstown, RI 02879
www.listentoyourheartmedicine.com

Library of Congress Cataloging-in-Publication Data

Sheff, Richard
    Hey, Doc! Are you listening to your heart?
    1.   Medicine   2. Humanistic medicine   3. Doctor patient relationship
            4. Physician patient relationship   5. Medicine and psychology
            6. Healing

ISBN: 978-0-6151-9767-8

This book is dedicated to the memory of Saul A. Silverman, Ed.D., JD (1923-2007). Thank you, Saul, for teaching me that love, courage, passion, and integrity are integral to living our values and expressing our humanity. Through your example you taught me that intellectual rigor can go hand-in-hand with an open heart. You were the first person to show me it was possible to put the pieces of a life of love and service together with critical thinking about what matters most. I hope I have honored your memory through my life and work.

"Wherever the art of medicine is loved, there is also a love of humanity."
Hippocrates

The events in this book took place during my training at the University of Pennsylvania School of Medicine in Philadelphia and the Brown University Family Medicine Residency Program at Memorial Hospital in Pawtucket, Rhode Island. The names of all individuals in this book, other than the author, have been changed. Occasional details have been modified to allow for a coherent telling of some of the stories. With these minor exceptions, all of the stories in the pages that follow are true.

Note on Medical Terminology

Throughout this book accurate medical terminology is used with the awareness that many lay readers will not readily understand all the medical terms used. Footnotes are provided explaining these medical terms and, when needed, the physiology or pathology of the condition experienced by a patient in a particular story. This formatting was chosen to allow those readers not interested in the underlying medical science to read the entire book without having to wade through the technical information. Other readers who may be interested in the medical science can read the footnotes, learning a great deal about medicine along the way.

Acknowledgments

Whatever is good in this book has come from the teachings of others. Some of my medical school and residency faculty were excellent role models of intellectual incisiveness, some of a compassionate heart, and a few combined both. I learned much from my fellow residents in training, and to them I owe a large debt of gratitude. The generous and wise guidance of Sam Horn through the publishing process has been invaluable. Sincere thanks go to Shirley Arendt for capturing everything I wanted in the cover photo and to John Chapin for his patient help with the cover design. And I am truly humbled by how much this book has benefited from Jane Bernstein's outstanding copy editing.

Hey, Doc! could only have been written because countless patients allowed me to join them in their deepest vulnerabilities. They entrusted their care to me. I took them into my heart, even when I was still learning and had not yet earned their trust. To all of these patients who will never know the gift they have given, I offer thanks for their being my most profound teachers of the art of medicine.

This book has lived in my heart for more than two decades, but it would never have become a reality in your hands today if it were not for the steadfast encouragement of my loving wife, Virginia. She has been my most ardent supporter and toughest critic. Her clear thinking and honesty in critiquing every page have made this a far better book than it would have been without her by my side. Thank you, Virginia. Thank you. Thank you.

Table of Contents

Prologue

This book is about unfinished truth. *Hey, Doc* is the true story of my training as a family physician in medical school and residency more than 20 years ago. I knew then, as now, there is something terribly lacking in how we take care of patients and train physicians. Even before completing training I felt compelled to write this book, to tell the stories that now unfold in the pages to come for doctors in training, for all who work in healthcare, and for everyone who ever has or will entrust their vulnerability to a doctor. In the ensuing years I have authored 12 books on medicine and healthcare, but the book I knew had to be written remained unfinished. Under the pressure of other responsibilities, the task I most wanted to complete kept being put off to some later date.

Yet ever since medical school my patients have taught me that bad things, catastrophic things, can happen to anyone at any age. I've cared for a beautiful baby girl born to young, loving, expectant parents. But she was born with Down syndrome, and her heart strained in her chest on the edge of failure because of an incurable congenital defect. Her parents sat helplessly by her bedside through the first nine months of her life, but she never went home to the nursery they'd prepared for her. I've cared for a father who struggled to overcome alcoholism so that he could be the kind of father he'd promised his five-year-old son he'd be. But this father died of testicular cancer at the age of 35, broken hearted with the knowledge he had failed his son because he was still drinking. I've cared for a seven-month-old boy who awoke from a nap just a little warm and two hours later lay dying, meningococcal bacteria teeming throughout his bloodstream.

Being a physician has taught me to cherish this fragile life we've been gifted, and not to take anything, or anyone, for granted. I've strived to live every day so that if an unforeseeable tragedy ended my life, I would be at peace with the choices I had made. I was at peace with my loving wife, my two sons, friends and family, and the choices I've made in my work.

I was at peace with all of my life, except this book.

Over the years since completing training, I had begun again and again to write the book you now hold, but could not find the words to express all that needed to be said. I dreamed of completing it, found the partially written manuscript calling to me from its home deep within a file cabinet drawer, felt guilty about not completing it, but still the book remained unfinished—until now.

And now it has become a different book than it would have been had I completed it immediately after my training. This book is different because I am different. More than two decades in medicine can't help but change you. The harshness of early judgments have tempered, perspectives have deepened. As a mature physician I have found the same stories I wanted to tell can now be told with insight I could not have articulated as clearly back then.

From the plaintive voice of Jim Schmidt, a frustrated, vulnerable man I met as a patient early in my internship, I have borrowed the title of this book. "Hey, Doc!" he called out as I entered his exam room for the first time. With those words he tried to get my attention, tried to let me know what he needed from me as his physician, tried to call me to action on his behalf. At the time he did so in anger and fear, a proud man trapped in a body experiencing inexorable decline. Over time, I came to realize that whenever I heard those words, "Hey, Doc," whether from Jim or anybody else, they held a multilayered meaning, calling upon me as a physician for something important but not well articulated. I now know that when I hear those words it is incumbent upon me to pay attention, for they announce a moment when a patient is about to tell me their truth, if only I would take the time to recognize it beyond their words. In the end, that is my hope as you read this book—that you will discern beyond its words something of the truths to which I have tried to give voice.

Though these truths are varied and complex, they center around a common theme: listening to the heart. At first, as a physician, I was taught to listen to the physical heart of my patients. This challenge, which consumed so much of my medical training, was to understand their physiology in all its intricacy. Then came the realization that to truly serve as a physician for my patients, I had to listen to their emotional heart, with its even more complex and challenging nuances. Finally, my patients taught me that if I wanted to become the physician I aspired to be, I had to listen to my own heart, which has proven to be the greatest challenge of all.

Over two decades in medicine has taught me that the most potent tool for healing we as physicians have is our heart. But each of us must learn to listen to our own heart before we can touch and heal the hearts of our patients.

As the 20-year labor that has finally produced this book drew to a close, I realized another motive for writing it which had remained unrecognized until the book was completed. As I relived the years of training, an awareness grew that the pressures and sacrifices of those years had left me wounded and scarred in ways I had only vaguely grasped until now. By giving voice to the stories of that time, I have experienced a gratifying healing of unseen scars not anticipated when first I embarked on this effort.

But this has been bittersweet, because even as I've experienced my own healing, I am aware that students of medicine who come after me will experience much the same wounds and scars. It is to you, current and future students of medicine, that I am speaking directly now. Know that your most potent tool of healing is not your knowledge but your compassion. You will come to medical training with this fledgling tool partially developed. During medical training, some of your teachers will show you the way to integrate your compassion, your capacity for offering healing that comes from your heart, with the knowledge and skills of medical practice. Others will ignore this aspect of the art of medicine, or worse yet, insist it is misdirected and belittle anyone who shows weakness of feelings and human frailty. The rigors of medical training will systematically harden you to the finest parts of yourself. At the end of your training, you will be wounded in your most sensitive and precious places. You can't help this. I write this book to share the path of one student of medicine who became at times lost and frightened on this path, only to find my way again, then to be lost, and again to find my way. It was for me, and can be for you, a path to becoming a skillful, caring physician, but not without a cost.

To physicians in practice, I hope these stories shed new light on your own journey through medical training and provide insights into your unseen scars that may get in the way of responding to patients, fellow healthcare workers, and perhaps your own family with the compassion and caring that first brought you to medicine.

To the many healers who are not physicians, I hope you come away with insight into the compassionate core of most physicians, and the very human experiences that sometimes get in the way of that part shining through.

And for any patient who has experienced the consequences of the scars your physician sustained during training and practice, who has been treated insensitively by your physician, causing you to come away angry and hurt when you longed to be heard and healed, I hope these stories provide much needed understanding. I wish for you and your physician that you find a way to meet each other in the healing place for which you both long.

This coalescing of desires has driven me to write the book you now hold. Would that I could truly share with you the richness of my inner world as I engaged my medical training, as I was taught by my teachers, and mostly by my patients. Please know my intention and my wish that, through reading this account of the journey of one student of medicine, you catch a glimpse of the honor I have had to serve in the noble but troubled profession of being a physician.

Part 1

Medical School

Chapter I

Gross Anatomy

That Final Moment When She Absolved Me of Guilt

Down we walked into the basement of the oldest building in the oldest medical school in America. Packed together so closely, we shuffled our feet to keep from tripping on the others around us. A bare bulb passed overhead, and then another. The lines between faded green tiles along the hallway walls, darkened by the years, seemed to tremble in the harsh light. Slowly snaking forward, we passed an open door, allowing a glimpse of the rat cages and microscopes within. The acrid smell of sawdust mixed with urine wafted from the door. We were silent, each consumed with personal thoughts swirling inside.

I'm not ready, a voice within me protested. They said we wouldn't go to the anatomy lab our first day. But bring your dissecting kit just in case, they'd said. And now this—15 minutes of the anatomy professor in an over-starched white coat laying down the rules for the gross anatomy lab and then sending us off to start dissecting a cadaver. I'm not ready. I need more time.

A new smell brought me back. It began faintly, almost sweetly at first. Then, as we moved deeper down the hallway, it pushed aside the other smells, increasing slowly until it pinched. Formaldehyde. The odor wouldn't stop growing, now piercing my nostrils, burning with every breath. Just then we stopped before two large, gray, metal doors. For the longest moment, nothing happened. Then, swinging back without a sound, the doors melted away, revealing a cavernous room filled with 40 black, stone slab tables, each with a harsh metal light hanging over it. On every table lay a still, silent figure wrapped in white linens. Death hung palpably in the air.

Suddenly, all was frenetic activity as 160 first-year medical students flooded the room, desperately releasing the tension that had been building. "The M's are over here," I heard one call out.

"Come to me, baby," another crooned. Nervous laughter filled the room as the dark humor began.

My fellow students streamed past, but I moved slowly, passing one table after another. Looking down at each still figure, I asked silently, who are you? How did you come to be here? Did you donate your body to science, only to find yourself here? Is being preserved in formaldehyde and cut apart by medical students the noble contribution you had intended to make? Silence was the only response. Yet a thought formed that even in their silence they were offering to me the first lesson of my medical training. What was it that coursed through me but not them? What separated me, alive and conscious, from these still, dead shrouds? Soon enough I would learn the mechanisms that caused a heart to beat or a brain to transmit electrical impulses, but would that knowledge answer this, the most fundamental question? Raised a devout atheist, trained as a scientist, I paused at a threshold I did not fully understand. Something told me in this moment that once I began the scientific study of the human body, I'd never again engage this mystery with the wonder it deserves.

My wandering finally stopped when I found "my cadaver." There were our names, crisply typed on a clean white paper taped to the black stone table: Schwartz, Sheff, Smith, Sullivan. My gaze rose from the paper to the figure lying on the table above it, wrapped in linens heavy with formaldehyde, the piercing odor now bringing tears to my eyes. The four of us surrounded the table. For a moment we shared something unnamed. The moment passed, and we began to unwrap the layers of linens on the arms and legs, then her torso, for we soon realized our cadaver was a woman. As I unrolled the cloth on her thigh, my gloved hand unexpectedly touched her skin. It felt surprisingly stiff and rubbery, with a cold that went beyond temperature.

Finally, she lay completely naked, except for the head which was still wrapped. The unnatural gray of her skin seemed to give off a dull, purple hue. Her breasts were shrunken, the abdomen concave, and her arms and legs extended out in unnatural angles. I reached for the linens on her head, but my hands quavered for a moment. Slowly, I unwrapped each layer, placing my hand under the back of her head to raise it with the unraveling motion, replacing it softly on the table each time. Why be so gentle, I wondered, but I couldn't do otherwise. Her nose appeared, and then her chin and forehead. Her eyes, clouded and sunken, stared blankly. Finally, her mouth emerged, pulled to one side, gaping open in a grimace that gave the impression she had died in pain, the agonizing moment of her passing forever frozen in her features. We could not bear to look at her face, so we rewrapped her head immediately.

"Come on, get out your dissecting guides. We've got work to do," I heard one of my lab partners say. At least one of us showed leadership in this moment. "Who else brought a dissecting kit today besides me?"

With a sense of relief, I responded instinctively to someone else's directive. I unrolled the kit that included a scalpel, scissors, forceps, and several dissecting needles and probes and opened the dissecting guide. At the start, we were told to split up so two of us would be dissecting the arm and two of us dissecting the leg. I found myself on her leg, working with Smith, or rather Mike. We turned to the page assigned by our lecturer this morning and read, "Incise the skin at a point half way between the anterior iliac crest and the pubic tubercle." I looked at Mike.

"Where the hell is that?" I asked.

"According to the picture here," he said, pointing to a diagram I hadn't noticed, "it looks like the anterior iliac crest is this spot on the hip and the pubic tubercle is right over there." Now I knew why the instructor had told us to work in pairs.

"So I should make the cut right here," I said, holding the scalpel above a point in the crease at the top of her thigh. The scalpel blade seemed suspended over the preserved skin. This is it, I thought to myself. I'm about to make the first cut into a cadaver. Then I will really have begun medical school. Nothing happened. My arm felt frozen as I scanned her body, taking in this stranger's nakedness. Instinctively, I wondered if she would feel it. Then, by an act of will, my hand and the scalpel descended into the rubbery flesh of her leg. She didn't move. My first cut was done. I let out a breath I hadn't realized I'd been holding.

As the skin peeled back, I paused, disoriented. According to the dissecting guide, I was supposed to find the femoral nerve and femoral artery in this exact spot, running above and through a muscle of the thigh. But the gray, featureless layers looked nothing like the picture in the anatomy book. How was I ever going to find a nerve or artery in here? How could I tell one layer from another when they were all matted together? Which of these stringy fibers was the muscle I was seeking, and which were just connective tissue in my way? Quickly I glanced around the anatomy lab. One hundred and fifty-eight students were bent over their cadavers, dissecting away and chatting lightly with each other. Was I the only one who struggled with this, the first and easiest part of the dissection? A feeling of panic rose in my chest. The words in the dissecting guide danced on the page. I didn't want to meet Mike's eyes. I can do this, I tried to say to myself. This shouldn't be that hard. Is everyone else in the class that much smarter than I am? What's wrong with me? I'm never going to get through this.

Just then, a voice said, "Do you have a question?" I looked up into the face of my anatomy laboratory instructor, Dr. Sanchez, his wrinkled eyes turning up at the corners with a look I couldn't read. Was he mocking me? Was he about to humiliate me in front of all my classmates?

"I...I'm not sure of what we're looking at here," I stammered.

"Let me see." Taking the scalpel and forceps from my hands he quickly scraped away most of the stringy connective tissue that had crisscrossed the dissecting field in which I was trying to work, and already it looked better. "You see this," he said in a thick Hispanic accent, grabbing a tangled piece of nondescript tissue. "These are veins." He quickly cut them out and tossed them aside. "We don't care about veins," he said with a sly grin. Then quickly, skillfully, he scraped the tissue away from two structures that appeared to emerge from the undifferentiated, gray, matted mess that had been my dissection. When he finished, I could finally make out two surprisingly slender structures, the femoral artery and femoral nerve.

He watched me dissect a little while longer. "I'm not sure surgery should be your specialty of choice," he said warmly, "but you'll get the hang of this. Just remember, get rid of the veins and everything will be much clearer." As he walked away, I knew I had just met my most important ally in the long, five-month ordeal of anatomy lab that had just begun.

The next two hours passed surprisingly quickly as Mike and I worked. Suddenly, it was time to finish. We wrapped our cadaver in the same linens, taking care to cover her well so she wouldn't dry out. I stood up, noticing for the first time an ache in my back that had crept up as I'd bent over her for several hours. I was asking Mike directions to our next class, biochemistry, as we passed the large, gray doors. Turning to look back, I felt odd. There were the same still figures on all the stone tables. The odor of formaldehyde seemed somehow less than when we'd entered, though in my mind I knew it had to be stronger from all our dissecting activities. I was aware again that each of these bodies was dead, but somehow that thought didn't hold the power it had when we'd first entered. Something had changed.

*   *   *

"You can do this," I muttered to myself, staring at the first sentence of the first paragraph.

> *The aponeurosis of the external oblique originates at the inferoanterior aspect of the costal margin and extends in a radial and caudad direction, inserting on the anterior superior iliac crest.*

What the hell does that mean? I glanced at the clock. Already 11, I sighed. Doing nothing but homework since classes had ended at 5, I'd waded through histology, physiology, and biochemistry. Now it was time for anatomy, and tonight's assignment was to read 10 pages in *Gray's Anatomy*, an imposing tome of some 1400 pages. I tried again, but my eyes drifted closed. I shook my head violently, trying to stay awake, but this was a losing battle. After all, throughout college I couldn't get by with less than seven hours sleep a night, and really needed eight hours not to fall asleep in lectures. How was I ever going to get through all this work tonight? How would I do it day in and day out? Worse yet, how was I ever going to manage the grueling hours I'd heard interns and residents were subjected to? Rather than give in to the rising sense of panic, I read the first sentence again.

*The aponeurosis of the external oblique originates at the
inferoanterior aspect of the costal margin and extends in a radial
and caudad direction, inserting on the anterior superior iliac crest.*

It made no more sense the second time. "You can do this," I repeated,
reaching for my newly purchased medical dictionary.

*Aponeurosis: A fibrous sheet or expanded tendon, giving
attachment to muscular fibers and serving as the means of origin or
insertion of a flat muscle.*

That's a start. Now what's the external oblique? I flipped through the
dictionary's pages looking for "external oblique." There was no such term.
Instead, under "external" it said:

Exterior; on the outside or farther from the center.

Under "oblique" it said:

Slanting; deviation from the perpendicular or the horizontal.

"So what's the 'external oblique'," I said out loud, venting no small
amount of frustration. I read the definitions again and was no further along.
All I knew was that some sort of fibrous sheet ran outside something and
slanted from it. I tried to move on, but "inferoanterior" wasn't even a word
in the dictionary. I had to settle for a definition of "inferior" that said:

Lower; below in relation to another structure.

And "anterior" was defined as:

*Before, in relation to time or space; in front of or in the front part
of. Sometimes used to designate a cranial and rostral position in
quadrupeds or certain embryos.*

I chose to ignore the second half of this definition, since I had no idea
what a cranial or rostral position was, and assumed I was dealing with a
sheet of something that was below and in front of something else. I plowed
on.

Costal: relating to a rib.

Now I was making progress. Whatever this thing was, it was below and in front of a rib. I thought I knew what margin meant, but maybe not how it was used medically, so I looked that up too.

Margin: the boundary or edge of any surface.

Alright, I thought, this fibrous sheet thing seemed to originate or start at the lower, front edge of a rib and extended somewhere. But where was that? It extended in a "radial and caudad direction." Here we go again, I thought.

Radial: relating to the radius (bone of the forearm), to any structures named from it, or to the radial and lateral aspect of the upper limb as compared to the ulnar and medial aspect.

Now I was really lost. What did a bone in the forearm have to do with a fibrous thing that was somehow connected to a rib? And, worse yet, the definition of "radial" even used the word "radial." Wasn't there a rule against that? There was a second definition of "radial" that said:

Radiating; diverging in all directions from any given center.

That made a little more sense. This fibrous thing started at the edge of a rib and "radiated" out in all directions. I couldn't picture any part of our anatomy that did such a thing, so I read on.

Caudad: In a direction toward the tail.

Did I miss something in biology back in college? I don't recall humans having a tail at this stage of our evolution. Then I thought of the tailbone, and made the assumption this meant a direction toward the tailbone. So now I had a fibrous sheet, starting at an edge of a rib, extending out in all directions, but also generally in the direction of the tailbone. I looked back at the picture in the textbook and realized the fibrous thing we were talking about was in the vicinity of the muscles of the abdomen. So this "aponeurosis" seemed to be part of the abdominal wall muscles. We had actually identified the "anterior superior iliac crest" in the lab today, so I finally figured out we were dealing with the fibrous sheet that surrounded something called the external oblique muscle, and that it attached to the lower edge of the last rib and ran across and down the abdomen to the top part of the pelvic bone. Now it made sense. "See, I knew I could do this," I said with some satisfaction.

I looked at the clock again. Reading the first sentence of my anatomy assignment had taken 15 minutes and had required me to look up nine words. How was I ever going to read 10 pages in this damn textbook tonight? I could feel panic start to rise again, but something in me steeled to the challenge. Anyone who had ever spent one evening having dinner with my parents would understand that giving up was not an option for me, so I plowed on. For the second sentence I only had to look up five words. One hour later I had gotten through two and a half paragraphs of the 10-page assignment. I finally lost the battle with my exhausted eyes.

"I can't do this," I moaned, and shut off the light to go to sleep.

*   *   *

Back in the anatomy lab, I bent over my cadaver. The harsh light of the lamp hanging over the dissecting table reflected an eerie glow off the linens covering her. I started unraveling the linens from her face, being careful to replace her head softly on the table with each pass. Looking around the cavernous room, I was surprised to notice no one else was in the lab tonight. Alone, surrounded by row upon row of still, shrouded figures on black, stone tables, my heart began to quicken. There's nothing to be afraid of. Just do what you came here for, and then you can leave. Slowly, ever so slowly, her features came into view. The wrinkles on her forehead seemed so deep tonight, deeper than I'd remembered them. Her nose, proud, prominent, stood straight out, almost too large for her face. As the next cloth came off, those cloudy eyes appeared, sunken, staring vaguely away. Finally her mouth appeared, still pulled to one side in that grimace, partly open, hinting at pain it almost seemed she still felt. Her matted, stringy hair stuck to the last of the sheets. I pulled, but the linen held tightly. Was that dried blood I saw causing the linen to be entangled in her hair? With one final tug, the linen tore off. As it did, I lost my grip on her head and it fell back onto the dissecting table with a dull thud. Suddenly her eyes turned on me and a low, unearthly moan erupted from that distorted mouth. Her eyes widened in a desperate panic, pleading with me. For what? "What do you want?" I shouted. "I'm sorry. I'm so very sorry. Please understand," I gasped, barely able to breathe.

I wrenched my eyes open. Staring into the blackness, struggling for breath, at first I saw nothing. I tried to move, but felt frozen. My muscles would not respond. I can't get enough air, I thought, the sense of panic deepening. Slowly, I became aware of my clothes clinging with sweat. Then my eyes made out the vague outline of a desk across the room. I sat bolt upright, chest heaving. It was just a dream. Just another one of those dreams.

*   *   *

C three, four, five keep the diaphragm alive
S two, three, four keep it off the floor

After a couple of months, I was getting the hang of this. Countless jingles, rhymes, and acronyms helped me learn and remember more anatomical facts than I had ever thought possible. The first half of this one meant that the nerves that came out of the spinal cord at levels three through five in the neck controlled the diaphragm. That's why someone could have a bad neck injury and be a quadriplegic, but still keep breathing. The second half meant that the nerves that came out of the spinal cord at levels two through four of the sacrum provided neurological input to the penis, controlling penile erectile function. At least that's how the textbook described it. In the anatomy lab we had lots of other names for this function.

It was one thing to read about genital anatomy in a textbook. It was quite another to dissect it, and that is what we were to start today. There was more than the usual level of humor and chatter as we entered the lab. Moving quickly, Mike and I unwrapped the cloths covering our cadaver's legs and pelvis. The dissecting guide showed a picture with a woman's legs bent at the hips and knees and spread wide apart, kind of like the position I'd heard women had to get into for a gynecological exam when they were up in stirrups, whatever those were. We tried to bend her legs at the hips, but the tough, rubbery skin and preserved fibrous tissue surrounding her joints made this impossible. Dr. Sanchez saw us struggling and came over.

"What do we have to do to get her legs like that?" Mike asked, pointing to the picture in the dissecting guide.

"Well," Dr. Sanchez began, with more than the usual twinkle in his eyes, "it is not my preferred method for persuading women, but in this case, I believe more force will do the trick. I'm sure you boys are up to it." With that, he walked away, chuckling softly to himself.

Mike and I tried again. He took one leg and I took the other. We raised her legs at the hips as far up and apart as they would go. "On my count," Mike said. "One...two... three...Now!" We each leaned our full weight into pulling her legs up and out. With a loud crack and a sickening, tearing sensation, her geriatric legs assumed the position. At the same time, a gush of formaldehyde poured from her vagina, forcing us to turn away to keep from choking.

We focused the light on her perineum, that part of her lower pelvic area which included her vulva and the surrounding tissues. Anxiety that had faded over the first few weeks of dissecting now returned. This was not just another anatomy session. As I cut away the skin of her labia to expose the vestibular bulb, the ichial cavernosus muscle, and the inferior rectal nerve, I was keenly aware that this woman's intimate sexual secrets were now splayed out before the two of us under a bright dissecting lamp. "Is that the urogenital diaphragm under this bulbcavernosus muscle?" I heard myself ask. But inside I was wondering if this woman, whose body I was now dissecting, had felt the warmth of tender embraces in her old age. What was it like for her to make love with a body as aged and shrunken as hers? Or had she lived alone, untouched and lonely? What was she like in the flower of her youth, when this vagina, this clitoris I had just cut in half, held deep mysteries for her eager, young lover? "I'm pretty sure this is the perineal nerve we've been looking for," I suggested, probing deep into an opening I had made alongside her vagina. "Let me just get rid of these veins that are blocking our view," I said, repeating the mantra I had learned from Dr. Sanchez.

All around us, my fellow students were cutting cross sections through penises, opening scrotums, bisecting testicles, and teasing out the delicate structures of the spermatic cord. When they'd completed their dissection, two women, Kate and Maria, working on the male cadaver next to ours, invited Mike and me over to review the anatomical structures they'd laid bare. It was all very clinical, at least that's how we tried to act as Kate used a metal probe to point out the structures inside the testicle she'd sliced open with her scalpel. Maria held the penis in one hand and showed us where she'd cut across it, completely removing half its length, and in monotonous tones described the corpus cavernosum and its role in erection. But as they spoke, neither Kate nor Maria ever looked up at us.

Then it was our turn. Mike and I showed Kate and Maria our dissection. We, too, used the appropriate clinical terms, but as I looked from Kate to Maria, I couldn't keep from stumbling over the names with which I'd been so fluid just minutes before. As I held the dorsal nerve of the clitoris with a forceps, my eyes met Maria's. For a moment we held each other's gaze. Then, as I glanced away, my eyes scanned down her body taking in her full breasts and the curve of her hips underneath the white coat. What would it be like, I wondered, to be with a woman again, alive, warm, and intimate? Would I feel the romance and mystery as I had before, or would I only see images of torn, dissected anatomy, each structure jumping out at me, shouting its Latinized name? Kate's voice brought me back to the moment. "So where does the superficial transverse perineal muscle run relative to the posterior labial nerves?" Fighting to come back from my own thoughts and focus on her question, I was relieved to hear Mike step in with the answer.

*   *   *

The formaldehyde smelled stronger today. That had to be because it was our first day back in the anatomy lab after Christmas vacation. I'd forgotten how much it stung the inside of my nose. The linens, no longer white but now streaked with shades of gray where they'd been smeared with strands of preserved fat and sinew, came off her body stiffly. I stared at the countless mangled structures we'd painstakingly teased apart hour after hour and whose names I had committed to memory. There was the artery that supplied blood to the ovary and upper part of the uterus, but what was its name? And wasn't there another artery that supplied blood to the uterus from below? Quickly my eyes scanned her body, pausing to try to name a nerve here, a tendon there. With rising panic, the realization grew that I couldn't remember the names of most of the structures I saw.

Hadn't I known every one of these names two weeks ago? Gone. Almost all of them had disappeared in only two weeks. I had known just as much about histology and all the details of every tissue type we'd studied. But now, trying to recall them again, I realized they, too, were mostly gone. And what of physiology and biochemistry? Equally a blur. Suddenly I could hear the voice of the biochemistry professor so clearly. "Remember this. It's really critical. Someday a patient's life may hinge on what you recall from today's lecture." Did this mean a time would come when I would kill a patient because the myriad facts and names I'd labored so hard to memorize were now fading beyond recall? I glanced furtively around the room. All the other students appeared engrossed in the day's dissection, chatting away as if nothing had happened. Was I the only one? Dr. Sanchez wandered by. I was ashamed to admit to him what was happening. Hoping to bring it all back, I pulled out my anatomy atlas, the one with the dramatic color pictures that made each anatomical structure appear so clear. But looking from the atlas to the grey mangle of dissected parts our cadaver had become made everything even more confusing.

I paused. It was clear an answer was not to be found in the atlas or in our cadaver. For a moment I closed my eyes and fragments of memories began to come back. But these were not names of anatomical structures. A memory from that first day of orientation rose in my mind. The dean was standing in front of the class. "You'll be asked to master a vast amount of knowledge this year," he was saying. He put one hand in his pocket. With the other he adjusted his heavy, black-rimmed glasses. Then he tried to smile, his transparent attempt at calming 160 anxious new students. "It might help you to know that if something is truly important, you'll see it again and again in the course of medical school here." At the time I had dismissed this as just part of his act, his scripted attempt to reassure the fledgling medical students before sending us to the slaughter. After all, hadn't it been only hours later that we had been ushered down to this anatomy lab to face our cadavers? But what if the dean was right? It made sense that no one could be expected to retain the extraordinary volume of details we'd been poring over throughout the fall. But every one of our lecturers had claimed that theirs was the most important course, theirs held the key knowledge we would require to be good physicians. Failure to memorize every fact thrown at me in every course this first year meant I would someday fail as a physician, fail a future patient, and cause irreparable injury or death to some innocent person who had entrusted their care to me.

Then another fragment of memory appeared. Sometime during my interviews for medical school a fatigued fourth-year medical student had pulled me aside and fervently said, "Medicine is a bottomless pit. You can pour all of yourself into it, seven days a week, 24 hours a day, and still not fill it up—still not do enough for your teachers or for your patients. Only *you* can decide when you've done enough. That's when you must walk away. Don't let anyone else decide for you. If you do, whatever effort you've given will never be enough." He paused, the exhaustion now coming through in his voice. "I wish someone had said these words to me sooner." And with that, he had turned and walked away, staring down at the floor with hunched shoulders and head hanging low. His words haunted me now. When was enough enough? Should I stay late in the anatomy lab tonight, trying to bring back the countless names that had seemed so familiar just weeks ago? Wouldn't they simply fade again within another few weeks? Perhaps I should take the dean at his word. Perhaps if something is important I will see it again and again in the course of my medical training.

I suddenly realized the half life of decay for everything I'd spent the last four months of my life feverishly memorizing was a mere two weeks. Something in me lightened as I accepted that I could do nothing to change this simple reality. I decided to trust the dean and heed the words of that tired, unnamed student who had gone before me and seemed so intent on my not making the same mistakes he'd made. I closed the atlas and walked out of the lab.

*   *   *

"That must be it over there. I'm sure of it," Mike said. I had come to learn that when he spoke with such authority, he was as lost as I was.

"How about that, over there," I said, pointing with a gloved hand, greasy from pulling aside mounds of preserved fat. "Dr. Sanchez," I called. "Could you help us here for a minute, please?"

Our lab instructor calmly approached and peered into the hole we had been digging deep in our cadaver's abdomen. "He thinks that's the celiac artery," I said, pointing to a slender, gray strand on a gray background, "and I think it's that thing right next to it."

The edges of his mouth curled up in that wry smile again. "May I borrow your forceps and scissors?" he asked quietly in that heavy accent. With several quick snips he lifted out both strands we had been arguing over. "Veins," he said. "Just more veins." With a few more deft cuts and scrapes, the celiac artery began to take shape.

I stood back while he cleaned up our dissection. "Our cadaver," as we came to call her, looked very different from that first day. The left leg on which I had begun my dissection was now open to the very bone. Long, gray muscles that ran the length of her thigh lay teased apart. The largest muscles had been cut through and pulled back to expose the deeper structures below. Her knee joint, completely open and disarticulated, provided an unobstructed view of the cartilage and ligaments within. Her right arm was just as dismantled. And her right hand, with its slender fingers punctuated by gnarled, arthritic joints, dangled from the side of the dissecting table. The palm was crisscrossed by incisions that caused flaps of skin to hang open, providing glimpses of the fine nerves and vessels nestled between the bones. Her breasts had been cut open and the nipple and aureole sliced through to reveal the glands and ducts within. But you couldn't see her breasts now, because we had sawed through the ribs and pried them back to dissect the chest organs. The right side of her chest cavity seemed hollow where we'd removed one lung to study the structures deep within. In the center of the splayed open chest lay her heart. It, too, had been sliced open to reveal the chambers and valves we had studied in such detail. In the center of her abdomen lay a large, cross-like incision. With the flaps of skin and muscle now pulled back where Dr. Sanchez was working, I could see each major organ where it had been teased away from the underlying structures, making the organs appear to hang by the arteries, veins, and ducts that connected them.

This was indeed "our cadaver." We had labored for more than four months over every inch of her body. She had taught us well. Yet, as much as we had dissected and worked, often late into the evenings, we had never touched the linens covering her face since that first day. Tomorrow that would change.

*   *   *

I unwrapped the first linen. Her nose and chin appeared, and then the mouth that had haunted my dreams since that first day. Still pulled to one side, contorted, again it seemed as if her last dying breath had been a moan of unbearable pain from which she had been finally released. Her cloudy, half opened eyes appeared for an unnerving moment to be looking directly at me, but then I realized they were unchanged, still gazing vaguely off somewhere else.

Slowly, my focus shifted to the task at hand. My first cut through her cheek was made with a slight hesitation, visible only to myself. Soon I was engrossed with Mike and my other lab partners in identifying the platysma muscle, the facial nerve, and the parotid gland. The moment I had dreaded since our first day in anatomy lab had passed.

*   *   *

I could not believe the words in the dissecting guide, so I read them again.

Hold a saw in the sagital plane at the forefront of the cranium. Then, with firm strokes, proceed to bifurcate the cranium in the midline. Take this incision down to the level of the cricothyroid cartilage.

Today we were to saw her, our cadaver, in half, right through her skull, right through her face, right through that contorted mouth. This was how we were to study the structures deep within her head. We all looked at each other. It seemed unthinkable to take a saw to her in this ghastly manner. For an interminable moment nobody moved. Then, slowly, I reached for the saw. Taking a position at the end of the stone dissecting table, I placed my left hand on the side of her head and placed the saw directly in the middle of her forehead. For a moment I could not bring myself to start. Though I was not a religious person, I found myself spontaneously whispering a prayer, a prayer to this woman who had given so much to me already. I asked her forgiveness for what I was about to do. I asked God to forgive me for what I was about to do, as I felt I was violating something deeper than I had ever violated before. With a sickening sound, I began to saw. First the saw pulled at her hair and her rubbery skin. Then came the harsh grinding of saw teeth into bone. Moist bone fragments flew out from the front and back of her head as the incision moved deeper. Now through her nose and then through her mouth, which pulled and twisted under the saw until finally her head parted in two.

I could not look at her any more. I could not examine and dissect. Instead, I put the saw down and walked out of the lab, a sensation of nausea deep in my stomach. As I walked through the university campus crowded with students laughing and chatting together, rushing to and from classes, I didn't want them to look at me for fear they would be able to see what I'd just done. What would they think? I felt monstrous and was afraid they'd think me so as well. Then I realized they couldn't tell, that my secret was safe. I wandered aimlessly, not knowing where to go or who to be with, wanting to be relieved of this feeling that I'd done something so very wrong. Gradually, the thought formed within me that students of medicine had been

performing this same ritual dissection for centuries. In that moment I
unwittingly entered into a deep, unspoken bond with all physicians
everywhere who had gone before me and who would come after me.

*   *   *

Preparing for the final anatomy exam had required one last, convulsive
effort. Now it was over. Our class held a celebratory party that evening. We
all came, to drink and dance, to laugh and release, and to do something
more. We told our stories—of the first day, the funniest moment, the most
botched dissection. But there was one story I didn't tell.

As soon as I had turned in my final exam paper, I had intended to join
my fellow students in a celebratory round at the local bar. Instead, almost
unconsciously, my steps had turned toward the anatomy laboratory as a
thought formed within me. It was time to say goodbye to her. As I had
approached those large gray doors, I'd sensed an eerie and unusual silence. I
had tried the handle, but found it locked. A research assistant had walked by
on the way to his lab. Upon my asking if he knew when the anatomy
laboratory would be open, he had remarked off handedly that he thought it
was shut for the semester. He'd seen the cadavers being removed that
morning, taking them to the crematorium.

The crematorium! So she was gone. As I stood alone before those gray
doors my thoughts swirled. That first day. The initial penetrating sting of
formaldehyde. Unwrapping her and seeing her face with the sunken eyes
and contorted mouth that had returned in my dreams again and again. The
rubbery feel of her skin. The sickening sensation of passing a scalpel
through her purple flesh. She had been so generous with me, had taught me
so much. Did she know how grateful I felt? All that I had hardened myself
to in order to dissect a dead human body crashed in upon me and a floodgate
opened. The gray doors became a palate upon which scene after scene
played out before me. The oily, clinging feel of her preserved fat as I dug
my hands into her abdomen, seeking some hidden corner of one of her
organs. The precious moments in which I had caught a glimpse of her life as
she might have lived it. And that final moment when she and I had become
inextricably linked as she absolved me from guilt when I had cleaved her in
two.

Finally the screen faded and became the gray doors once again. I became
aware of the taste of salt and the wetness on my lips. Suddenly self-
conscious, afraid what others would think if they found me here, I spun
around and began walking briskly down the hall. But the thoughts wouldn't
stop. Did she know? Would she understand?

Now, beer in hand, laughing with my fellow students, I wondered if they
would understand. I looked around the party and realized that each of us, in

our own way, had completed this initiatory rite of passage we called gross anatomy. Behind the laughter, how much were they each keeping to themselves? I wondered if they, too, had been haunted in their dreams. They must have.

Chapter II

Psychiatry

"Just treat them like people. They'd like that."

I put on my white coat and rang the buzzer. Through the small window near the top of the door, a pair of eyes peered down. Quickly the door opened, and a firm grip pulled me in as the door slammed shut, bolted behind me.

"We have an acute runaway risk today." Sue, the head nurse, looked me over. Had I been able to look through her eyes I would have seen a gangly young man whose wide blue paisley tie didn't quite match the yellow shirt that was a size too big for his thin neck. The short white coat, wrinkled by the 20-minute bicycle ride through the potholed city streets to the psychiatric hospital on my old, black, three speed, displayed the green nametag which had allowed me entrance.

"I don't need anyone to tell me you're the new medical student this month. Take that thing off," she said, pointing to my white coat. "We don't wear that stuff in here." As I took it off I looked around, and she was right. Unlike the other medical units I'd been on during my first year, everyone here was in street clothes. But that meant I couldn't tell who the staff were and who the patients were. "The drill is that we all get together for rounds with Dr. Wilson, chief psychiatrist on the West Unit, at 9 AM every day. Until then, he told me you're to talk to some of the patients."

"About what?" I asked.

"Whatever you want. They'll pretty much tell you anything. Just treat them like people. They'd like that."

As she turned and left me standing there, I looked around the unit. The day room was bright, clean, with simple orange and yellow vinyl chairs and a couch. Two adolescent boys played ping pong. They began to argue, getting louder with each point. Suddenly the game exploded into open threats, and a man in his early twenties intervened, obviously a staff member. A hint of urine odor rose as a disheveled man in his fifties shuffled past, his eyes staring blankly. Feeling very uncomfortable, I searched for someone to talk. Finally my eyes landed on a tall, muscular man standing to one side of the room looking away from me down a long corridor. With a deep breath I approached him and stuck out my hand.

"Hi. I'm Rick, a medical student. What's your name?"

His hand reached out to grasp mine, the taught grip tightening painfully around my hand. "I'm Max," he said, turning to face me. My breath caught. The right half of his face was gone, replaced by a mangle of scar tissue, a lifeless eye gazing off to one side from the right socket. His one good eye stared down at me. He held onto my hand a little too long before releasing me.

"What happened to your face," I managed to ask.

"Oh that," he said matter-of-factly, his voice disarmingly calm. "I tried to kill myself. Pointed the gun right here." He pretended to hold a gun to his mouth. "But I flinched as I pulled the trigger. Missed my brain, but took off half my face. I've had three surgeries already, and I'll need a few more."

What to say now? My mind froze. He sounded so normal. Should I ask him why he attempted suicide? No, that didn't seem right. I wanted to say something—needed to say something. It looked so painful. I didn't know what to say. He seemed so damn calm about the whole thing.

"Sounds like it's been tough," I said feebly.

"Yeah, man." He turned to resume staring back down the corridor.

Pulling away from Max, I looked anxiously around the room. "Just talk to them, they'd like that," I repeated to myself, hoping to feel less disoriented. I migrated to the couch in the center of the day room, drawn to the most "normal-looking" person I could find. A girl in her late teens, dressed in jeans and a T-shirt, sat alone.

"Mind if I join you?" She looked up with a friendly smile and motioned for me to sit down. I began to relax. "My name's Rick. I'm a new medical student starting on the unit today."

"My name's Patty. You seem nice. You've got good eyes."

"Thanks," I replied, suddenly not sure where this conversation was going.

"Relax. I'm not coming on to you," she said, reading my thoughts. "If I want to do that, there are lots of guys on the unit I could fuck."

"How long have you been here?" I asked to change the subject.

"Seven months." Glancing down I noticed odd shaped scars on both forearms.

Hesitating, I pointed to her arms. "What happened there?"

"Oh, I do that sometimes, just cut myself with whatever I can get a hold of." She saw my eyes focusing on several round scabs on her upper arms. "You're wondering about those, too, I'll bet." I nodded. "I did that last night with my cigarette."

"But why?"

"Don't know," she said hollowly. "Sometimes it's a way of feeling something, anything, I guess."

"Did you want to kill yourself?"

"No, not this time. Sure I've tried to do that before, but I'm not feeling that way right now. Thanks for asking," she said flatly.

For the first time, I noticed she wasn't looking at me. Instead, her eyes appeared to move furtively around the room, taking in every detail.

"What are you looking at?"

"I'm watching the door, and the staff, too. You see, I'm the reason we're locked today. They think I'm gonna run away. I am, you know. They think they can stop me, but they haven't stopped me yet."

"You mean you've run away before?"

"Yeah. I get just so far, and they find me and bring me back. This'll be my fourth time."

"Why do you keep running away?" Now I got the look of "how stupid can you be." "I guess I can see why you'd want to get out of here," I recovered.

"Yeah." She paused, looking right at me. "I was right. You are nice."

"Thanks for talking to me," I said, as I got up. "I'll talk to you some more later." This was starting to feel more comfortable. Looking around the room, I saw a young woman standing over near the door. I walked up and introduced myself. "I'm Rick, a med student starting here today. What's your name?"

"Martha," came the clipped reply.

"How long have you been here?" I asked, learning this was a good lead in question.

"Six months."

"That's a pretty long time. What are you in here for?"

Just then I noticed a teenage girl pulling at Martha's skirt and giggling. Martha was telling her to be quiet, but smiling at the same time. Something felt very wrong. "You're not a patient here, are you," I stammered.

"No."

"What do you do here?" I asked, turning quite red.

"I'm the assistant head nurse."

*   *   *

Entering the meeting room a few minutes before 9 AM, I was surprised to find myself the first to arrive for "rounds." As this was my first clinical rotation in a hospital as a beginning second year medical student, I didn't know what rounds meant, but I was soon to find out. The room was furnished as sparsely as the rest of the unit, even though it lay just beyond the locked door, a few steps closer to the elevator—the elevator which, I had already learned, symbolized freedom to the patients of this locked psychiatric ward. If patients behaved well and showed an ability to follow the rules of the unit, they could earn their "grounds" privileges, which meant they would be allowed to walk around the hospital grounds without staff supervision. Individual chairs, some folding and some made of that same orange and yellow vinyl, crammed along every inch of the outer perimeter of this 12 foot by 15 foot room. I took my place in one chair in a corner as far from the door as possible so I could watch everyone else come in.

It didn't take long. At one minute before 9, the door swung open and in rushed almost a dozen people, some with clipboards or patient charts, each grabbing the nearest chair. At precisely 9 AM, as I was to learn he did every Monday, Wednesday, and Friday, Dr. Wilson, medical director of the West Unit, calmly strode in. A man near 50, he wore a well tailored tweed sport coat and carefully selected tie. He took his place in one of the two remaining empty chairs which magically now formed the center of the room's focus. Facing him was a single empty chair that everyone but me seemed to know was to be left empty. Dr. Wilson paused, looking around the room as if acknowledging each of us individually. When his eyes met mine, he looked directly at me, or, more accurately, directly into me. Then he moved on to the person seated next to me, continuing his silent greeting to the group.

Finally he turned to Sue and said, "What do we have going on today?"

"Well, it was a busy weekend. We had two Code 7's involving Max, and Patty's back on runaway risk again. She made it as far as the front gate of the hospital Saturday night. We've been on acute runaway risk status ever since. By the way, she's got some fresh cigarette burns this morning. Jimmy's been more cooperative lately, and he seems ready to earn grounds with supervision at today's group meeting like we discussed last week. We've got one new admission for you this morning. That's about it." She paused. "Oh yeah, and we've got a new med student starting today," she added, pointing her clipboard in my direction.

Dr. Wilson turned to face me. Simultaneously everyone in the room did the same. Instinctively, I tightened, prepared for…prepared for…for what I did not know. And then Dr. Wilson's measured voice, with more invitation than I'd expected, said, "Welcome, Rick." He then introduced all the other team members. "This is Sue, our head nurse, and Mary, assistant head nurse." Mary informed Dr. Wilson that we'd met, and smiled at me. From the stifled chuckles around the room, I realized the story of our meeting had already spread rapidly. Tom McLaren was one of the two second-year residents on the unit, along with Andy Schmidt.[1]   Bethany was introduced

---

[1] Medical training is broken up into different stages, with a strict hierarchy among them. Medical students are at the bottom of the pile. Once graduated from medical school they have earned a doctorate of medicine degree and hence have earned the right to be addressed as "doctor." After medical school physicians start a residency training program. During the first year of residency they are usually referred to as interns. This internship year may include some study of the physician's specialty field, but usually also includes experiences rotating through a number of different clinical areas or services in the hospital. It is during this internship year that physicians in training experience some of the most extended and severe periods of sleep deprivation, often working 36-hour shifts without any sleep, though recent

as the head of PT/OT. She had short hair and a pert, engaging way about her. I interrupted Dr. Wilson to ask what PT/OT stood for and was informed it meant physical therapy and occupational therapy.

Dr. Wilson then asked, "Rick, you could have completed your required psychiatry rotation with just one month at one of the other hospitals. Why did you choose to come to the Institute here, which is a two month rotation?"

"I'm seriously considering going into psychiatry and wanted to get as much exposure to it as I could early on," I replied.

Dr. Wilson paused for a moment and then responded, "We hope you have a good experience here with us. Please speak up with questions any time you don't understand something." He appeared to mean it.

With the introductions completed and after a few further questions to update Dr. Wilson on some of the patients, he turned to Tom and said, "It's pretty clear from how tired you look that you were on call last night. Why don't you present our new admission?"

Tom straightened up, took a deep breath, and began the following remarkably rapid torrent of words. *"This-is-the-first-PPI-admission-for-this-24-year-old-female-presenting-with-depression-and-a-suicide-attempt.-She-has-a-benign-past-medical-history-except-for-a-left-salpingo-oophorectomy-for-a-dermoid-cyst-at-age-17.-Her-medications-include-birth-control-pills-and-Tylenol-PRN-for-headaches.-Family-history-is-positive-for-a-mother-and-maternal-grandfather-with-depression-both-of-whom-attempted-suicide-the-grandfather-apparently-successfully.-Social-history-is-significant-for-the-recent-breakup-of-a-three-year-relationship-with-a-man-to-whom-she-was-engaged.-She-found-him-in-bed-with-another-woman-and-took-an-overdose-of-Tylenol-that-night.-A-friend-found-her-and-brought-her-to-the-Presbyterian-emergency-room.-She-underwent-the-Tylenol-overdose-protcol-including-induction-of-emesis-and-nasogastric-lavage.-After-24-hours-in-the-ICU-without-a-bump-in-her-LFTs-she-was-deemed-stable-enough-for-transfer-to-the-Institute-where-*

---

regulations place restrictions on the maximum period of duty without sleep for physicians in training. Depending upon the specialty selected, residency training can last from three to five years, which includes the first or internship year. For example, family practice, pediatrics, and internal medicine require three years, obstetrics and gynecology requires four, and general surgery requires five. At the completion of residency training, many physicians begin practicing. However, many also go on to fellowship training in a subspecialty. For example, cardiology requires a three-year residency in internal medicine followed by an additional two to three years of fellowship training specifically in cardiology. A physician who has completed the required training in their field is frequently referred to as an attending physician.

she-arrived-last-evening.-Her-physical-exam-is-benign.-Her-mental-status-exam-reveals-a-tearful-withdrawn-somewhat-distraught-female-who-makes-poor-eye-contact.-Mild-psychomotor-retardation-was-present.-Her-thought-content-appeared-obsessively-focused-upon-her-relationship-and-the-event-of-walking-in-on-her-partner-in-bed-with-another-woman-without-evidence-of-other-thought-disorder.-She's-still-got-pretty-serious-suicidal-ideation.-No-clear-plan,-but-said-something-about-more-pills-if-she-got-a-chance.-She-remembered-zero-out-of-three-objects-at-five-minutes-though-this-appeared-to-be-effort-related.-Similarly-she-would-not-cooperate-with-serial-calculations-or-number-recollection.-Her-response-to-aphorisms-was-hostile.-In-summary-we-have-a-24-year-old-female-with-a-family-history-of-depression-and-suicide-attempt-who-presents-with-depression-and-her-own-serious-suicide-attempt-following-an-acute-emotional-trauma.-She-required-sedation-with-20-milligrams-of-Valium-last-evening-and-was-started-on-100-milligrams-QHS-of-nortriptylline-and-50-mg-QHS-of-trazadone."

I could not believe Tom could produce all that without any notes while pausing for remarkably few breaths. He looked like he'd been up all night, yet, as tired as he appeared, his recitation had been crisp and focused. It was only later that I was to learn that "LFTs" meant liver function tests, "emesis" meant vomiting, and that "QHS" meant at bedtime. Still, I could piece together enough to understand that this poor woman had just walked in on her fiancé cheating on her, got depressed, and tried to kill herself with an overdose of Tylenol. It sounded like she'd been through hell in the emergency room and the hospital before even getting here.

Dr. Wilson nodded a kind of approval to Tom. "That's a nice summary. What's the plan?"

"Antidepressants, sedation as needed, milieu therapy. I'll do some one-on-one sessions with her."

"How's she been doing on the unit?" he asked, turning to Sue and Mary.

"She was quite upset last night. Didn't want to talk to anyone. Nights said she was crying off and on during 11 to 7. This morning we had some trouble getting her out of bed, but she eventually came into the day room with a lot of prompting from our psych tech, Bobby. It seems she's opening up some to Bobby, so I've asked him to spend some extra time with her. She's still pretty suicidal, so we've got her on precautions."

"What's her insurance situation?"

Lorraine, the social worker, jumped in. "She's got pretty good insurance, so at least the first 30 days should be covered. I'm looking into the private funds situation for possible long term inpatient treatment." I didn't realize it at the time, but this was my first introduction to the biggest determinant of the care all patients receive. If a service is covered by insurance, patients get it. If it's not covered, they don't. The specialty of the prescribing physician doesn't matter—the money does. In psychiatry, I would soon learn, when the money runs out, it's amazing how the patient is suddenly ready for discharge.

"Bring her in," Dr. Wilson concluded. "By the way, what's her name?"

"Maria. Maria Alphonzo."

Mary placed a call to the unit. A few moments later the door opened, and the man I had seen breaking up the boys in the ping pong game, who I presumed was Bobby, led in a dark-haired young woman with red, swollen eyes. I would have described her as pretty, except that her stooped shoulders, rumpled clothes, and tear stained face did all they could to hide her attractiveness. She paused at the door, staring at the faces around the room. Her eyes widened and she looked pleadingly at Bobby.

"I didn't realize there would be so many," she whispered.

"It's alright," Bobby responded. "They're good people, and they're all here to help you. Trust me." With a gentle pull of her hand, Bobby coaxed her into the room and into the one empty seat across from Dr. Wilson. Bobby started to go.

"You can't leave," she pleaded. "I don't know any of these people."

Sue said quietly, "It's OK to stay, Bobby. Mary will cover the unit for you." Mary nodded and left, and Bobby took his seat.

Dr. Wilson now took over. "Maria, I'm Dr. Wilson, medical director of the West Unit here at the Institute."

"I...I know. Bobby told me about you."

"Really? What did he tell you?"

"He said I had to talk to you. Everybody new here does. I told him I didn't want to."

"That's understandable, Maria. I hear you've had a rough time recently." Maria just stared at the floor in front of Dr. Wilson. In a soft but firm voice he said, "Tell me what happened that brought you here."

Almost against her will Maria began to speak. "I don't know. I have this boyfriend—my fiancé—at least he was my fiancé." She paused. "It's really hard to talk," she said thickly. "They gave me so much medication I can't think straight, and my mouth is so dry."

"I'm not surprised. Those are pretty typical side effects from the medications you received last night. Would a glass of water help?" Maria nodded. Sue got up and left, returning in a moment with a glass of water she gently placed in Maria's hand. Maria mindlessly raised it to her lips and sipped. "You started to tell me about your fiancé," Dr. Wilson prodded.

"Yeah. My fiancé," she said bitterly. "He's a shit head."

"Why is he a shit head?"

Maria looked up at Dr. Wilson, surprised to hear him use her own profanity. It seemed to make her soften toward him a little. "I'm sure they told you what he did. Everybody here seems to know everything about me. There are no secrets here."

"Do you have secrets, Maria?"

Again, she was taken a little aback, but this time her eyes met his. "Everybody has secrets." For a moment there was life back in her eyes as they blazed at Dr. Wilson. "Anthony had secrets," she hissed.

"Yes. I know about at least one of Anthony's secrets. It must have hurt very much."

"You wouldn't know," she bit off the words.

"You're right. I can't know just what you're feeling. Nobody can." His eyes were locked with hers. Why didn't he say something more comforting at this moment? Her heart was splayed open so publicly in front of all of us. He could have said something about wanting to hear how she felt about it. He could have talked about how his own painful experiences, or at least his professional experiences, gave him some insight into what she must be going through. That's what I would have done, I thought. Instead, the silence deepened. The room and all of us seemed to dissolve around the two of them. Slowly, her eyes filled, and one tear spilled over, running down her cheek. "It's lonely, isn't it," he said. She nodded a deep ascent as another tear ran down. Then he added, "What secret do you carry, Maria?"

She stiffened. Her eyes narrowed. "How did you know?"

"You told me, when you said everyone has secrets."

"It was last year. I met this guy. Even though Anthony and I were engaged, I let it happen. But it was different. I knew after the first few times it was a mistake. That's why I broke it off. I knew I wanted Anthony." The tears started again.

"So you're not just a victim in this, are you?" Silence again as she looked directly at him. Shifting gears and using a somewhat more authoritative tone, Dr. Wilson then said, "There are still a few more questions I have to ask you, questions I ask everyone. Please do your best to answer them. I'm going to begin by mentioning three objects. Please try to remember them, as I'll ask you to recall them in a few minutes. The three objects are a book, red shoes, and 35 Broad Street. Will you try to remember them?"

"Yeah. Sure." Maria's tone also shifted.

"Now, I'd like you to start at 100 and subtract 7. What do you get?"

"I don't know. You guys have me so doped up my mind is awfully fuzzy."

"Try your best anyway."

"Take 7 from 100," she said slowly, as if in a haze. "That's 94. No, it's 93, I guess."

"Now subtract 7 from 93 and keep going, subtracting 7 from each number."

"What's this got to do with why I'm here?" she said, struggling to clear her head. "I tried to knock myself off. For what, so I can take a math test?"

"Just try. It's important for us to understand how your mind is functioning at this time."

"My mind is functioning like crap because of how drugged you've got me. But it's functioning well enough to know this has shit to do with me right now. I just want to put an end to this whole thing."

"Meaning you want to kill yourself?"

"No shit! Isn't that what landed me in here?"

"So if you were to leave now, you'd try to kill yourself again."

"I don't know, but my mind, as you call it, has figured out that if what it takes to get me out of here is to tell you I won't kill myself, then that's what I'm telling you."

"I understand." Shifting gears again, Dr. Wilson said, "I'm now going to share a few sayings. I'd like you to tell me what they mean to you. The first is 'A rolling stone gathers no moss.'"

"What are you talking about? I can barely think straight, between the drugs and looking at all of you." She was staring at me at that moment.

"Just try. What does 'A rolling stone gathers no moss' mean to you?"

She continued staring at me, imploring me with her eyes, as if looking desperately for a place to hide. "I guess," her faltering voice finally began, "it means nothing's for sure. Anthony is kind of like a rolling stone. I thought I could count on him." She started to sob quietly.

Dr. Wilson leaned forward. "Thank you, Maria. You did very well. We're finished now. Just one more question. Do you recall those three things I asked you to remember?"

"I can't remember what they were," she said in a defeated voice through her tears.

"That's OK," Dr. Wilson responded.

Bobby got up and took Maria back to the unit.

As soon as the door closed, Dr. Wilson looked at Tom and Andy and asked, "So what did you see?"

"She had a cloudy sensorium, consistent with the medications she's received, and appeared to have no sign of a thought disorder," Tom volunteered.

"Good, Tom. So, Andy, what would be your diagnosis?"

"Axis I would be depression with suicidal ideation.[2] We don't have enough information on her premorbid condition to identify an axis II

---

[2] Psychiatric diagnoses have been standardized through the *Diagnostic and Statistical Manual*, published by the American Psychiatric Association. This standardization includes looking at psychiatric diagnoses as occurring on several

diagnosis, but I believe there is a good chance, from what we saw here, that she may have an underlying histrionic personality disorder. Axis III is status post acetaminophen overdose without sequelae."[3]

"Excellent, Andy. Rick, since you think you want to become a psychiatrist, let's get you involved on your first day." I was suddenly sitting bolt upright. "What do you make of her resistance to answering the mental status exam questions I posed to her?"

"Which questions were those, sir," I responded, trying to buy time to clear my thoughts.

"That's right. It's just your first day. Tom, make sure you take him through the formal mental status examination format and questions today. The mental status exam included the specific questions I asked beginning with remembering the three objects, though, as all of the rest of you know, I was assessing her mood, affect, speech, and thought processes from the moment she entered the room."

I decided to be truthful. "Well, honestly, I was a little embarrassed for her. It seems like it would be really hard for someone to answer questions like that in front of so many people, especially strangers. And she did look pretty drugged. I thought it was kind of a normal response."

"That's an understandable answer for your first day. But we work very hard to create a therapeutic milieu here. You could see how every member of the team did everything they could to make her feel at ease and supported." I did have to agree with this, except for their insisting she discuss her profoundly personal feelings in front of a group of complete strangers. This was the first time, but certainly not the last, I learned how much the practice of medicine assumes we should talk about our patients' most intimate and personal details in a public, humiliating way, even in their presence. "She is resistant to the help we're offering her," Dr. Wilson went on. Her "resistance" looked to me like normal behavior in a crazy-making situation. "She's certainly suffered a major betrayal, but until she's ready to open up and let us help her, I'm afraid she shouldn't be allowed to go home.

---

different axes. Axis I represents clinical disorders of thought, mood, learning, or development, such as depression, schizophrenia, or attention deficit hyperactivity disorder (ADHD). Axis II represents a personality disorder, if one exists, such as borderline or narcissistic personality disorder. Axis III represents acute medical conditions and physical disorders, such as substance abuse or diabetes. Axis IV represents psychosocial and environmental factors contributing to the disorder. Axis V represents a global assessment of functioning.

[3] Sequelae are the impact or results of an injury, illness or, as it is often referred to medically, an insult. The Tylenol overdose was the injury, and Maria appeared to not have suffered any permanent medical damage from her overdose, even though a Tylenol overdose can be lethal due to severe liver damage.

I agree with the staff's assessment that she is still suicidal and should remain on suicide precautions. Is she on a voluntary or involuntary admission?"

"Voluntary for now," Lorraine answered.

"Good. But if she starts to make noises about leaving, especially without signs of opening up and letting the staff really help her, be prepared to go for an involuntary commitment for as long as it takes for her to be safe." So she was caught in a catch 22. If she wants to get out, she has to bare her soul to every staff member or doctor who asks her to. If she doesn't, it's interpreted as resistance to help and a sign she should be held against her will in a locked psychiatric ward. But she's just been betrayed and humiliated by the man she loved and trusted. Demanding her to confide in a group of strangers the next day, under threat of keeping her locked up if she does not, felt not just crazy-making, but abusive. I could already see the whole team had been working in this system for so long nobody questioned it. Maria did. And now, thanks to the protest she managed to get out through her drugged fog, so did I.

Dr. Wilson's voice interrupted my thoughts. "Rick, this being your first day, do you have any questions before we break?"

"I have lots of them, but I guess that's pretty usual for a first day." Several people chuckled. "But I'd like to just ask one of them now. What's the Code 7 Sue talked about Max having over the weekend?"

"A Code 7 is an episode of violent behavior that requires restraint of the patient. Calling a Code 7 brings in what we affectionately refer to as the 'goon squad,' at least four of the strongest males on duty right then. Max can get surprisingly violent with little provocation. He's sent at least one psych tech to the hospital already. Why do you ask?"

"Just curious," I said, silently shuddering with the memory of Max's all too tight handshake and the look in his one good eye.

*   *   *

Standing outside room 107, I reviewed the steps of a patient workup[4] in my head one more time: history of present illness, past medical history, social history, review of systems, and then the physical exam. I knocked. No answer. I knocked again and opened the door a crack. On the bed, staring

---

[4] Workup is a term used to describe the process of medically evaluating a patient to arrive at a diagnosis. It usually involves taking a medical history, performing a physical examination, and ordering appropriate laboratory and imaging studies. It can be used as a noun, as it is here, or as a verb, as in, "I need to workup the new fever for Ms. Jones, the patient in 243B."

directly at me with wide eyes, sat Peter McDonough, my first psychiatric admission workup.

"Hi. I'm Rick Sheff, a medical student here. Dr. Schmidt has asked me to talk to you to get some information about you and do a medical exam to make sure you're OK physically."

"Haloo. Haloo. Haloo." His eyes suddenly darted around the room, searching for something. "Yes. Yes. He said you'd be coming."

"Who said I'd be coming? Dr. Wilson?"

"He did. He did."

"Who is he?"

"He's right over there," Peter said, pointing to an empty corner of the room. The patient then leaned forward, and, in a conspiratorial whisper said, "He doesn't like you, you know, so we'd both better be nice to him or it may mean trouble."

"Trouble? What kind of trouble?" I asked warily.

"Trouble. You know. Trouble." His eyes continued to dart furiously around the room.

"OK. I'll be nice to him," I said with a glance over my shoulder to the empty spot. "Now, why are you here at the Institute?" A good place to start my history of present illness, I thought, though the reason was already perfectly obvious.

"Dunno. Dunno. Dunno. They just put me here."

"Who's they?"

"They. Them. They're friends with him," he said pointing again to the empty corner. It was clear I was getting nowhere with this line of questions so I tried a different approach.

"What medicines are you taking?"

"They give 'em to me. But," again the conspiratorial whisper as he leaned forward, "he doesn't like 'em," he said, pointing again to the empty corner. "So sometimes I pretend to take 'em."

I was quickly learning that I'd have to get his medical history from another source. But at least I'd learned he'd been pretending to take his medications, so I could report that to the nurses and Dr. Schmidt.

"Alright. I have to examine you now." With that, I started by taking both his hands in mine and looking at them. His cigarette-stained fingers already told me more than I'd learned from my aborted attempt to take his medical history. This trick of taking a patient's hands at the start of an exam, learned from my physical diagnosis class during the first year, allowed me to begin the exam by touching the body part we already extend to others in a routine symbol of greeting, a handshake. Starting an examination with the least threatening touch possible allowed me to work from the outside in—from the hands, to the face, down the body, finally examining the genitals and rectum. We'd learned this last year by practicing examining each other. But of course we didn't do genital and rectal exams on each other, or breast exams either. We read about them, but they said we'd learn how to do these during our clinical rotations. Here I was on my first clinical rotation, doing my first workup, and nobody was telling me what to do with the genital and rectal exam. I didn't know if it was for better or worse that the first patient I'd be doing these on was psychotic.

Working through the exam, I looked in his ears, which were completely blocked with hard, black ear wax. Then I tried to look in his eyes, but they continued to dart around in spite of my instructions for him to stare at a single point. Next I asked him to open his mouth. As I put the tongue depressor in, he bit on it, making it impossible for me to see the back of his throat. Three times I tried, and three times he bit down on the tongue depressor, each time with a weird grin, like he was enjoying frustrating me. I gave up trying to see his throat as well.

Then I asked him to pull up his shirt so I could listen to his chest. He looked at me and said, "You're not gonna hurt me with that thing, are you? He wouldn't like it," once again glancing towards the empty corner.

"No. I'll be very gentle," I said. But as I placed the stethoscope on his back, he jumped.

"That's really cold, Doc," he said.

As much as he jumped at the touch of a cold stethoscope, I jumped to hear myself addressed as "Doc." Just a beginning second-year medical student, I didn't see myself as a doctor yet, not nearly. "Sorry. I'll warm it up," was the only response he heard as I held the stethoscope in my hands, which were cold themselves from my being so nervous. Eventually the stethoscope warmed up enough for me to continue. He let me examine his lungs and heart, but when he laid down for me to examine his abdomen, with every touch he burst out laughing, tightening up his stomach muscles and making it impossible for me to feel anything in his abdomen. He did let me examine his legs and feet. Then I took a deep breath and said, "Lower your underpants please," trying to sound as nonchalant and professional as possible.

"Sure Doc. Hey, you ain't queer, are ya?"

"No. I'm not," I responded. But I wasn't prepared for what happened next. As I put on my exam glove and started examining his scrotum and testicles, he developed a large erection, pointed straight at me. They never told us to expect this during physical diagnosis class, and they certainly never told us what to do if it happened. I did my best to cover up my discomfort, but it didn't seem to help.

"I guess he likes you," the patient said, pointing to his now throbbing erection. Quickly I finished examining the contents of his scrotum and his penis. I put my finger into his inguinal canal, that place up in the groin doctors had always prodded as they told their male patients to cough, and I heard myself say the same words I'd heard from doctors so many times, "Turn your head and cough." At least now I knew what they had been looking for all those years.

"Roll over on your side. I need to examine your rectum," I heard myself say, my own quickening pulse now pounding in my ears.

"I thought you said you wasn't queer. Ya know only queer guys put things up other guys' butts."

"This is just part of a routine physical exam," I said in the most professional voice I could muster at that moment. I placed a small amount of KY Jelly on my gloved finger, spread his buttocks and slowly worked my finger into his rectum. I wasn't sure who was more uncomfortable at that moment, him or me. I moved my finger around, trying to visualize the prostate and rectal anatomy inside to figure out what I was feeling, but none of it made sense. Feeling something I thought was firm, like his prostate, I suddenly realized I was pushing my finger up instead of down, so I had to have been pressing on his sacrum. I then turned my finger around, feeling the KY Jelly squish inside his anal canal as he squirmed, trying to get away from my examining finger. To great relief for both of us, I finally pulled my finger out. There was brown material on my glove, which I smeared on the card I'd brought with me to check his stool for trace amounts of blood. "You can get dressed now," I said with extraordinary relief.

As I stood up to leave, he said, "Can I ask you something?"

"Sure," I responded.

"Can I suck your Joe?"

"Excuse me?"

"Can I suck your Joe?"

"Uh, no, you may not. Have a good day."

*   *   *

"Fifteen-12, my favor," I said.

"You were just lucky on that last one," Jimmy taunted.

I'd been on this psychiatric rotation for only 10 days, and already my ping pong game had much improved. Funny how I was supposed to be learning psychiatry, but found myself spending a good part of each day playing ping pong. I was coming to understand how this kind of "play therapy" could actually help build a relationship with a patient like Jimmy. You certainly couldn't talk to Jimmy directly about his feelings. He'd shut down in a second. But, while playing ping pong, we'd already had some great exchanges.

"This has nothing to do with luck," I responded.

"Yeah, bad luck is what got me in here in the first place."

"What happened?" I casually asked just after serving.

"My parents split up," Jimmy said, returning my serve with a little bit of a slice. "I was with my mom, and she was drinking, as usual." Was it my imagination, or did he hit the last shot back a little harder as he said that? "She tied a pretty good one on that night. She didn't even notice when I left. It was easy for me to take her wallet. At least she had a little money in it that time, so I was able to buy some booze for me and my friends. We just shouldn't have stolen that Mustang."

"Why not?" I asked as he started serving.

"I know cops are out looking for kids driving cars like that. We should have stolen something different, like a Chevy. Then the cop wouldn't have been interested. Besides, the Mustang had shitty traction, so if we were in a Chevy we wouldn't have hit that pole. That's what I meant by bad luck."

So this is what a budding sociopath looks like, I said to myself. Already I could see that he had no sense of guilt and no inner prohibitions against things the rest of us would condemn. Still, for some reason I liked Jimmy. He was different from Smitty, the other boy I'd seen Jimmy starting to argue with that first day. Smitty's problem was impulse control. He hadn't developed any of the usual inner controls you'd expect, even in a 15-year-old. If he felt it, he said it. If he wanted it, he took it. And he had a hair-pin trigger for going into a rage. I'd seen them call the goon squad on him twice already. It took all four guys to hold him down and get him into a seclusion room the last time.

I tried playing ping pong with Smitty a few times, but each time he'd broken out in a rage at something. He couldn't follow the rules, and he absolutely hated to lose a single point. As a result, I tended to avoid him on the unit. Most of my time, when I wasn't working up new patients, I spent talking with Patty or playing ping pong with Jimmy. I'd even had a few more conversations with Max, but always from a wary distance. And there were now regular meetings for me with the patients I had worked up. Though I was told to keep these to 15 minutes, as the residents did (Dr. Wilson's patient meetings were even shorter), my meetings always ran much longer. I couldn't just talk about the patient's response to their medications and any medication side effects, which seemed to be the primary focus of Dr. Wilson's and the residents' meetings with patients. I liked to talk about their responses to what was happening in the "therapeutic milieu of the unit,"[5] as the staff called it. I was even more fascinated by their

---

[5] Milieu therapy is the term that applies to the process of using the hospital environment and relationships with both staff and other patients (all of which together are referred to as the therapeutic milieu) as part of the therapy to help patients improve while in a psychiatric hospital.

personal and family histories. This allowed me to apply some of the skills I had learned leading therapy groups before starting medical school, hopefully helping these patients gain insight and commit to new behaviors, even during their inpatient psychiatric stay. The medical model of psychiatry, focused on drugs as the intervention of choice, is not what sparked my initial interest in the field, so I struggled against this approach while simultaneously working hard to master its basics.

The routine on the unit became more familiar. Though rounds still made me uncomfortable due to the public display of such personal information, I always learned something from Dr. Wilson there. One time he was interviewing an anxious, hysterical woman in front of the group. Under the stress of the interview and feeling exposed in front of our large group, she had started to escalate and become uncontrollably anxious. Dr. Wilson leaned forward at that moment and took her wrist, assuming the position for taking her pulse. Instantly she calmed down and was able to complete the interview. After she left, Dr. Wilson explained this was "playing the medical card." This meant that when a patient was upset, doing something familiar that doctors routinely do can have a calming effect. It had been remarkably effective, I noted.

I began to participate in discussions at community meetings as if I was actually a member of the team. Kudos to the staff, the residents, and Dr. Wilson for encouraging me to participate in this way, I thought. It also fit my own tendency to take risks and get involved. My training as a group therapy facilitator before medical school clearly paid off here, allowing me to be comfortable jumping into the community meetings, articulately expressing insights and comforting patients, as well as sometimes sharing my own personal feelings. The term therapeutic milieu began to make more sense to me, even though it still seemed "crazy making" to expect patients with such diverse problems (psychopathology, as I was learning to call it), patients who barely knew each other or the staff, to talk about their most intimate feelings and fears so publicly. In the therapy groups I had led, such trust had had to be built over many months, and in some cases, years. Here the doctors and staff demanded it the first day.

I was also actively participating in patient conferences and educational sessions designed to cover the core content of the field of psychiatry. Medications I had paid only fleeting attention too just months earlier during my pharmacology course now became familiar as I learned their indications, side effects, and drug interactions. Terms like "psychomotor retardation" became more than just another textbook phrase when a profoundly depressed woman was admitted to the unit. I watched as, in response to a question, it would take her almost a full minute to begin answering, and then she'd speak only in slow, monotone phrases. Her hands barely moved from her lap during an entire interview. And I got to meet my first high-flying manic. He stormed all over the unit, calling out colorful, sexually explicit comments, often in clang or rhyming associations. Then I watched as, under the effects of lithium and other medications, over a 10-day period his bizarre, outrageous behavior moderated so he became less dangerous to himself and others. In short, I was getting a thorough introduction to the field of psychiatry through this clinical rotation. I enjoyed much of what I did, liked much of what I saw, but felt critical of some of it. Would I be happy doing this for an entire career? I wasn't sure.

*   *   *

One morning, a new patient appeared on the unit. Holga was a private patient, so I did not perform a workup on her. This meant I didn't know much of the details of her background. All I knew that first day is that she didn't want to come out of her room for community meeting, but finally arrived, wearing her calf-length, faded, flannel nightgown, clinging to Bobby's arm. Her long, gray, stringy hair hung lifelessly down her back as she shuffled slowly to her seat. Another case of depression with psychomotor retardation, I heard myself saying silently. The terms were coming more easily now.

At rounds that day, Sue mentioned the new admission to Dr. Wilson. She finished by saying, "And the plan is to initiate ECT immediately." ECT? Did that mean what I thought it meant—electroconvulsive therapy? Shock treatment? My only picture of shock treatment came from reading Ken Kesey's book, *One Flew Over the Cuckoo's Nest*. The main character had gone through repeated shock treatments. He hated the shock treatments so much. He had to be held down by a goon squad and finally tied down in four-point restraints with a padded tongue blade in his mouth. The paddles were placed on either side of his head, and, when the "shock" was applied, his body violently convulsed, seizing repeatedly. Is that what this frail, elderly woman was to be subjected to?

My fears were further heightened when she returned from her first "treatment." She appeared to be in a daze, sleeping off and on the rest of the day. The following morning, as I entered the unit I could see her pacing slowly back and forth across the day room screaming in an Eastern European accent, "I don vant shock treatments!" This pattern repeated itself throughout her first week on the unit. Both because I thought it should be part of my medical education and out of sheer curiosity, I finally approached Dr. Wilson asking if I could observe one of her shock treatments. The arrangements were made.

The morning arrived. I accompanied Holga as the nurse brought her in a wheelchair down into the basement of the hospital to the ECT unit. All the way she called out to everyone we passed, "I don vant shock treatments!" As we paused outside the ECT unit, she looked into my eyes, grabbed my hand, and implored me, "I don vant shock treatments! Don't let them do this to me. Stop them." The nurse pulled her away from me until our outstretched hands finally parted, and she disappeared into the ECT unit, calling back to me all the time to "stop them." I waited outside the door as her muffled cries moved further away. Eventually, there was silence. Time passed. Finally the door opened and the same nurse invited me in. I entered some sort of holding area with several stretchers along the walls, only flimsy curtains separating them. Most were empty, but a middle-aged man lay sleeping quietly, even peacefully, on one of the stretchers. A nurse periodically checked him. He had just finished a treatment, I was told.

Another door opened before me, and I entered the innermost room of the ECT unit. There lay Holga on a table, an IV in her arm, electrodes on her chest, sleeping soundly. Two men stood at the head of the table, and both looked up as I entered. "Welcome. I'm Dr. Wyman, Holga's psychiatrist. We understand you are interested in what we do down here," he began. "You heard how upset Holga was when she came in. We sedated her, as we do all patients prior to ECT therapy. In fact, we perform ECT under general anesthesia. That's why Dr. Martin here, our anesthesiologist, attends every ECT treatment. It's not like in the old days when patients were awake for their treatments. Today, not only will Holga be asleep, but we'll inject succinylcholine, a neuromuscular blocking agent, just before the treatment. This means her muscles will be temporarily paralyzed, so she will not convulse during the treatment."

My mind raced. Here was Holga, peaceful, asleep. I watched as Dr. Wyman placed two small cylinders attached to wires on Holga's temples. He nodded to Dr. Martin, who injected the succinylcholine. In seconds her entire body appeared to collapse into the table as every muscle went limp. Dr. Martin placed a mask over Holga's face to maintain her airway as Dr. Wyman said to the nurse, "Now!" It was over in no more than three seconds. Dr. Wyman stood up, as Dr. Martin squeezed a bag that forced air in and out of the mask, and I watched Holga's chest rise up and down. I looked incredulously at Dr. Wyman. This was it? It seemed so benign, so medicalized. No drama, no restraints. So why was Holga so upset about these treatments?

"Holga seemed very upset to be undergoing ECT," I said to Dr. Wyman. "Why would she find this so objectionable?"

"That's a common response we see when we mention ECT to patients. They have images of dark rooms, restraints, and seizures. Nothing like it is today. Patients do experience some memory loss and disorientation during the series of treatments, which they find unpleasant. But for Holga, it's her only hope. She has been profoundly depressed and unable to function in her life. She hasn't responded to or hasn't been able to tolerate any of the antidepressant medications we've tried for her. ECT is remarkably effective in such refractory cases of depression. It's truly a life saver for Holga and patients like her."

I looked down at Holga, now starting to breathe on her own again as the succinylcholine wore off quickly. She appeared calm, more peace in her face than I'd seen since she arrived. I thought of what I'd learned in neuroanatomy during my first year. The large number of new chemical messengers called neurotransmitters that were being discovered regularly, the billions of cells that made up the brain, the unimaginable complexity of microscopic connections among these cells, all of which together produced minute electrical signals any one of which was too fine for our most delicate instruments to measure. Into this, the most refined creation in all of biology, Dr. Wyman had just pumped countless volts of raw electricity. It didn't cause convulsions only because her muscles had been paralyzed. But what was actually happening in her brain? Her brain had to be seizing, but silently, with no outward expression of the chaos echoing inside. Her haunting eyes still implored me to "stop them" and her voice echoed inside me, "I don vant shock treatments!" Who was right?

*   *   *

One day, another new patient arrived on the unit. Paul, as he was called, was catatonic. I'd never seen a catatonic patient before, and he fascinated me. Sitting in one corner of the day room, hunched over, with five days beard growth and greasy, black hair, he never moved. Always staring in the same direction, he did not acknowledge anyone who passed or anyone who tried to speak to him. Tom invited me over for a demonstration of plasticity, in which he raised Paul's hand and left it there. It stayed up over his head for over 15 minutes, until Tom eventually returned and lowered it back into Paul's lap.

But there was something in his eyes that told me he was taking it all in. Though he looked away from everything and everyone, something appeared to be getting inside. I was discussing Paul one day with Bethany, the head of PT/OT. (I was getting comfortable with the incessant abbreviations everyone used). She, too, had a sense that he was taking in more than he appeared to let on. I mentioned to her that prior to medical school I had spent time studying both dance and drama. I'd been particularly interested in modern and jazz dance, as well as improvisation for dance and drama. As we talked, we simultaneously got an idea. Since Paul certainly wouldn't engage anyone with words, could we engage him through movement? It was worth a try.

The next day, Bethany and I entered Paul's room. He was seated in his usual hunched posture, staring at the wall. Bethany and I sat on his bed on either side of him and assumed the same body position he held. We sat there for five minutes, got up and left, never saying a word. The next day we did the same thing. This went on for almost a week. During this time, Bethany and I both had the sense that he was coming to expect our visits, and his body appeared to relax a little during our presence.

One day, as we sat, Bethany gently leaned into Paul. Then she sat up, and I did the same. After a while we were "passing" Paul back and forth between our shoulders, while he maintained his seated, hunched position. The next day we did the same thing. It was on the third day that, in the midst of this "game," the edges of his mouth came up in a slight smile. It was the first change in his facial expression we'd seen since his arrival on the unit. That day Bethany and I left his room high-fiving each other for this first hint of progress.

Each day, we returned to work with Paul. Eventually we began moving his limbs. Two weeks into our experiment we succeeded in getting him to stand up with us. By the end of the next week he held our hands and walked with us. As we worked with him, he now appeared to steal little glances at each of us, telling us we were getting in. It was time to take him to the physical therapy area.

Over the next few weeks, we expanded his movements. One day we handed him a large ball. He just held it. Bethany took it back, and she and I demonstrated having a bouncing catch with it. When we bounced it to him, it just glanced off his hands as he made no effort to catch it. We tried this again and again. Finally, on the fourth day of our efforts to have a catch with him, he miraculously held onto the ball. A few days later, we were having an active game of catch among the three of us.

We carried out these activities in silence for the first two weeks. Then we began speaking. At first we spoke about Paul in his presence, commenting on whatever he was doing at that moment. Eventually we directed our comments and questions directly to Paul. Always the sly stealing of a glance followed one of these efforts, but never a single word. Bethany and I were convinced we were getting through to Paul, that he was actually "interacting" with us. The depth of his catatonic state appeared to be lifting. Was it just a change in his medications, or was our physical engagement with him having an effect? Both Bethany and I felt what we were doing went beyond "having an effect." We were developing a relationship with Paul, at least that's what we told each other. He appeared to recognize us when we entered his room, his slight smile told us when we "touched" something inside him, as did the glances he came to know we saw, commented upon, and appreciated.

This became more than a job for Bethany. And it became more than just a clinical rotation for me. Both of us grew to care about Paul very deeply. Both of us wanted so much for him to break out of his catatonic state. We cheered any small advances. We convinced each other that what we were doing was working, even if only a little at a time. But for me, the end of my time at the Institute was approaching. With two weeks left, I began to focus more on getting Paul to speak. Even if he could say only one word, I would feel we had truly made a breakthrough. I spoke directly to him, asked him questions, made comments about him, and eventually personally asked him to speak. Each time I asked him to say something out loud, he would throw one of those glances at me and the edges of his mouth would go up, but still not a single word came out.

My last week at the Institute began. To my surprise, yet true to the precepts of the psychiatric field, Dr. Wilson began to engineer "closure" to my experience at the Institute. He introduced the timing of my leaving to the rounds group, as well as to the general community meeting of patients and staff. With two days left, he invited comments from both of these groups about my time at the Institute as well as how people felt about me and about my leaving. My last day, Dr. Wilson took 20 minutes out of rounds to go around the room, inviting each person to say whatever they wanted to as a public goodbye to me in front of the group. The same occurred in community meeting. In each case, I heard appreciative comments about my willingness to get involved, about how available I'd made myself personally to patients and staff, and about my openness to taking risks. These comments, and the feelings expressed with them, caused a lump to rise in my throat. I hadn't realized how much I'd come to care about these people, both the patients and the staff. I had worked so hard initially to figure out how to fit in, and to learn what it took to carry out my responsibilities as a medical student on the unit. Now it all felt so comfortable. Yet I had to leave. Was this what the rest of medical school would be like—starting a new rotation every one to two months, making the effort to fit in, beginning from scratch each time to learn the necessary clinical information for a new field, starting relationships, and then moving on? I didn't want to go.

After the final community meeting, I made my way around the unit, saying a personal goodbye to each of the patients I'd come to spend time with. Patty said how much she didn't want me to leave, that everyone she ever cared about left her, and that I was just doing the same. No amount of rational discussion of rotations and schedules could undo the hurt she felt. She eventually went with a staff member to her room, crying. I played one last ping pong game with Jimmy. I let him win, but he was beaming with pride after the final point.

Then it was time to say goodbye to Paul and Bethany. I found them in the physical therapy area, Paul with the large ball in his hands and Bethany encouraging him one more time. I came directly up to Paul, took both his hands in mine, and told him directly how proud I was of him, how much I had come to care for him, and how much I wished he would recover enough to go home. One last time I asked if he would speak at least one word to me. His eyes rose briefly to meet mine and then flashed away again. The edges of his mouth curled up, but no words came. I hugged him and Bethany and walked away, fighting back tears.

I returned to the unit one final time. To my surprise, Smitty came up to me and asked for a last ping pong game with him. He had just earned his grounds privileges with supervision, so he asked if we could go to the ping pong table in the hospital's recreation center. I agreed. On the way there, and all during our game, he kept up a rapid conversation ranging from how proud he was to have earned his grounds privileges, to what his favorite paddle was for ping pong, to his plans for when he left the Institute. As we finished the last of three games, I said, "Time to get back to the unit."

"Aw no. Just one more game. Commawn, just one more?"

"OK," I answered. "Just one more. But you have to promise to go straight back to the unit as soon as the game is over."

"I promise."

This game took on a new intensity. Smitty tried to smash every point, but most of his shots careened wildly out of control. The game ended quickly. "Alright," I said. "Time to head back."

"You have to catch me first," he said, something wild flashing in his eyes.

"No. I won't do that," I said, walking deliberately around the ping pong table towards him. It was at that moment I noticed we were alone in the rec room. "You promised to head straight back if I played one more game."

"Yeah, but I'm not going back until you catch me." And with that, he darted away from my outstretched hand and raced around to the other side of the table.

I paused. Do I chase him or just call the goon squad? My instinct told me not to call right now. Something was at work between us I didn't yet understand, but it felt important to play it out, at least for a while. I made it a point to stay between him and the door out of the rec room so he couldn't bolt away from me. He recognized what I was doing and appeared content to play the game like this. As I lunged for him around the table, he quickly pulled away. If I came over the table directly at him, he scooted under the table to the other side. Though I was fit and in my mid twenties, I was no match for a wiry 15-year-old bent on staying out of my reach. Our chase went on for a full 10 minutes, both of us sweating, at times laughing and at times focused in sheer determination. I realized I was never going to catch him this way. Time was running out, and I knew I would need help.

"You know we can't continue this," I said breathlessly. "I'm going to have to call the goon squad."

"No. Don't do that. I hate those guys!" he snapped. "Just try to catch me a little longer."

"I won't keep doing this. I'm setting a limit. If I haven't caught you in the next two minutes, I'm calling the goon squad."

He gave no response, but appeared to redouble his efforts to elude me. Except for the sound of grunts and our heavy breathing, we continued our frantic chase in silence. "Thirty seconds left," I said, glancing at my watch as I got up from a split-second lunge I thought would nab him. He said nothing, but just kept leaping and diving around the table, eluding my grasp by inches each time. "Ten more seconds." He slid under the table as time ran out. I anticipated this move and met him under the table, grabbing his leg and wrapping my other arm around his waist tightly. It was clear he had let me catch him at the last second.

We lay there for a moment, our chests heaving, wrapped tightly together. "Will you come with me to the unit now without a struggle?"

"I promise."

We stood up together. With one hand I held his upper arm tightly and with the other I kept a firm grip on his belt at the back. He couldn't get away and didn't try. We made our way out of the rec room, down two corridors to the elevator. Up the three floors it went. But as the doors opened just outside the unit, he suddenly grabbed the edge of the elevator door, refusing to come out of the elevator. All heads at the nurses' station turned toward us at the sound of the struggle. "You need help?" Mary's voice called out tensely.

"No. Not right now," I hissed, holding him tightly and sensing this needed to be finished between us. The nurses watched anxiously as I slowly wrestled him out of the elevator and, step by step, closer to the locked unit door. They opened it for us as we fell inside, landing on the floor of the day room. Safely inside the unit, I breathed a sigh of relief and tried to stand up. But Smitty wouldn't let me go. Desperately he clung to my leg, refusing to release me, dragging me back down.

As we struggled together in the middle of the day room, all eyes on us, a flood of words came, and with them his tears. "Why didn't you spend more time with me?" he wailed. "You only spent time with Jimmy, not me. I wanted you to be with me! You liked him more, I could tell. And Patty, too. But it should have been me." He was sobbing now. "I don't want you to go. I won't let you go. You have to stay." With these final words he gave expression to my own inner voice which had been ceaselessly bombarding me with the same thoughts silently all day. I did not want to leave. I cared so deeply for everyone here—for Patty and Jimmy, for Mike, Tom and Dr. Wilson, for Sue, Mary, and Bethany who had been so kind to me, and so very much for Paul. I didn't want to leave. And now Smitty, with the rawness and honesty his impulsiveness produced, gave voice to it all. But it had to come to an end. We remained in each other's tight grip on the floor as I told him how I, too, felt, how much I truly did not want to leave. I apologized for not having recognized how much he wanted me to spend time with him. But I also told him that my time at the Institute was coming to an end and that nothing he or I could do would change that. He asked me if I'd come back to visit, but with honesty I told him I didn't know and couldn't promise to visit. This was Friday, and I was beginning my gynecology rotation at another hospital on Monday on the other side of the city, not knowing what demands would be placed on me there. He let out another wail. Finally, I told him that if he didn't release me I would have to call the goon squad. Again he refused, but I counted down from 10, telling him when I got to zero I'd call the goon squad. Just as I reached zero, he released me and ran sobbing to his room. Bobby followed to comfort him and make sure he did not do anything to hurt himself or anyone else.

Slowly, I stood up from the floor. Shaking with my own tears the struggle and Smitty's raw words had released, I turned to the door. With a wave to Sue and Mary, and a silent acknowledgement to Patty and Jimmy, I left. In spite of my intentions otherwise, I never returned to the Institute.

Chapter III

Gynecology

The Woman I Knew So Intimately but Would Never Meet

"Gentlemen, and, uh, uh, ladies," our lecturer began, "welcome to Gynecology 200." As I looked around the darkened lecture hall, I could already sense how different this would be from my psychiatry rotation. Thirty of us arrived together, all wearing those short, rumpled white coats that identified us as medical students. The residents and attending physicians[1] wore full-length white coats which, particularly for the attendings, always appeared starched and fresh. You could identify the more senior attending physicians because they had their names monogrammed into their white coats. Instead of ones and twos participating in small conferences and discussions, as in psychiatry, now we were all to meet in this lecture hall three days a week as they poured the core content of the field of gynecology into us. Today, we began with the basics.

"Let's begin with a review of the anatomy," he said, as a huge picture of a woman's vulva, spread eagle with her legs up in stirrups, filled the 20-foot-high screen. His review of external and internal female anatomy was thorough and helpful, since I'd completed anatomy almost a year earlier and had forgotten most of it already. However, in that darkened room the conditioned response I'd developed during the first year of basic science lecture courses took over, and my eyes became heavy as I began to drift off.

"It's one thing to examine this anatomy in a textbook," he continued, "but it's quite another to examine it in a patient. Today you will learn how to conduct an examination of the female genitalia." I was suddenly sitting bolt upright. So this is how it was to begin, I thought. My heart quickened.

"This is a speculum," the white-haired gentleman said, holding up an awkward looking duck-billed contraption. "It's a gynecologist's best friend. Learn how to use it well, and you'll be doing right by your patients. Don't learn how to use it well, and a lady who comes to see you once will never return. I will now teach you the art of the seductive pelvic exam." He didn't just use the word "seductive," did he? "Your lady must place her feet in the stirrups and slide her vulva down to the edge of the examining table, like this," he said, showing us a picture of a woman in the appropriate position.

---

[1] Every patient admitted to a hospital must have one physician in charge of the patient's care. This person is referred to as the attending physician. Residents and medical students assessed and treated patients, but they did so under the supervision of the patient's attending physician who was ultimately responsible for the patient's medical care.

"This is what you'll see." A 12-foot-tall vagina and labia filled the screen.
"You seat yourself at the end of the table between her legs." It looked
remarkably awkward and embarrassing for the woman, but there was no
doubt the examiner had a bird's eye view of everything they were supposed
to be examining. "It's helpful to keep up a running explanation to the patient
of what you are doing, or more precisely, what you're about to do. Telling
her what to expect goes a long way to helping her relax. Now here's the
seductive part." I couldn't wait to hear this. "If you immediately place the
speculum in her vagina, the patient's bulbcavernosus and pelvic floor
muscles are very likely to go into spasm, sealing the vagina and making the
rest of the exam difficult for you and painful for her. To avoid this, start by
holding the speculum in your right hand, like this, but then gently touching
her inner thigh with the outside of your right hand, at least halfway down to
her knee. Then move up a few inches and touch her again. She'll now know
you're coming." An unfortunate choice of words, I thought. "Then touch her
a final time just a few inches away from the vaginal introitus, the opening of
the vagina. Resting your hand that holds the speculum there, with the other
hand place two fingers in the posterior portion of the introitus and slightly
spread the labia minora, the inner lips. If you've done this right, with the
appropriate seductive technique, she should be able to relax and allow you
to enter." Did he not realize what he was saying? Apparently not. "Then
insert the tip of the speculum into the vagina aiming down at a 30-degree
angle. Insert it all the way in before opening it. Once fully inserted, spread
the blades of the speculum." Again a poor choice of words as the image of
blades opening within the vagina momentarily made me squirm in my seat.
"The cervix, the bottom part of her uterus, will literally pop into view." A
picture of a woman's cervix as seen through the speculum now filled the
screen.

   He went on to describe how to obtain a Pap smear specimen and how to
collect samples of vaginal fluid to examine under the microscope for any
possible vaginal infection. He ended with demonstrating how to test for
gonorrhea and chlamydia, two common venereal diseases. "Once you've
done this, carefully remove the speculum. I said 'carefully' because if you
let the blades close prematurely, you'll pinch skin from the vaginal opening
between them, and you will never see this woman again. So be sure to
maintain gentle pressure on the lever that controls the blade opening as you
remove the speculum. Too much pressure, however, and you'll be stretching
the vagina excessively, pressing on the tender urethra anteriorly, and this,
too, will be uncomfortable. Remember, too little pressure and you'll pinch
her skin, too much pressure and you'll be stretching her uncomfortably."
How the hell do you know how much pressure is too much and how much is
too little, I wondered almost out loud? I pictured myself awkwardly trying to

remove a speculum from a woman and could vividly see the blades close on her, hear her scream with pain, and watch her recoil from me. I shuddered.

"One more thing," he added. "After removing the speculum from your patient, you perform a 'whiff test.'" This didn't mean what I thought it meant. "Have your assistant place a drop of potassium hydroxide 10% solution directly on the tip of the speculum immediately after you've removed it. Then take a whiff, meaning smell the speculum for a fishy odor. If you smell that fishy odor, then she is likely to have bacterial vaginitis." I could just picture the expression on her face as a woman watched me remove the speculum and literally put it up to my nose and smell it. What would she think?

"Next you perform the bimanual exam." What was this all about? "Step up between your patient's legs and place your second and third fingers into her vagina, again angled 30 degrees posteriorly, aiming for her cervix. Place your other hand on her lower abdomen, about halfway between her mons and her umbilicus."[2] This looked awfully intimate. "Then, use the fingers in her vagina to identify and lift the cervix up toward your other hand. With the hand on her abdomen, feel for the contour of her uterus. It should feel like a firm, inverted pear. You are literally trying to hold her uterus between your two examining hands, one inside her vagina and one on her abdomen." What a bizarre thought. "Next feel for her ovaries on either side of the uterus. Do this by compressing your two examining hands just above where you expect the ovaries to be. Then slide your hands posteriorly, and you will feel her ovary slip between your fingers. Be careful, though, because the ovary can be as sensitive to compression as a testicle." I pictured someone squeezing my testicle between their two hands and squirmed again.

"The final step is the bimanual vaginal-rectal exam." The what? "Remove your fingers from the vagina and reinsert one finger in the vagina and one finger in the rectum." How weird. "You may need to reapply some more KY Jelly for this step." I bet she'd appreciate that. "Now palpate the wall between the vagina and rectum as well as the cul de sac area.[3] Then repeat the bimanual exam steps, feeling for the posterior aspect of the uterus and ovaries in particular." So having performed more than 10 rectal exams during my psychiatry rotation, I was finally going to be taught how to do a rectal exam. Of course, during psychiatry I'd only done these on the men I'd examined, not the women. This would definitely be different.

"Any questions?"

---

[2] Umbilicus is the medical term for the belly button, related to its function as the point of connection for the umbilical cord.
[3] The cul de sac is the area above the innermost portion of the vagina and behind the uterus and ovaries.

"My girlfriend said she hates it when the speculum is cold," one of the male students commented. "Is there some way to warm up the speculum before putting it in?"[4] Many of the female students cheered.

"I'm aware of this issue. But I must tell you, in my experience, this is an overrated concern. Most women get used to this procedure fairly quickly, so I wouldn't worry about it." Some of the female students hissed, but it didn't seem to bother the lecturer. "Any other questions?"

"This looks a little more involved than learning to use a tongue depressor," said one student to muffled laughter, "but we got to practice that before we had to do it on real patients. How do we practice this?" I listened carefully.

"We've thought of that. So we've arranged for you to practice performing pelvic exams this morning on 'surrogate patients.' These are undergraduate women who have 'volunteered' to allow you to practice on them. We've broken you up into groups of three. The groups are listed on the back wall of the lecture hall, along with the locations you'll be meeting in. Please find your group and we'll begin the practice sessions in 20 minutes. Good luck."

*   *   *

The three of us huddled together, waiting for Dr. Hershowitz to return. "We'll be ready for you in just a minute," he had said, his white hair and starched, white, monogrammed coat carrying an air of competence and experience I so clearly yearned for at that moment. The other two male students were in the year ahead of me, so they chattered away with a familiarity I didn't share. Their jokes and nervous laughter reminded me of the gross anatomy lab. Left with my own thoughts, over and over again I pictured what I was about to do. I would walk in, calmly sit down, instruct the "patient" to slide down to the end of the table, take the speculum in one hand and subtly, unobtrusively, touch her inner thigh progressively closer to the vaginal opening, smoothly place the speculum, open it for a clear view of the cervix, obtain the specimens, remove the speculum carefully, and perform the bimanual examination. My thoughts were interrupted by Dr. Hershowitz's return. "Alright, who's first?" he asked.

There was a long pause as we looked at each other. Should I volunteer? Was I ready? One of the other students stood up, and I breathed a sigh of relief. Not yet, I thought. He disappeared into the examining room with Dr.

---

[4] When these events took place in the late 1970's, virtually all speculums were made of metal. Today, most speculums used for exams are made of plastic which doesn't feel as cold to the touch, and this significantly reduces, but doesn't eliminate, the long standing problem of cold speculums.

Hershowitz. I returned to vividly imagining every step of the pelvic exam procedure. I noticed my foot anxiously tapping up and down as I did. Sometime later, the first student emerged, a smile on his face. Was that a smile of relief? A remnant of his nervous laughter?

"How was it?" his friend asked.

"Piece of cake," came the reply. So his smile was one of success and satisfaction I thought. Good. If he can do it, so can I. The first student left as the second student went in. I was alone now. I tried again to visualize the steps of the exam procedure, but instead my mind drifted to thoughts about the "patient." What would make a college coed volunteer to undergo repeated pelvic exams by clumsy medical students? I figured they must pay these "surrogate patients" pretty well, at least by student standards. Why else would they do it if it weren't for the money? My thoughts were interrupted as the second student emerged. He wasn't smiling. He just headed for the door, like he couldn't get there fast enough. Just as he left he called out over his shoulder, "Good luck." What did he mean "good luck"? Why wasn't he smiling? What had happened in there?

"Come on in." I looked up into Dr. Hershowitz's face. It was kind, wizened, with a soft smile that made the wrinkles at the corners of his eyes turn up. I stood up and walked stiffly into the exam room.

As I entered the room, I glanced quickly at the "patient." She was beautiful! I hadn't expected this. Even lying on the exam table, her long blond hair fell flowingly over her shoulders. Her features were soft and attractive. Her breasts were full and prominent, even lying supine. My heart began to race. "Michele, this is Dr. Sheff, the last of the medical students for today. Dr. Sheff, this is Michele." Did he introduce me as doctor? But I'm not a doctor yet.

"Hi, Dr. Sheff," she said sweetly.

"Hi, Michele." Did she hear the quiver in my voice? I looked briefly into her face. It was so open and friendly. She lay on the exam table, naked from the waist down, I knew. Her bare legs extended out from under the sheet covering her lap, a hint of the deeper beauty that lay beneath. I felt an unwanted stirring. Oh God, please don't let me have an erection now.

"Don't you think you should sit down," I heard Dr. Hershowitz say. I tried to remember the steps to follow. Right, I thought, I should sit down. As I did, Michele raised her knees, placed her feet in the stirrups on either side of my ears, and slid down just to the end of the table. A musty smell of fresh vaginal secretions arose as I stared directly at her vulva, a small tuft of blond hair rising on her mons. Her vaginal opening was still wet from the KY Jelly of previous examiners, the wetness making her look all the more arousing.

"Here is the speculum," I heard a strange voice say. I hadn't realized a nurse was in the room with us. She was clearly impatient. I started to take the oddly shaped instrument. "Don't you think you should put your gloves on first," she said firmly.

"Thank you," I managed to get out. I picked up the gloves, my hands trembling slightly. Then I took the speculum in my right hand. What to do now? Oh yes, use the fingers of my left hand to open her vagina. I reached forward. As I touched the lips, her vagina spasmed shut for a moment as her buttocks instinctively rose off the table.

"You probably should have told her you were about to begin," Dr. Hershowitz prompted. Oh yeah, I'm supposed to tell her what I'm going to do *before* I do it. And her thigh. I forgot to touch her thigh first. OK, let's try this again.

"I'm going to touch the inside of your thigh first. Then you'll feel my fingers at the opening of your vagina. Then you'll feel the speculum going in." That was better, my full focus finally coming to bear on what I was doing. The fingers of my left hand spread her vaginal opening slightly, and I placed the tip of the speculum into her vagina. It slid in easily. Now just press open the speculum with my thumb on this lever. "Oh shit!" I said half out loud as my thumb slipped off the lever, causing the blades to collapse and pinch her vaginal skin.

"Ouch!" she said as her buttocks jumped off the table again.

"I'm so very sorry, Michele," I said. "Do you mind if I try it again?"

"No. I know you need to practice," she said, after taking a deep breath.

"Thank you for being so understanding," I said. "Alright. I'm opening the speculum again. You should feel a mild stretching inside." I looked in. No cervix.

"You probably needed to place it more posteriorly," Dr. Hershowitz suggested. "Michele, I'm going to ask Dr. Sheff to remove the speculum and insert it again."

"Sure," came the remarkably calm reply.

This time Dr. Hershowitz put his hand over mine on the speculum and guided it in, pointing much more back and down than I had done initially. I opened the speculum, and her cervix popped into view. I started to take my hand away, but suddenly the speculum began to close, and I certainly didn't want that to happen again. "What do I do to keep the speculum open?" I asked Dr. Hershowitz.

"Use your forefinger to tighten this screw here," he said, pointing to a round-headed prominence on the speculum. I tightened it awkwardly, but the speculum stayed open. I took the necessary samples. This time, I made very sure to keep gentle pressure on the speculum as it came out. "Excellent," I heard Dr. Hershowitz remark, the first positive comment I'd heard from him since we had started.

I stood up and suddenly saw Michele's face again. Instantly I felt the stirring one more time. I had been too lost in the mechanics of what I was trying to learn to even notice that I had been probing and manipulating the innermost recesses of this beautiful woman's sex organs. The nurse placed some KY Jelly on the tips of the gloved forefinger and middle finger of my right hand. As I rubbed the jelly over my fingers I could feel its lubrication spreading, my mind associating this sensation with intimate memories I'd rather not be recalling at this moment. My fingers then entered her vagina smoothly and easily, reaching in for…for…what was I supposed to feel? Was that it? I looked up at Dr. Hershowitz with a quizzical look.

"You are trying to feel for her cervix which has the consistency of the tip of your nose," he replied to my silent question. "The rest of her vaginal wall has the consistency of the inner portion of your lower lip." Did he have to add this graphic detail? But the difference was helpful as my fingers recognized the firm round nub of her cervix.

Next I placed my left hand on her abdomen just below her belly button. Suddenly, I felt her hand on mine. For a moment the feeling was electrifying, but then she firmly lowered my left hand to just above her pubic hair. "There. That's where you should feel it," she quietly prompted. Feel what? I tried to picture her uterus, tubes, and ovaries, where each was located, and to feel deliberately in each region, but it all felt the same.

"Michele is thin enough that you should be able to feel the contour of her uterus clearly," Dr. Hershowitz commented, adding to the pressure I was already feeling. Silently, she placed her hand over mine again and pushed in much harder and deeper than I had felt comfortable doing. She was showing me how hard I could press without hurting her. There, was that it, I thought. It was terribly subtle, but I had the distinct impression of firmness under my left hand that I could imagine was her uterus. I felt nothing in the area I was supposed to feel her ovaries.

"It would be unusual for you to feel her ovaries on your first exam," Dr. Hershowitz reassured me. "That will come with practice. Now for the vaginal-rectal exam."

I glanced momentarily at Michele's face, but she was looking away, perhaps trying to relax in the face of what was about to happen. The nurse placed more jelly on my fingers. I slid the forefinger in her vagina and pushed the middle finger just up to the opening of her anus. With a silent pop, my finger entered. I advanced both fingers, feeling the delicate wall between her vagina and rectum. "Now repeat the bimanual exam," Dr. Hershowitz prompted. Yes, that's it, I thought as my hands went through the same maneuvers. I felt again for her uterus and again felt only the vaguest sense of fullness. When I felt for her ovaries, I noticed her wince slightly, though I felt nothing but indistinct irregularities on either side of her uterus. I then withdrew my fingers with an unconscious sigh of relief. "That was fine for your first time," I vaguely heard Dr. Hershowitz say encouragingly.

My eyes met Michele's. "Mind if I ask you a question?" I heard myself say.

"Not at all," she replied, sitting up and facing me directly.

"Why are you doing this, being a surrogate patient and all?"

"Well, my friends told me about it 'cause they thought it was good money. While I could certainly use the money, I actually wanted to help you new doctors learn how to give a sensitive exam. I hated it when I got an exam from a doctor who was rough. So if I can help new doctors like you learn how to do it better, then that's really important."

"Thank you, Michele. You really did help me." I left the room smiling.

*   *   *

"So you're the sorry lot I've got to whip into shape this month." With this welcome Bertha Smith began her orientation to the operating room for me and my fellow students. Bertha was a short, wiry black woman in her mid fifties who had obviously been doing this for a long time. The deep lines around her eyes turned up as she addressed us, belying the tough tone she used as she spoke. "Each of you put on a pair of scrubs and meet me at the scrub sink over there. Come on now, move fast. I don't have all day. I run the whole gynecology operating suite, all five rooms, so I've got better things to do than babysit the lot of you." She huffed off quickly.

I stared at the closet that housed the scrub suits. It was a jumble of tops and bottoms of what looked like light blue pajamas. I looked for a medium or even a large, but it was already after 10 o'clock because we'd had another lecture this morning, and the only sizes that remained were extra large and small. In the locker room, I put on the extra large top, and its V-neck opening stretched almost to my shoulders. But the pants were comical. As I stood in them, they stretched out to more than twice the size of my waist. Pulling on the drawstring, I could barely get it tight enough to stay up over my boney hips. Sure that any minute my pants would fall down, I walked out to the scrub sink.

Standing around the scrub sink listening to Bertha, I muffled a laugh as we all looked like we'd come from a coed slumber party and were just getting up for breakfast in our rumpled pajamas. Bertha looked right at me. "You think this is pretty funny, don't you." Her dark eyes burned into me. "We're just getting started. The most important thing in the operating room is maintaining sterile technique at all times. Me and my nursing team will be watching each of you like a hawk for any break in sterile technique. Mess this up, and you'll be personally responsible for infecting one of my patients, and believe me, you don't want that to happen to you in *my* OR." Still staring straight at me, she bit off these last words.

Next she showed us how to put on hats designed to cover our hair and to keep it from falling into the "sterile field," the area where the surgeons actually worked on an open patient and the instruments were passed between nurses and surgeons. These "hats" looked just like the silly nets the lunch ladies wore back in my elementary school cafeteria. For those of us with beards, we had to put on a contraption that covered both our hair and our beards, leaving our faces bound like a mummy's. The booties we had to wear over our shoes turned up at the toes, making it look like we were all wearing elf shoes. And we finished it all off with face masks that left only our eyes showing. I looked around. Standing at the other scrub sinks were residents and fellows and attending physicians, all wearing the same oversized pajamas, the same silly hairnets, and the same pointed booties. How could surgeons take themselves so seriously, I thought, when they walked around in these get-ups all day?

"Now, to start with, you have to get sterile, and that begins with a good surgical scrub. Turn on the water, like this," she said, reaching for a large faucet handle with her hand. "But remember, once you start scrubbing, you can't touch that handle or anything else that isn't sterile with anything closer to your hand than your elbow." She went on to teach us how to open a prepackaged scrub brush embedded with an orange, germ-killing soap called Betadine and use it to scrub. This meant washing from our elbows to the tips of our fingers continuously for five minutes, then from the mid forearm to the fingertips for another five minutes, and finally from our wrists to the ends of our fingertips for the final five minutes. If at any time while scrubbing we accidentally touched anything, like the faucet, the sink, a fellow student, or the wall, we had to start all over again. When we finished, we were to hold our hands higher than our elbows so the water, and any accompanying germs, would roll down our arms and fall off at the elbows, the look made famous in countless movies and TV shows.

Once we completed our scrubbing, Bertha and a team of nurses showed us how to put on a surgical gown and gloves in a sterile fashion. "No matter what we do," Bertha had said, "hold your hands between your nipples and your waist. Once your hands move out of that zone, it's back to scrubbing for you." A nurse held up a sterile gown in front of me, and I put my hands through it carefully. She pulled it back over my shoulders, tied the waist belt behind me, and pulled down on the back of the gown to extend it fully. I put surgical gloves on just as we'd been taught. Watching us with a careful eye, Bertha corrected each of us the moment we deviated from the technique she demanded we follow exactly. Finally, she was satisfied. "Alright, you're ready to enter my OR. You've each been assigned a room for the day. Check the list outside the OR for your rooms. Now take everything off and start scrubbing from scratch."

As I finished scrubbing, hands held high, water dripping from my elbows, I entered operating room number 5. Bertha was waiting for me inside. "This is Dr. Stanley's room," she whispered softly. "He likes it real quiet in here, so mind that. He's already started the procedure, but you can watch from over there once you're gowned and gloved." She presented me the gown in just the way she'd taught us. I placed my hands through the sleeves, keeping them held at waist height as she pulled the gown back over my shoulders. I could feel her tie the belt at the back. Suddenly, I felt her grab hold of something low behind me and yank down hard. I had the distinct sensation of those big baggy pants coming down, and let out a loud "Yeow!" Dr. Stanley and everyone else in the OR stopped and stared at me. For a moment, all was silence as I blushed instantly. Then Bertha started to laugh. Her deep, throaty laugh filled the operating room.

"You thought I was pullin' down your pants, didn't you." She was laughing harder. "Well, boy, your pants are still standing, I can assure you of that. What would I want with you with your pants down?" Everyone started laughing, including me. Bertha loved that I could laugh at myself. "You know what? You're alright, boy. You're alright."

         *   *   *

Dr. Stanley finished his case just then, so I didn't get to see much. Excited and a little nervous to finally be in an operating room, I hoped the next case would be different. I wasn't disappointed because it was to be done by "my team." Composed of a chief resident, a second-year resident, and an intern, this is the team I was to work with when not in lectures, going on work rounds,[5] working up new admissions, scrubbing on their cases in the OR, and joining them when they had shifts in the gynecology clinic. On rounds this morning with the team, I had already heard about the case they were about to do, a 48-year-old woman who had had two children, but now had an abnormal Pap smear[6] that suggested early cancer. She was

---

[5] As I had learned in psychiatry, rounds is a generic term used to describe any event when a group of doctors gathers together for a purpose. Work rounds, usually just referred to as rounds, was the process of our team stopping outside of the room for each patient, hearing either a presentation of the new patient or an update on an ongoing patient, going into the room as a team and speaking with and examining the patient, and making an action plan for that patient's care for the rest of the day. This was when we carried out much of the important work of the day (hence the name "work rounds"), deciding the plan of care for the patient for the rest of the day. We did this every morning from 6:30 AM to7:30 AM before starting in the OR.

[6] A Pap smear is part of a gynecological exam in which the examiner takes a small sample of cells from the vagina and cervix and places them on a slide for microscopic examination. The test, developed by Dr. Papanicolaou (hence the

undergoing a total hysterectomy, meaning they were going to take out her ovaries and tubes as well her uterus, for reasons I didn't fully understand. I found my team at the scrub sink, already up to their elbows in orange lather, scrubbing for this case.

"Come on, Rick, get scrubbing," Martin, the chief resident, said as I approached. "You know we can't do this case without you," he joked with a smile. I liked Martin, all five feet six inches of him. The other members of the team were civil to me, but as a medical student, I came to understand I represented more work for them, which was not welcome when they were already overworked and sleep deprived. Martin seemed to take his responsibilities to teach me, as well as the other team members, quite seriously. "Did you get to do an EUA on Dr. Stanley's case?" he asked.

"What's an EUA?"

"It stands for 'evaluation under anesthesia'."[7]

"I guess I still don't know what that means," I said a little sheepishly.

"Well, you already told me you didn't think you'd be going into OB/Gyn[8] as a specialty. So whatever else you go into, the most valuable thing you can learn during this rotation is how to do a good pelvic exam."

"Don't remind me," I said. "My first one isn't something I'd care to relive."

"Yeah, mine wasn't much better. Was the speculum a challenge for you too?" I nodded. "But the really hard part is figuring out what you're feeling on the bimanual exam, and the EUA is the best way to learn this."

"Why is that?"

"When they put these women to sleep, and especially once they're paralyzed with succinylcholine, everything relaxes, including their abdominal muscles. So if you do a bimanual exam on a woman once she is asleep and in stirrups, you can really feel everything much more clearly."

"And she won't be looking at me to make me nervous?"

---

abbreviation Pap), is an effective screening procedure to identify early cellular changes likely to progress to cancer of the cervix long before any symptoms develop.

[7] The years of my training coincided with early use of ultrasound and CT scans as new imaging modalities. Because these weren't readily available, nor were the interpretation of these studies standardized, evaluation under anesthesia (EUA) was an important tool in diagnosing pathology in patients, especially in the Fallopian tubes and ovaries which were difficult to clearly feel in awake women. Today, ultrasound and pelvic CT scans are routinely performed and exquisitely accurate in identifying pathology in the pelvic organs, making EUAs obsolete for diagnostic purposes.

[8] OB/Gyn is a standard abbreviation for the field of obstetrics and gynecology.

"No, she won't," he smiled, "but you get used to that. Besides, once you've done the exam, if it's an open case, you get to see the anatomy and see exactly what you were feeling. It's the best way to learn a really good bimanual exam."

I hesitated to tell him psychiatry was still my first choice for a specialty, in case he might not continue to be so helpful. Also, I was beginning to find the thought of becoming competent at performing the bimanual exam intriguing. "How do I get to do these EUA's?"

"Every morning check the schedule for which cases involve EUA's. Then, just as the case is about to begin, find a way to be there and ask the attending surgeon if you can do an EUA as well. Some will say yes, and some won't, but you won't know until you ask."

"Thanks. I'll try that."

"Hey," Martin said suddenly, "let's do one now. Let's break scrub and go do an EUA on our next case. I'll show you how it's done." Uncertain of what would happen next, I rinsed and dried off, tagging along behind Martin. We went back into room 5, where the patient had just been put to sleep. "Before you do the vaginal scrub, I'm gonna show our new student, Rick, here, how to do an EUA."

"I don't think EUA was on her informed consent form," the anesthesiologist interrupted.

"I know, but she's already asleep. She'll never know. Besides, this is for teaching purposes, and she did sign permission for us to involve students and residents in her care for teaching purposes." I felt a little uncomfortable with what all that meant, but Martin seemed so sure, and I was getting more and more interested in what I could learn from doing EUAs. The patient, a woman I had never met, was asleep under anesthesia, her legs up in stirrups, her thin abdomen completely relaxed. The anesthesiologist was busy at the head of the table, and scrub nurses scurried around the room getting instruments ready. Martin stepped between her legs, putting on a pair of gloves. "Alright. I'll do the exam first so I can tell you what you should be feeling." He did a quick bimanual exam. "Your turn. Tell me what you feel."

I too put on gloves and stepped between her legs. With my left hand on her abdomen I placed two fingers of my right hand into her vagina, moving slowly down and backward until, sure enough, I could feel her cervix. I then lifted her cervix up and pushed down with my left hand, just as Michele had instructed me to do. Was that it? Yes, I could actually feel something firm under my left hand.

"Describe what you feel," Martin said, encouragingly.

"I definitely feel something under my left hand."

"Go beyond 'something'. Describe it."

"I feel something firm, kind of rounded."

"Like the round part of a pear," he prompted.

"Yeah, kind of like that." I felt a certain thrill to be actually feeling this woman's uterus.

"What do you feel at the top right most part?"

"I'm not sure, but it seems a little more full than the other side."

"Exactly. She has a small fibroid in that part of her uterus. We'll see it when we do her hysterectomy. Now feel for the ovaries."

"I don't feel anything," I said, a little disappointed.

"Push the fingers of both hands together harder. Remember, she's asleep and you can't hurt her."

"Hey. I think I felt something." A vague, rubbery fullness passed between the tips of my left and right hands.

"Now, try to feel how big it is."

"You've got to be kidding."

"No. Try."

I replaced my fingers above the area I'd felt that rubbery thing and slid them down and towards each other again. I did feel something rubbery there. "I don't know. It feels about four or five centimeters long." I was partly guessing, but somehow did have a sense that what I felt was approximately that long.

"Exactly. Now try the other side. This one will be a little harder because her left ovary is tucked a little behind the top of her uterus."

I couldn't feel anything on the left side. Martin said that was good enough for now. He hurried out and resumed scrubbing. I walked more slowly, something I couldn't name rising up within me. A mystery had somehow been revealed in that moment, and with it came a sense of discovery, even excitement.

We finished scrubbing and reentered room 5. As I entered I could hear Bertha say to the other nurses in a loud whisper, "That's him. That's the one. You should have seen his face when he thought I had his pants down around his ankles," and they all giggled with her. Bertha came up to me with a broad smile. "Now don't you fret any, son. I'm not planning to do anything with those pants of yours." And with that she laughed out loud as she helped me on with my gown and gloves. "Now go on and find your place over there," she said helpfully, pointing to the operating table around which Martin and other members of the team had already assembled.

I approached the table and took my place to the left of Elaine, the second-year resident. Martin and Suzie, the intern, were on the other side. "Scalpel," Martin barked. Suddenly, the handle of a scalpel landed with a smack in his open palm, expertly placed there by a scrub nurse I hadn't noticed until then. Past the scrub nurse, I could see a tray with a dizzying array of instruments. She knew just where to find everything Martin asked for, even before he asked for it. "Are we ready?" Martin asked the anesthesiologist. With his nod, Martin brought the scalpel down. The blade effortlessly parted the skin almost from one hip to the other, a Pfannenstiel or "bikini cut" incision, they explained. Instantly blood oozed from the glistening tissue, filling the wound. With surprising speed and skill, Elaine used a white gauze to dab the fatty tissue, identify each specific site of bleeding, and apply one clamp after another to control the bleeding. Hemostats, I heard them called. In moments, the incision was dry, but it was filled with almost 20 hemostats. One by one, Elaine lifted them in Martin's direction. He applied a "Bovie," as it was called, which gave a small electric current into the hemostat, and I watched as the fatty tissue to which it was attached sizzled, charred, and smoked for a moment. As the smoke arose, it smelled just like…like a barbecue—the smell of fat burning. It was sweet and pleasant, until I realized it was human fat, this woman's fat, that was burning. My stomach turned for a moment. One by one Elaine removed each hemostat, checking to make sure none of the sites bled further.

Some of the vessels that had bled were too large to be "barbecued," so they had to be tied, Martin explained. He showed me how he tied a perfect square knot with the absorbable suture. As soon as it was tied, Suzie cut the ends of the suture off, leaving a neatly tied knot with its ends consistently a quarter of a centimeter in length. Martin asked if I'd like to cut some of the sutures, and I instantly said yes. He showed me how to hold the scissors, how to guide the scissors down the ends of the suture to the knot, how to turn the scissors to see the knot before cutting, and how to cut the ends so they'd be the right length. "Now you try it," he prompted. The next suture he tied I held out the scissors, guided them down to the knot, and cut the suture. "No. No. That's too short. That knot's going to untie before enough scaring, so it could rebleed after we've closed her up. Now I've got to tie it again." He placed another suture. "Try it again." This time my hand shook a little as I tried to cut the suture to the right length. "No. This time you cut it too long. Those ends can stick into her tissue and cause pain in the healing process." With that, he trimmed the edges a little shorter. I had just learned how much I, and every medical student, desperately wants to hear the words "That's just right" in the operating room. I would find this to be a rare occurrence.

Martin then used the scalpel to delicately open the next layer. "What's that?" he asked me.

"Isn't that the fascial sheath of the rectus muscle?" I heard myself respond. Where did that answer come from?

"Not bad," came Martin's reply. The muscles of the abdominal wall came into full view. They were a deep, dark red. "And that?" he asked, pointing to a line of white tissue running up and down in the middle of her abdominal muscles.

"Isn't that the linea alba?" I replied.

"Exactly. And that is what we will now open to enter the abdominal cavity." With that, he made a small hole in the linea alba and literally pulled the abdominal muscles apart. A thin, translucent layer of tissue covered everything below. "I'm now entering the parietal peritoneum," Martin said to the entire team. Lifting the delicate tissue with a forceps, he cut a small hole in the translucent tissue and extended the opening the full length of the incision.

I stared into her abdomen. The small intestines glistened a yellowish pink as they contracted and relaxed, appearing to subtly slither around within her belly with peristaltic action. The liver was deep maroon with a smooth, glistening edge, and the gall bladder peaked out below it, a rich, dark, blue-green. The colors surprised me with their vibrance, everything so smooth and perfectly formed. To my eyes, it was beautiful. This human body, in its aliveness, was breathtaking—so different from the dull, gray, preserved organs I had dissected in my cadaver.

But instantly Martin and Elaine were "packing" the abdomen, placing large white gauze pads strategically to push the small intestines out of the way. Didn't they see it? Didn't they see how beautiful it all was? No, they didn't. To them, everything we saw was something to be moved out of the way so they could "visualize" this woman's pelvic organs, their real goal. Martin handed me a large, wide metal instrument that extended downward into the incision. I could not see where it stopped, but it took considerable strength to keep it in place. "Hold onto this, and, whatever you do, don't let go. This retractor's holding the small intestines up against the liver and out of the way."

"How long should I hold it for?"

"We should be done in about an hour and a half."

Oh. So I was to stand here, holding tightly to the handle of this retractor, for the next hour and a half. What if I got tired? What if it slipped? These thoughts I kept to myself, for, just at that moment, the uterus and ovaries, the very ones I'd just examined, came into view. Her uterus was a deep pink, smooth, carefully nestled deep in her pelvis. The tubes arose gracefully from either side with extraordinarily delicate, lacey fimbria, or fingers, draped over the ovaries where they ended. The ovaries were even a darker pink, almost brown, with an irregular surface. The left ovary sat tucked behind the uterus, just as Martin had predicted. And there, on the upper right corner of her uterus, I could see a bulge, approximately four centimeters across—the fibroid I'd felt just minutes before. My heart raced. A drama was unfolding before my eyes. But nobody else appeared to notice. Martin and the scrub nurse carried on a playful banter that bordered on the flirtatious. Elaine and even Suzie appeared to work with a dry precision.

Didn't they feel it? Didn't they appreciate entering the sanctity of another's body? As I looked around that operating room, I felt very alone.

"Tell me about the blood supply to the uterus," Martin's voice intruded.

Think, Rick. Think. You used to know this. "I believe there is an artery that comes from the top and one that comes from the bottom."

"You've just proved my theory right," he bellowed. "This business of spending only one year in the basic sciences turns out students who don't know squat about anatomy or any of the other important subjects like pathophysiology or pharmacology." Then he softened slightly. "Of course I'll work with you just the same, Rick. But I feel for you, not having enough of the basic sciences under your belt and being thrown into clinical rotations like this. It's nothing but stress on you and extra work for us residents." I had to agree that tackling the clinical rotations would have been easier with more basic science preparation. But I knew in that moment I'd take this stress any day for the privilege of standing here in this operating room, gazing into the pulsating abdomen of a live woman, rather than sitting in a deadly classroom lecture for another minute or spending one more anatomy session with a cadaver. To graduate from this medical school, I would eventually have to go back for four more months of electives in the basic sciences. That was the design of the school's controversial curriculum. Its architects recognized that no one can ever fully prepare to assume the care of patients through classroom work in the basic sciences. And students can't retain the basic science information they will need to practice until they have cared for patients and have a clinical context for such voluminous information. Medical knowledge was of a whole. A medical student had to start somewhere, but any starting point would have its challenges. This required plunging in somewhere and slowly filling in the gaps of what you didn't know over time. Only at the completion of training would a student of medicine be able to grasp the whole. As a result, medical students and residents lived in this state of incompleteness, of dissonance, which fueled their anxieties and made them easy fodder for older residents and attending physicians seeking to humiliate them.

The architects of my medical school's curriculum understood this, and knew that it would be better to err on the side of more and earlier clinical contact with patients and less basic sciences classroom time. Already I knew they were correct. Yet every basic science professor believed otherwise. Each one thought his or her subject was the most important, contained the essential nuggets that would keep every student of medicine from killing a patient, and deserved more time in the medical school curriculum. I wasn't sure another pathology lecture on tubular necrosis of the kidney would keep me from killing a patient. But thanks to constant reminders from basic science professors, the thought kept turning over in my mind. Someday I would be in a position to potentially kill a patient with my personal error. I pushed the thought aside, finding it too overwhelming to contemplate. But as I did, I uttered a silent prayer that I would never make such an error. I'd never be able to live with myself if I did.

Again Martin's voice intruded. "You've got the general idea, but this here is the ovarian artery," he said, pointing to a slim red structure, "the 'one from above,' as you called it, and this is the inferior uterine artery, your 'one from below'." A mocking tone was in his voice. "Go home and study this anatomy tonight and I'll quiz you on it tomorrow. You need to know it because in order to take out the uterus and ovaries, we need to isolate and tie off each of these arteries on both sides, or this poor woman would bleed to death. You wouldn't want to be responsible for that?" Suzie and Elaine let out a snicker, as did the anesthesiologist.

This was clearly meant as a put down, an attempt to humiliate me for not knowing this anatomical "detail." For a moment, I felt blood rise in my cheeks. But then I remembered what I'd learned after returning from Christmas vacation last year. I had known the names of these two arteries, at one point known them cold, but I'd forgotten them. If I ever were to find myself operating on anybody's uterus, I sure would make a point of knowing about its blood supply. But I don't expect I'll be doing that in my future. It's clear how much Elaine and Suzie need to know this information. OB/Gyn[9] is their career. Whatever my career choice, I reassured myself, I'd make a point of knowing everything I'd need to practice it competently. At least I'd try with all my heart. Until then, I would have to accept what I didn't know and couldn't do, and not let anybody else humiliate me for it. And with that, the color subsided from my cheeks.

An hour later the blood had drained from my hands. They ached from holding that retractor. The uterine and ovarian arteries had been isolated and clamped, and Martin and Elaine were almost finished dissecting everything that needed to come out. There. It was done. They lifted the uterus, tubes, and ovaries into a steel bowl to be sent to the pathology lab for examination. After what seemed like an interminable time, Martin took the retractor from my hands. With relief, I shook them out. Realizing there was nothing more for me to do as they were closing up, I asked if I could break scrub. "Sure," Martin said hollowly, as he focused on placing the next suture.

I stepped back from the operating table and approached the steel bowl that had been set aside. "Do you mind if I look at the 'specimen'?" I asked the scrub nurse.

"Go ahead," she said, barely paying any attention to me as she worked quickly, exchanging instruments with Martin. "But don't mess it up. They like to get it in pathology just like we find it here."

---

[9] OB/Gyn is a common abbreviation for the specialty of obstetrics and gynecology.

With that I took the steel bowl to one side. I glanced momentarily at this sleeping woman, the one I'd just examined without her knowledge. Turning back to the bowl, I lifted her uterus in my hands. She had had children, I knew. So this was her womb. This was the site of the miracle of conception, gestation, and birth. This little, muscular pear had nurtured new life and eventually propelled it out into the world. What life-giving force was I holding in my hand? Finally, reluctantly, I placed it back in the bowl and started to leave. At the door to the operating room I paused. Turning, I bowed slightly to this woman I knew so intimately but would never meet, and then I left.

*   *   *

"Alright, Rick, you write the post op orders and I'll start scrubbing for the next case," Martin said.

"What post op orders?"

Martin paused, gazing at me with a sigh. "You've never written post op orders?" I shook my head. "What rotations have you done before this one?"

"Just psychiatry."

"Well, that explains it. Here's what you do to write post op orders," he said patiently. "Think of everything you have to do to take control of every aspect of the patient's life." What an overwhelming responsibility, I thought. "The way I was taught, and it's always worked for me, is to use 'ADC Vandissl'."

"AD what?"

"ADC Vandissl. It stands for admit, diagnosis, condition, vital signs, activity, nursing orders, diet, IV, specific meds, symptomatic meds, and labs. Get it? ADC Vandissl."

Quickly I wrote this down in my little black book. The book was a four-by-six-inch black loose leaf notebook I'd seen other medical students carry around. Their "ectopic brain,"[10] they'd called it. It's where they wrote down countless facts they would otherwise forget. I'd started doing this in psychiatry, and was amazed how many facts I could write down, learn, and then forget. It was definitely helpful to keep them with me in this book.

"Admit: unit or ward to which the patient was to be admitted," I scribbled, as Martin explained each of the items.

"Diagnosis: patient's admitting diagnosis and any other important diagnoses.

---

[10] Ectopic means out of place, as an ectopic pregnancy is in the Fallopian tube rather than inside the uterus. So an ectopic brain is a brain out of place, meaning in my back pocket rather than in my head.

"Condition: patient's condition, such as stable, unstable, critical, or terminal." Terminal? The thought struck me that someday I would be taking care of a dying patient. But Martin was rattling on.

"Vital Signs: how often should they be taken?

"Activity: what can the patient do, such as bed rest, out of bed to a chair, ambulate ad lib, etc.

"Nursing orders: special nursing activities such as dressing changes.

"Diet: how many calories and any restrictions such as low sodium or antidiabetic.

"IV: intravenous fluid order, including what should be hung in the IV bag and what rate it should drip in.

"Specific meds: medications for the patient's medical condition(s) such as antibiotics, insulin, oxygen, etc.

"Symptomatic meds: medications to keep the patient comfortable such as Tylenol, sleep medication, and stool softeners." Being responsible for the functioning of a patient's bowels in the hospital was an important activity, Martin emphasized. Failure to anticipate the patient's needs in this area was certain to lead to problems. "Besides," Martin noted, "have you ever tried to shit into a bedpan?" Neither of us had, but the image made it clear that anything that could be done to make this easier would be appreciated by the patient.

"Labs: blood tests, radiology studies, consultations, and any other diagnostic testing."

"Now go ahead and write the post op orders, and I'll check them," Martin concluded.

Haltingly, and with constant references to the notes in my ectopic brain, I completed her post op orders. I paused before calling Martin over to review them. It felt odd, in a way I couldn't name, to assume this much control over another person. Slowly the word formed in my mind. This was power.

*   *   *

The next day I peered at the OR schedule. Bertha noticed, as she noticed everything that happened in *her* OR suite, and approached me. "What you lookin' for?" she asked.

"I wanted to figure out which cases had EUA's so I could ask the attendings if I could examine their patients to better learn how to do bimanual exams."

"You actually want to learn that much?" She paused, looking me over. "Heck, since you were such a good sport yesterday, I'll see to it that you participate in as many EUA's as possible. You just follow my lead." With that she stepped right up to my side and surveyed the schedule with me.

"Now you remember Dr. Stanley from yesterday. I told you he likes it quiet. He's very particular about what he lets students do, but if you approach him just right, at the moment he's going in to do an EUA, I bet he'd be willing to let you do it with him. He's a good teacher too. Now Dr. Hershowitz, he's a real kind gentleman, so anyway you approach him should work out OK." On she went, describing the unique qualities and quirks of each attending gynecologist, and doing her best to give me the right cues so I could do as many EUA's as possible. "Here comes Dr. Stanley now for his 10:15 case. Remember what I said."

I watched Dr. Stanley carefully. He spoke briefly to a resident I gathered would be scrubbing in with him today, checked the schedule, and approached the door to room 3. Just before he got there, I came up to him. "Excuse me, Dr. Stanley, my name is Rick. I'm a med student on the gynecology rotation this month. I noticed you had an EUA followed by a D and C for this patient. I'm trying to learn to do a really good bimanual exam. Do you think I could participate with you in this EUA?"

Dr. Stanley looked me over as I held my breath. "Weren't you in my operating room yesterday?" he asked, scrutinizing me more closely.

"Yes, sir. Unfortunately I was the one who thought my pants had come down, so I made that loud noise. I'm terribly sorry, sir."

The corners of his mouth turned up. "Well, you gave us all a good laugh yesterday." He paused. "Sure, come on in on this EUA."

And so began my pursuit of participating in as many EUA's as possible. Bertha became my co-conspirator. She knew just the right moment for me to approach each attending, or when to just slip into the OR and, as the nurses were prepping the patient, ask the attending if I could perform an EUA before the procedure began. At first, each exam still had that sense of uncertainty, with me never really knowing exactly what I was feeling. But slowly, over the course of the month, I came to more consistently be able to feel the contour of the uterus. I also came to be able to feel the ovaries in more than half of the cases, and when I couldn't feel them, it was usually because the woman was overweight which made feeling anything more difficult.

This training in EUA's paid off in the gynecology clinic as well. This was where, once a week, I joined my team of residents seeing outpatients with gynecological complaints. Some were just in for annual Pap smears, but in this racially mixed, inner city clinic, more often we were seeing women with multiple medical problems who had no other physician. Virtually all were on Medicaid. In addition to their medical problems, most had staggering social problems ranging from having no money for transportation, to showing up for an exam with four or more children who had to fend for themselves while mom was up in stirrups, to drug and alcohol abuse, to being victims of physical abuse. Often they couldn't speak

English, so we had to work with translators, when they were available.
When they weren't, we did our best to make do with sign language for
symptoms, but trying to obtain any medical history was hopeless. The
residents described this as practicing veterinary medicine, since we had no
way of talking to the patients.

My first day in the clinic happened to come right after a lecture we'd had
on vaginal infections that morning. One of the patients, a very overweight
black woman named Mulanda, described symptoms of vaginal itching with
a watery, greenish vaginal discharge. "Any pain with intercourse," I heard
myself asking in the natural course of taking the history.

"Yeah, it burns like hell, but he don't care," came the response. I was
surprised at how easily she answered such a personal question, even though
I was just a medical student. I'd seen this before. During our first year, we'd
had a course in taking a medical history. Once they had taught us the basics,
each week we went into the hospital and talked to a patient to take their
medical history so we could practice. The second patient I ever did this with
was a man undergoing chemotherapy for prostate cancer. I got to the part of
the history called a review of systems. That's when we were supposed to ask
a long list of symptoms the patient might have, but which they hadn't told us
yet. The number of symptoms was so large and the terms so unfamiliar, I
had to read them from a written list until I could memorize all of them. The
symptoms were organized by organ system. I was going through the urinary
symptoms, reading them off my list, when suddenly I got to "impotence." I
paused, not sure if I should ask this. After all, I was only a medical student,
just four weeks into my first year. Why did I have a right to pry into such a
personal area? But our instructors had told us to ask about all the symptoms,
both for thoroughness and so we would become comfortable asking. I
plowed ahead.

"Any problem with impotence?" I managed to get out.

"Oh that. That's been gone a couple of years, ever since they took my
balls off," he answered matter-of-factly.

Now what? What was I to say to this response? He seemed so
comfortable with it, even though I wasn't. I just moved to the next symptom
on my list. "How about any back pain," I was asking next. That's when I
learned that patients will answer just about any question I asked, at least
when I was wearing a white coat bearing a name tag. I was the one who had
to become comfortable asking these very personal questions, and hearing the
answers, whatever they may be.

So, sitting with this unhappy woman in gynecology clinic, I was about to
move on with the rest of my history when I realized she had given me an
opening about herself. I took it. "He doesn't care?" I repeated. "You mean if
you're uncomfortable having sex, he doesn't care about that?"

"He could care less, long as he's gettin' it. My man, he don't care 'bout nothin' when it comes to me."

"That must make you pretty mad."

"Damn straight it does."

"I'm sorry he treats you that way."

"Ain't nothin' for you to be sorry 'bout. But, thanks for actin' like ya care," she said, meeting my eyes for a moment.

"I do care, and you don't deserve to be treated that way." She seemed genuinely appreciative for a moment of kindness in her hard life, and I was glad to offer it. "Now let's move onto the exam," I said.

She put her feet up in the stirrups and slid her bottom down to the end of the exam table. Suddenly I was assaulted with a foul, acrid smell of sweat, vaginal odor, and something else that made me want to pull away. Get used to it, I told myself. The nursing assistant handed me the speculum. "You'll feel my hand on your leg here." I touched her inner thigh just above her knee, then again closer to her vagina. "Now you'll feel a stretching of your vaginal opening." I placed two fingers of my left hand in the posterior portion of her vaginal introitus. "Next you'll feel the speculum as I put it in."

She flinched as the speculum touched and her vagina spasmed shut. "That's really cold," she said.

"Sorry. I'll warm it up first." I held the speculum in my right hand, warming it as best I could. At that moment I swore to always warm the speculum whenever I did a pelvic exam. I repositioned the speculum. It went in easily as I guided it back and down. The blades opened, and her cervix popped into view. I let out a sigh of relief. Her vagina was filled with a watery green discharge, just as she'd described it. Suddenly I realized this looked exactly like one of the slides we'd seen that morning. This is what trichomonas[11] looks like, I said to myself excitedly. I took the necessary samples to examine under the microscope. As I took out the speculum, the nursing assistant held a dropper with the potassium hydroxide solution in it over the speculum. I paused. How could I perform a "whiff test" without her seeing it? I nodded to the assistant who let a drop of the solution fall on the tip of the speculum. I turned my back to Mulanda and took a quick sniff. I don't think she noticed. It smelled pretty foul, but it didn't smell like fish. I assumed that made this a negative whiff test.

I took the slides of her vaginal discharge to the room at the back of the clinic where the residents met and the microscopes were kept. Martin was there, and he asked me to "present" the patient to him. This meant describing her history and the results of her physical exam, as well as my

---

[11] Trichomonas is a single cell organism that can infect the vagina, causing symptoms of discharge, burning, and pain. It is a common cause of vaginitis.

working diagnosis, just as I'd watched the residents do over and over again during my psychiatry rotation. Haltingly, awkwardly, I described her symptoms and exam. Martin kept correcting me on the order of things and how I should present them. Finally he said, "So what do you think she's got?"

"I think she may have trichomonas."

"You're probably right," he replied. A feeling of pride and satisfaction rose within me. "Now let's look at those slides under the microscope to confirm it."

I placed the first slide on the microscope. As I'd done countless times throughout my first year, I slowly turned the knob to bring the field unto focus. At first I couldn't make out what I was looking at. Then the irregular shapes of cells came into focus. Those must be vaginal lining cells, I thought. But something was moving. Single cells that looked quite different were making their way around the field. "What are those things that are moving around?" I asked Martin.

He looked in. "Those, my dear friend, are trichomonads." I'd made the right diagnosis! The feeling was electrifying. "She'll need metronidazole five hundred milligrams twice daily for a week. Of course, you'll have to treat her partner too."

"Her partner? Why?"

"We now believe trichomonas is a sexually transmitted disease. Either she or her partner has probably been having sex with someone else."

My heart sank. The elation I'd felt at making my first correct medical diagnosis was replaced by the anticipation of telling Mulanda her "man" had been unfaithful.

"That lying shit!" she let out. "I sure as hell know I ain' been doin' it with nobody else. I don't care if his thang shrivels up and falls off, he ain' gettin' no more of it from me." I explained that her partner needed to be treated for a full week, as did she, in order to make sure it didn't come back. "Oh he'll take it, alright," she assured me. If he continued to be unfaithful, I explained, it was likely to come back even with treatment. "You don't have to worry 'bout that. He be done gettin' it from me!"

*   *   *

The rest of the month went by quickly. Each day I rode to the hospital on that old, black three speed bicycle. On many days I could be found ducking in and out of operating rooms under Bertha's guidance, stealing chances to perform EUA's whenever possible. If a procedure I hadn't seen was scheduled for one of the operating rooms, I tried to observe at least some of it. This way I got to see a wide range of procedures from routine hysterectomies to microsurgery of the fallopian tubes for infertility. One day Martin called us all in to demonstrate what a full blown case of PID, pelvic inflammatory disease, looked like. This woman, he explained, had a large abscess behind her uterus that needed draining. He placed an operating speculum in her vagina. "Stand back," he told us, plunging a scalpel into the soft tissue just behind her cervix. A large stream of green, foul smelling pus poured out. This left a lasting impression on me far greater than the lecture we'd had about PID.

When not in the operating room, lecture hall, or clinic, I worked with my team on the gynecology inpatient unit, performing workups on new admissions and participating in rounds. When Suzie, the intern, was on call, I stayed at the hospital working with her until 9 or 10 at night. This is when I did many of the workups. I had the option of staying later, even all night, but it wasn't required. The time would come soon enough, I knew, when I'd be expected to stay in the hospital all night, often working through the whole night without sleep, and I had no desire to do this any sooner than was absolutely necessary. So I happily left at night when I could. This meant I didn't participate in much scut work. Scut work is the term for the busy work medical students and interns are expected to carry out, such as drawing blood samples, starting intravenous lines, and tracking down laboratory test results. My friends at other medical schools, and even some at my school who were the most eager, told me they stayed in the hospital at night during their clinical rotations, working through the night performing endless scut work.

Since I still assumed I would become a psychiatrist, I saw no reason to spend my time doing scut work. Time at night in the hospital would be spent at the expense of reading up on the conditions we learned about in lectures or saw in the OR. Besides, reading and studying would help me pass the final exam at the end of the course in just a few weeks, I rationalized. Even stealing this time out of the hospital, I was always behind in writing up the history and physical exams I had performed as these write ups were many pages long and often took hours to complete. There would be time enough for scut work ahead, I knew. But most importantly, I feared that I would not be able to handle the sleep deprivation that came with working through the night, catching an occasional nap, and then working all through the next day. My body still needed eight hours sleep a night, seven when I pushed it, just to function. Any less than that and I became physically ill. Deep inside I felt a fear and foreboding as I anticipated the days, weeks, and months of severe sleep deprivation looming ahead of me.

The last week of my gynecology rotation began. On rounds there was no discussion of "closure" from Martin or the rest of our team. The work continued as it had all month—rounds, the OR, workups, and the clinic. Mulanda came back into the clinic during that last week. Her symptoms had resolved with the medication we'd prescribed, but now they were back again. Her "man" swore he'd taken the same medication. He also swore he hadn't been with another woman again. Mulanda knew better. Her angry eyes and bitter voice told me all I needed to know. Though I tried to comfort her, Mulanda and I both knew nothing would change. Making the right diagnosis and knowing what to prescribe for her had given me a thrill. But her return visit taught me that it would take considerably more than diagnostic acumen and knowledge of pharmacology to really "treat" her effectively. Already I was realizing her relationship, her community, and ultimately her sense of self worth would have to change for her vaginal infection to resolve. Though I couldn't know it at the time, Mulanda was teaching me a powerful lesson about what the responsible practice of medicine would require if I was truly committed not just to treating patients, but to healing people.

On Friday I mentioned to Martin during rounds that it was my last day. He seemed surprised, but that was all. Suzie and Elaine said goodbye politely. Martin said I had done pretty well. I took, and passed, the final exam that morning. We then went to the OR as usual. At the end of the day, I looked for Martin to thank him.

"I appreciate everything you did for me this month," I said.

"Sure. Glad it helped," came the clipped reply.

"I guess I'll be going now."

"See ya around."

"Yeah. See ya around sometime."

With that, it was over. I walked out of the hospital. A subtle, lingering emptiness followed me home.

Chapter IV

Obstetrics

"It's wrong. It's all wrong."

Sitting in the student on call room just off the labor and delivery floor, I stared down at the two objects that were supposed to be helping me. In my left hand I held an actual female pelvic bone. In my right hand, a doll the size of a newborn baby.

Flexion…internal rotation…descent…extension…external rotation…

I recited these over and over again, trying to visualize the path a baby takes as it comes down through the pelvis, through the vagina, and is born. It was all very confusing. Our lecturer this morning had made it sound so clear. Now, trying to visualize it in three dimensions using the models in my hands, I couldn't make any sense of it.

I looked around the room. So this is what an on call room looks like, I thought. Sparse walls painted institutional green, a single bed—a cot really—a bright overhead light, and one small reading lamp on a nightstand next to the bed. The bathroom was down the hall. Oh yes, and the phone. Sitting on that nightstand was the phone that, when it rang, would tell me a new patient had arrived, my patient, and my obstetrics rotation would begin in earnest. Here I sat and waited. My first day—and soon to be night, I reminded myself—on call had been uneventful, at least so far, but it was still only 10 AM.

Yesterday, bright and early Monday morning, we'd arrived to start our obstetrics rotation, all six of us. After the usual introductions and explanation of the schedule of lectures, the course director, an affable, large man dressed in those comical scrubs, looked at us and said, "Two of you won't be going home tonight." We all sat still, not daring to look at each other or him. "We go through this every month," he continued. "We need two of you to volunteer to be on call tonight or we'll just draw lots." I was about to volunteer when I thought…no toothbrush…no change of underwear. "You'll each be on call every third night, so it's just a matter of setting the order," he continued. If I didn't have to be on call tonight, I'd just as soon delay it by a day or two. Besides, the sooner I was on call, the sooner I wouldn't be sleeping. Obstetrics was my first rotation that required me to work all through the night, and I honestly didn't know if I could do it. With no volunteers we drew lots. I got the second night. This morning I'd shown up with a toothbrush and a change of underwear, prepared as best I could be for my first night on call.

Suddenly, the phone rang. "A new admission just arrived. Jerry wants you to work her up," the ward clerk's voice intoned. Jerry was the intern I'd been assigned to work with. Once our on call rotation was established, this paired us with whichever intern was on call the same nights. For me, that meant working with Jerry. He seemed nice enough. Friendly, easygoing, a muscular build which, unlike most of the attending physicians, was enhanced by the scrub suit he always wore. But he looked tired. It was December, which meant he'd been an intern for six months already, on call every third night, and up working most of those nights. His eyes were perennially half closed. Looking at him made me wonder how I would ever be able to do it. Yet he seemed to be patient with me, willing to teach me. And, as it turned out, it was my good luck to be doing obstetrics in December. The OB/Gyn interns did six months of obstetrics and gynecology and six months of internal medicine. When the interns first started their OB/Gyn half year, they were hungry to do as many deliveries as they could. But, after five months, Jerry had done many routine deliveries, and he had become more interested in the difficult or complicated deliveries. This meant he didn't mind allowing me to "catch the baby" for the routine deliveries. If this had been July or August, or even January when the next group of interns would switch from medicine to OB/Gyn, I would have gotten to "catch" a lot fewer babies than I was going to be able to in December. Timing, I was beginning to learn, had a lot to do with the quality of my training experience.

I stood up, looked at the bed, wondering when I'd see it again, took a deep breath, and walked out of the on call room. It was beginning.

"She'll need an IV and bloods drawn," Jerry started saying even before I'd reached the nurses' station. "She's in room 214," he added, pointing without looking up. Yesterday Jerry had shown me how to start an IV for the first time. "Obstetrics," he'd said, "is the perfect rotation for learning how to start an IV." The women were all young and, unless they were addicts, had great veins. "Piece of cake," he'd said. But it didn't prove to be a piece of cake for me. "The best vein is this one here on the radial side, the thumb side, of the wrist," he'd explained. "It's plenty big, and, if you put the IV here, it doesn't interfere with her moving around and pushing during labor." He showed me how to wash off the skin in a sterile manner and hold down the wrist to stabilize the vein. "You gotta watch out for veins that role," he'd said. "Otherwise you'll be poking all day and won't get nothing." Holding the wrist down in just the right way was supposed to keep the vein from rolling away from the tip of the needle. In my hands, it would prove to be less than effective.

I watched Jerry do an IV for a patient yesterday. Once the vein was stabilized, he showed me how to hold the specially designed needle that had a catheter surrounding it. I watched as he pierced the skin with the needle and advanced it just a few millimeters, easily puncturing the vein. Dark blood appeared in the clear plastic holder at the base of the needle. "This is the tricky part," he explained, as he slid the catheter over the needle, advancing it inside the vein without going through the far side of the vein. Then he quickly withdrew the needle, pressed on the vein just ahead of the catheter tip to stop blood from back-flowing and leaking out. He deftly attached a syringe to the exposed hub of the catheter, withdrew a full syringe of blood so she wouldn't need another needle stick to draw her blood for lab tests, removed the syringe, and attached the end of an IV tube filled with fluid to the hub. He adjusted the IV so it was running smoothly, drop by drop. He'd already torn several pieces of tape which he'd placed just next to where her arm hung. With a few quick movements, he carefully placed the pieces of tape so that the IV was securely taped down. I was in awe. "You'll do the next one," he'd said matter-of-factly.

Today it was my turn. I gathered the materials I'd need: syringe, tubes for the blood samples, tape, IV bag and tubing, and the IV needle/catheter contraption. It took several trips to the supply room for me to remember everything I'd need. Jerry had said an 18 gauge catheter was best because it was big enough to pour in blood or IV fluid quickly if it was ever needed, but that a smaller, 21 gauge catheter, the size usually used by labs to draw blood, would probably be adequate unless the patient got into trouble. I now had everything I'd need. I knocked on the door of room 214. A tired female voice said, "Come in."

"Hi. My name's Rick Sheff, medical student here on labor and delivery. Jerry, uh, I mean Dr. Kosky asked me to examine you and start your IV.

"Do whatever you gots to do," she said, barely looking at me. "I done been in labor for hours, so jes' be quick about it. I's exhausted."

The chart revealed her name was Lorraine Johnson, she was 22 years old, and this was her first baby. She'd had two abortions before. I wondered to myself if this was a pregnancy she wanted or not. Doing her history and physical exam went quickly, especially because as a medical student new to obstetrics, I didn't know what to look for. She was overweight but, aside from her two abortions, had a negative past medical history. I noticed a large scar on her upper right arm which she said came from an accident as a child. I listened to her heart and lungs, but that was about all Jerry told me to do for a physical exam. "What about looking in her ears, her eyes, and her throat," I'd asked.

"The ears, eyes, and throat are not pelvic organs," he had said. "Here on labor and delivery, just listen to her heart and lungs and check out her belly. That's what counts. I'll do the pelvic exam with you."

Check out her belly meant measuring how high the top of her uterus was from her pubic bone and then performing what were called Leopold maneuvers. These were four different ways to feel for the position of the baby in her uterus and to estimate its size. Since this was only my second day on obstetrics, I couldn't really perform them properly. I just felt her abdomen as best I could, trying to figure out if I could tell if the baby's head was up or down, which I couldn't. Besides, every few minutes she started tossing and turning, moaning with the pain of another contraction. I'd never been with a woman in labor and didn't know what to do during those times, so I just awkwardly waited for the contraction to end and then resumed my evaluation. After the first few contractions, I realized they came on gradually, built to a peak at which she was calling out and writhing in pain, and then subsided. With the next one, rather than stand awkwardly by, I took her hand and said, "Now you know this one's going to build. Here comes the peak. You can get through it," I said, trying to help her in any way I could. "Now the peak is passing. It's going away now. The worst is over." She seemed to calm down a little as I talked to her in this way.

Then it was time to start the IV. After preparing the pieces of tape and attaching the tubing to the IV bag, I took a cue from my training in doing pelvic exams, and began explaining each step before doing it. "First I'm going to wash off your skin here on your wrist so the area will be sterile." That was a good start, I thought to myself. "Then I'm going to hold your wrist down, like this, in order to stabilize the vein." I wasn't sure she was even paying attention to what I was saying. I could see another contraction starting to build, so I stopped trying to put in her IV and talked her through the contraction. Hurrying, I resumed trying to put in her IV. "Now as I hold your wrist down, you'll feel a sharp needle stick, like a bee sting." That's what I'd heard Jerry say. It seemed to be what the nurses said every time they put in an IV as well. The needle went in and she winced and tried to pull back her hand, but I held it as best I could. I paused, not wanting to hurt her. I could see the bulge of the needle tip just under her skin. It was right next to the vein Jerry had pointed out to me. I advanced the needle toward the vein, but to my horror, it "rolled" out of the way. She winced again. I pulled the needle back a little and tried advancing it again. Again the vein rolled out of the way as she called out in pain from the needle. I could sense another contraction coming, so I pulled the needle back a little and quickly advanced it a third time. This time I felt a subtle pop as it entered the vein, and I could see blood appear at the base of the needle hub, just as it should. I was elated. Her contraction was starting to build. "Try to hold your wrist still for just a little longer," I pleaded. Now, advance the catheter over the needle tip, I said to myself. I tried to do this, but the catheter seemed somehow stuck on the needle. It wouldn't advance. Jerry had made that part seem so easy. With my left hand I was still holding her wrist down as her contraction built to a peak. I wanted to use my left hand to hold the needle base as I advanced the catheter over it, but I couldn't let go of her wrist, so I had to do it all with my right hand. She was moaning loudly now. It was everything I could do to keep her wrist from flailing around. In a desperate attempt I pushed the catheter harder with my right hand. Finally it slid off the needle and moved forward under her skin. With a sigh of relief I pulled out the needle. As I did, no blood flowed back out of the catheter, which was fortunate as I suddenly realized I had forgotten to compress the vein to prevent the blood from flowing out. I connected the syringe a little awkwardly, as at first I didn't realize I had to twist it to get it to attach. Once it was firmly attached, I drew back on the plunger of the syringe as I'd seen Jerry do. Nothing happened. No blood filled the syringe. I pulled back a second time, but again, nothing happened. She was moaning again, but this time she was complaining that her wrist hurt.

"What you doin? You'se hurtin' me." I looked at her wrist. To my horror, I saw that her wrist was swelling around the catheter, creating a lump at least half an inch high and two inches wide. I could only imagine it was filled with blood oozing from her vein. I pulled the catheter out and pressed on the area with a gauze pad. She cried out again when I pressed.

"I…I'm so sorry," I stammered. "I'll get Dr. Kosky to put your IV in. Just put pressure on this gauze here, and I'll get him." As I got up to leave, I could tell another contraction was starting. "Just hold the gauze," I pleaded as she started to writhe, her arm flailing as the gauze fell to the floor. I ran out to get Jerry.

*   *   *

Several hours later, Lorraine lay in the labor room. This was more like a ward than a room, since it had four beds, separated only by curtains that hung from the ceiling. Each bed now held a woman in labor. They all tossed and turned with every contraction, just as Lorraine had done and was still doing. The farther along in labor they were, the more they called out, moaned, and sometimes outright screamed.[1] The steady "beep, beep, beep" of fetal heart monitors could be heard in the background. All the women were black. No one sat by their bedsides. No one held their hands. No one comforted them. With the same instinct that had caused me to take Lorraine's hand and talk her through her contractions, I wanted to reach out to each of these women, to ease their fear and loneliness. In what little time

---

[1] These events predate the consumer movement in birthing. Driven in part by competition among hospitals for the baby boomers entering the age of pregnancy and childbirth and in part by the women's movement, hospitals have mostly eliminated labor wards and some, but certainly not all, of the other medicalized, insensitive, and abusive labor practices I witnessed and was actually trained to perform.

they had, I watched the nurses try to teach them basic Lamaze breathing techniques to help them relax and get through each contraction.

"Now look at me and breathe just like me," one nurse was saying to Lorraine. "Hee…Hee…Hee…Hewww…Hee…Hee…Hee…Hewww…"

"I cain' do that," Lorraine responded, tossing her head and writhing as another contraction gripped her.

"Didn't you go to childbirth classes?" the nurse asked, scowling.

"I didn' go to no classes," Lorraine moaned, her head flailing back and forth.

"Well, do your best, honey," came the reply as the nurse quickly looked at the long, continuous graph paper that rolled out of the fetal monitor and ran off to respond to another patient. I looked at the paper with no idea how to interpret the seemingly endless trail of wavy lines and countless dots, knowing only that they represented every beat of the tiny heart in Lorraine's womb.

I watched Lorraine go through one more contraction, tossing and turning, calling out to some unnamed source for help. Something inside caused me to step up to her bedside, take her hand firmly in mine and get right in her face, our noses inches from each other. "Listen to me, Lorraine," I said firmly. Her eyes came into focus on mine. "I'm gonna breathe with you. You do exactly what I do," I said with authority.

"Yas. Yas. I'll try." The contraction started to build.

"Do everything I do, and breathe with me. Hee…Hee…Hee…Hewww… Hee…Hee…Hee…Hewww…"

"I cain. I cain." Her head started to toss.

"Lorraine, look right at me!" I barked with an authority I didn't know I possessed. Suddenly her eyes were locked on mine again.

"Hee…Hee…Hee…Hewww…" She started to breathe as I was. "Hee…Hee…Hee…Hewww…" she continued. The contraction was subsiding. "You did it, Lorraine! Now you know you can do it for the next one." We were both smiling. The next contraction came. This time she held her gaze on me, and we breathed through it together. With each contraction she relaxed a little more, and grew with the confidence that she could get through the next one. Then she looked at the clock.

"How long I been doin' this? How long I been here?"

"A few hours," came my reply. I didn't want to be too specific.

"When cain I get some medicine, somethin' for this pain? I cain keep doin this. I need you to gimme somethin' now."

"I'll ask your doctor."

"But I thought you'se my doctor."

"No, Lorraine, like I said before, I'm a medical student here. But I'll do whatever I can to help."

"I'd like you to be my doctor. When you're here, I ain' feelin' so afraid."

"I'll stay with you through your labor as much as I can. Right now I'll get Dr. Kosky to check you." That's what I heard the nurses say to other patients in the room, but I didn't really know what "checking her" meant. Jerry came back with the nurse. He pulled the curtain two-thirds of the way around, leaving a gap at the end of the bed. Privacy didn't seem to matter here.

"Bend your knees and put your feet together," he said to Lorraine. She put her feet in a frog leg position. In her scant hospital Johnny, this left her vulva fully exposed to Jerry, the nurse, and me. Jerry put on one glove and the nurse put some of the orange, bacteria-killing soap on it. He leaned over, his face near her belly, and put his forefinger and middle finger into her vagina.

"Oww! Owww!" Lorraine cried out as he examined her. I looked at her face and moved to the head of the bed to take her hand. She squeezed so tightly it hurt.

"She's only 3 centimeters and -2 two station, but she's now 80% effaced,"[2] Jerry said to the nurse. "Rick, come on in here and examine her. You've got to learn how to evaluate a cervix during labor."

I released Lorraine's hand, but she didn't want to let mine go. "I'll be back," I reassured her. Moving down her bed the nurse held out an examining glove for me. I put it on, and she squirted some of that orange liquid on it. I placed my fingers in her vagina. It felt different from other bimanual exams I'd done. Everything felt soft and swollen. I couldn't tell at all what I was feeling. Jerry read my mind.

"Everything feels pretty soft in there," he began. "That's from the hormonal changes of pregnancy that will allow the vagina to stretch during birth. But reach way back down her vagina to find her cervix. It won't feel

---

[2] The 3 centimeters refers to how far the cervix, the bottom opening of the uterus, has stretched. It needs to stretch to 10 centimeters for the baby to pass through it. The 80% refers to how much the cervix has thinned out, which is called effacement. The non-pregnant cervix is approximately 2 centimeters thick or just a little less than an inch. During late pregnancy and the first stage of labor, it becomes increasingly thinner. The degree to which it has thinned is expressed as a percentage of the original 2 centimeter length. So an 80% percent effaced cervix is about 0.4 centimeters long, or less than a quarter of an inch. The station refers to how far down the baby has descended through the birth canal. Zero station is a location even with the tip of two small boney prominences on either side of the pelvic bone called the pelvic spines that can be felt during a vaginal exam. If the most descended portion of the baby (usually the head) is 2 centimeters higher up the canal than the tip of these spines, it is referred to as -2 station. If the most descended portion of the baby is 1centimeter beyond the tip of the spines, it is referred to as +1 station. A baby at +4 station is at the opening of the vagina and on the verge of delivering.

at all like it usually does. You're feeling for the anterior or front lip of her cervix. It's pretty thin now, only about four millimeters thick. You'll feel the baby's head behind that." I pushed my fingers in as far as I could. Everything still felt so soft and confusing. Suddenly my fingers felt something hard, bone hard. I looked up at Jerry.

"I think I feel the baby's head," I said excitedly.

"Now feel for the edges of her cervix around that." I felt some vague softness rising on either side of the hard part I was feeling. "Put one finger on one edge of her cervix and one on the other. How far apart are your finger tips now?"

"I don't know. I'd guess about 3 centimeters," I said, counting heavily on the information Jerry had just shared with the nurse.

"That's it. That's what you're feeling for. The cervix has got to get to 10 centimeters to let the baby's head through it, and she's only 3 centimeters now." I took my examining hand out of the vagina.

"What's that all mean," Lorraine looked from me to Jerry pleadingly. I'd almost forgotten about Lorraine in my focus on what I was feeling.

"Your cervix has dilated some," Jerry began. "It's become pretty thin, but it has to dilate some more before we can give you anything for the pain. If we gave it to you now, it would slow your labor down."

"Well, what do I has to do to get somethin' for this pain? I cain keep doin' this."

"When you get to four or five centimeters, we'll be able to give you something for the pain," Jerry responded. Just then his beeper went off. "I've got to get this," he said and walked out the door.

Lorraine looked at me, her eyes wide with bewilderment and fear. "What did he jes say? What do I gotta do to get somethin' for this pain?" Another contraction started.

"Breathe through this one with me, and then I'll explain. Look at me. Hee...Hee... Hee...Hewww... Hee...Hee...Hee...Hewww..." The contraction finally subsided. "Here's what's going on," I explained to Lorraine. "With each contraction, your uterus, your womb here," I said, laying a hand lightly on her protruding belly, "is pulling the bottom opening of your uterus, your cervix, open. It opens a little more with each contraction. Right now you've pulled it open this far." I held up my fingers 3 centimeters apart, which was just over an inch. "You need it to stretch this far for the baby's head to fit through." I held up my fingers 10 centimeters apart, which was about 4 inches. She let out a wail.

"If I been in labor all this long, and I's only gotten to that little bit, I ain' never gonna make it all the way."

"But if you can hang in there until your cervix gets just this open," I said, holding my fingers up about 4 centimeters apart, "Then you'll be able to get some medicine for the pain." Her face brightened a little.

"How long is that gonna take?"

"I don't know. But my suggestion is that you take it one contraction at time. That's the best way to get there."

"Alright. I'll do that, long as I know you'll stay here with me."

"I'll stay right here as much as I can."

*   *   *

All afternoon I was in and out of her labor room. I had another chance to start an IV, this time with Jerry watching. On my first try the wrist swelled up just like Lorraine's had, but Jerry insisted I try again. The second try worked. I felt the pop, pushed the catheter forward, and this time it slid easily. When I removed the needle, blood flowed out everywhere from the hub of the catheter, making a bloody pool on the floor and on the bed sheets, until Jerry applied pressure to her vein. "You'll get the hang of it," he had said encouragingly.

At one point, a nurse came to find me in Lorraine's labor room. "Jerry said he's about to do a delivery if you want to watch." I jumped at the chance. "It's a private patient, so you can't do anything during the delivery. He said you should step just inside the room and stay down near the far end of the delivery table where he'll be. But stay clear of the sterile area." I certainly knew how to do that from my gynecology rotation.

I entered the delivery room greeted by a chorus of voices calling out, "Push! Come on, push!" The room was filled with harsh light. At the head of the table sat an anesthesiologist, checking the woman's blood pressure. Next to him sat a bewildered looking man I presumed was the woman's husband, or at least the father of the baby. He chimed in with the calls for the woman to push, but with a little less certainty. A green sterile drape covered the woman's abdomen. She lay supine, with her legs raised almost vertically off the delivery table, strapped into widely spread stirrups that supported both her legs and feet. Her legs, along with the stirrups, were covered in that same green sterile material. Her vulva, which hung exposed just at the edge of the table, was the only part poking out. The small pucker of her anus was swollen and distorted by large hemorrhoids that bulged even larger with each push. A steel bucket had been wheeled between her legs and sat on the floor, just beneath her vagina. Jerry straddled the bucket. He was dressed in a sterile surgical gown, with the head covering, face mask, and gloves I'd learned to wear in the operating room. Only his eyes showed. With the next push, I watched as her vagina stretched partially open, revealing a patch of dark baby hair that receded between contractions.

So this was the mystery of childbirth unfolding before me. One of the allures of becoming a doctor was to pull away the veil of secrecy we have constructed around the major events of our lives—birth, death, suffering, and illness. All of these very human events that used to take place within our homes and with our families now take place within the walls of hospitals with strangers. At this moment, I sensed I was being initiated into another of the great human mysteries.

Her vagina stretched and strained to the chorus of "PUSH!" Suddenly Jerry grabbed a large syringe with an even larger needle on its end. Placing his fingers inside her vagina just to one side of the bottom of the opening, he plunged the needle at least 4 inches deep into the outer edge of her vagina. She let out a scream as he injected the full contents of the syringe into the thinned tissue. Grabbing a large scissors, he cut through where he'd just injected. Blood poured out of her vagina and down into the waiting bucket. On the next contraction, everyone screamed "Push!" again, and out popped a wet, blue, swollen and distorted head. It seemed so odd, the head poking out of this woman's vagina, as if just resting there. "Stop pushing," Jerry called. Quickly he grabbed a suction bulb and pulled slimy mucous and fluid out of the baby's nostrils and mouth. "Now push gently," he ordered. She pushed as he pressed down on the baby's head. Suddenly a wet arm emerged, and then the rest of the baby followed, a huge gush of blood mixed with clear amniotic fluid exploding out behind it. "It's a girl!" Jerry exclaimed as he flipped the slippery wet baby around onto his arm, suctioned some more, clamped the umbilical cord in two places, and cut the cord between the clamps. Instantly he handed the baby to a waiting nurse as its first cry filled the room. The nurse took the baby to a warmer and dried it off. I went over to watch. She took an imprint of the baby's hands and feet for the birth certificate and put some chemical in the baby's eyes. The baby fussed and cried during all of this. "Silver nitrate drops," she'd said the chemical was. "The babies seem to hate this stuff, but it has to be done to prevent transmission of gonorrhea from the mom to the baby through her eyes." She finished by giving a shot into the newborn's thigh. The infant wailed. "Vitamin K. Needed to prevent bleeding," she'd said.

I came back to the end of the table where Jerry was working, pulling steadily on the clamp attached to the umbilical cord. More blood than I'd ever seen continued to pour out of her vagina and down into the bucket. "We've got some bleeding here," Jerry called out to the anesthesiologist, a slight edge in his voice. "Open up her IV." Jerry instantly reached up and massaged her belly to sounds of protest from the woman. At the same time he pulled firmly and steadily on the clamp remaining on the umbilical cord. For a while the flow of blood slowed considerably. "Here comes the placenta," Jerry called out. With another extraordinary gush of blood that splashed all over Jerry and beyond the bucket, a large slimy, irregular, bloody mass emerged from her vagina. He threw the placenta into a metal bowl. "10 and 10 now," Jerry ordered.

"I've just given 10 of Pitocin IV push and 10 in the bottle," came the anesthesiologist's reply. "I'm way ahead of you."

The bleeding slowed. Jerry used his foot to pull over a stool and sat down between the woman's spread legs, straddling the blood-filled bucket. He picked up a long metal instrument with two rings on the end that came together like a mouth, a ring forceps he called it. Reaching one end of the instrument high up in her vagina, he closed the rings together on something as the woman let out a yell. "I'm just checking inside for any tears," Jerry called to the woman after the fact. He pulled the instrument toward him and a swollen, bloody, thick, maroon-colored object came into view within the vaginal opening, caught tightly between the metal rings. "Her cervix," Jerry said in an aside to me. "I'm looking for any tears. With that much bleeding, you have to rule out a cervical tear, but it looks fine." Next he used the instrument to stretch and probe her vagina, again to protests from the woman. "No other lacerations seen," he said half to himself and half to me. He surveyed the torn, distorted, and bloody area that had recently been this woman's vagina. Next he lifted a long needle holder from the instrument table and attached a suture with a needle to it. Reaching high up into the vagina, he dug the needle in, pulling it out again with the needle holder. The woman's legs tightened as he did, but they were tied to the stirrups, so she couldn't move away. "Just sewing you up," he called out over the drapes that separated him from this woman. Deftly he tied a knot with one hand, pulling it tight. In a few minutes I could see the edges of this woman's vagina coming together again. "I cut a mediolateral episiotomy, meaning I cut diagonally down at the seven o'clock position," he was now explaining. "I could have cut a medial or straight down episiotomy at the six o'clock position. It would have been smaller and healed with less scar tissue, but if that kind of episiotomy tears, it tears down toward the rectum, sometimes right through the anal sphincter muscle and into the rectum itself, and that's a mess to put back together. A mediolateral episiotomy is more painful as it heals, but it's definitely safer. And keeping moms and babies safe is our primary job." He was now placing stitches deep within the bloody tissue that gaped on the lower, outer edge of her vagina. "It's also our job to sew her up good and tight for her husband," he said in a hoarse whisper, "so sex will be better for him." I glanced up uncomfortably, concerned she'd heard him. The sewing was finished, and remarkably, her vagina had come back together again in a semblance of the normal anatomy with which it had started. Jerry stood up. "All set," he called out to the anesthesiologist. Then, as if as an afterthought, he turned to the couple, "Congratulations. Your baby looks great," and he walked out of the room.

*   *   *

Lorraine slept between contractions now. That was new. An hour ago, she was still writhing in pain, struggling to breathe with me through the contractions. Jerry had checked her and said she had gone from 3 to 4 centimeters. "Doesn't that mean she can get some medication now," I'd asked.

"She could," came the reply. "But I like to hold out until 5 to make sure she's in a good labor pattern."

"I thought you said she could get medicated when she reaches 4 to 5?" I asked, upset at the thought of Lorraine having to wait much longer for some relief.

"That's always a judgment call, and I like to wait until 5." That's because you haven't been sitting back there in the labor room with Lorraine through every contraction she's had for the last four hours, I thought. If you had, you'd sure feel otherwise.

"If it's a judgment call, given how much pain she's been in and her lack of childbirth preparation, wouldn't it seem reasonable to come down on the side of giving her some medication now," I suggested hopefully.

"Look, you've got to get used to the fact that women in labor are in pain. You just can't let it get to you. Wait until you've got to handle a woman laboring with a baby who's OP."

"What's OP?"

"It means the baby's occiput, the back part of the head, is facing posteriorly, towards the sacrum. Pounds the hell out of the mom, and makes for a long and difficult labor. I know you think I'm being hard, but it's just that pain is what childbirth is about. If you responded to every woman in pain, you'd make yourself crazy." Not responding to someone in pain would make me crazy, I thought. "But, alright, since you seem to feel so strongly, we can order the Demerol and Vistaril now." I ran back to give Lorraine the good news.

She had relaxed considerably once she'd received some medication. When she wasn't sleeping between contractions, she seemed ready to talk. "Wonder where Rad is," she said at one point.

"Who's Rad?" I asked.

"Oh, he's the baby's daddy," she said brightly.

"Rad is an interesting name," I said. "Is that his real name?"

"No, it's the nickname he got from his…his…friends." Something dark came over her face, but then it was gone. "I wish he was here, doin' what you're doin."

"That could happen if he'd come," I suggested.

"Ya think so? I'd really like that."

"Do you know how to reach him now?"

"I thinks so. I could call my girlfriend and as' her to find him."

"It's a little hard for you to make a call, but maybe the nurses could call for you. What's your girlfriend's name and number?"

Two hours later, a short, heavyset, uncomfortable-looking black man approached the nurses' station. He was escorted to the labor room. "Rad! I's so glad you'se here." They embraced, though Rad appeared stiff. His eyes quickly scanned the room, uneasy with the moans of pain rising from behind the three other curtains in the labor room. "Sit down, Rad. Sit down here," she said, pointing to the chair I'd just vacated next to her. "Dr. Sheff here done been helpin' me with my contractions. They gave me somethin' and they ain' so bad now. I still needs to breathe through them. One's comin' on right now. Dr. Sheff, where are you?"

"I'm right here, I said, moving in beside her and taking her hand. "Hee...Hee...Hee... Hewww... Hee...Hee...Hee...Hewww..." The contraction passed. Rad's eyes were wide. "You wanna try it?" I suggested.

"Come on, Rad. You can do it. I knows you can. Jes' try it. Try it for me." Rad shifted in the chair uncomfortably.

"It's not hard," I said encouragingly. The next contraction was beginning. "Do just what I do. Breathe like this. Hee...Hee...Hee... Hewww...That's it. Hee...Hee...Hee... Hewww...Now take her hand in yours, here." I stepped away, putting her hand in his. Self conscious at first, Rad slowly relaxed into the breathing. The next contraction I did the breathing from one side of the bed while Rad held Lorraine's hand and did it from the other. A few minutes later, I left them alone.

*   *   *

A little while later I came back to the labor room to see how Lorraine and Rad were doing. As I peaked around the curtain, they were breathing together through the end of a contraction. I smiled. Just then I heard talking behind the curtain next to Lorraine's. Since it wasn't fully closed, I could partially see what was going on.

"We've decided you need an epidural," a tired-looking man in scrubs was saying to a young black woman. She couldn't have been more than 18, and appeared more exhausted than the man standing over her. "We need you to sign this consent form for it."

"A wha?"

"An epidural. It's for your pain."

"I don want no more pain. You do wha ever you gots to to stop me hurtin'."

"That's why we think you need an epidural. But we can't do it until you give informed consent."

"I done been in labor for 20 hours. What you wanna do to me now?"

"I'm going to put a needle in your back, at L3."

"A needle? I don't want no more needles." Just then a contraction started.

The man kept talking. "I have to put this needle in your back to make the pain go away. I'm going to push it in until it is up against the dura in your spine. Then I'll put a catheter through the needle. Once the catheter is in place I can inject the lidocaine. You just need to sign this informed consent form."

All the while she writhed and twisted, trying to get away from the pain. "I cain. I cain," she kept saying.

As the contraction subsided, the man, who by now I had figured out was an anesthesiology resident, said impatiently, "So will you sign this form?"

"Say what?"

"Will you sign the consent form?"

"What form? I don' know nothin' bout no form."

Exasperated, the resident continued, "I've just explained it to you. This is for your epidural. I said I was going to put a needle in your back, push it up against the dura in your spine, put a catheter through the needle, and inject lidocaine through the catheter. You won't be able to move your legs after that. We'll let it wear off later so you can push. You're supposed to give consent before I can do it." She looked blankly at him. "Oh yeah," he added, "it's possible the needle could accidentally go through the dura. If that happens you could develop a really bad headache and you'll suddenly get numb below your waist. Your blood pressure might drop suddenly. The level of numbness could go too high and you could stop breathing, but we would just put a tube in your throat and breathe for you."

Just then another contraction started. "I'll sign. I'll sign. Jus' gimme the pen and I'll sign. Jus' so you make this pain go away." With one hand she scribbled her name at the bottom of a paper attached to a clipboard the resident held in front of her while the other hand clutched her belly. The moment she finished writing, the pen dropped to the floor and she resumed writhing.

The anesthesiology resident, relieved to have the paperwork done, left to get the set-up tray and instruments for placing the epidural. Was he that blind, I thought? Did he not see that he used medical terms and jargon the woman couldn't possibly have understood? And the insensitivity he showed by continuing to "explain" the procedure to her during a contraction. I fervently hoped never to be that out of touch with the experience of any patient I ever had the chance to treat.

*   *   *

"I think it's time," the nurse called down to Jerry. He ran in and checked her. Nobody bothered to pull the curtain this time.

"Yup. She's 10 centimeters. It's time for her to push. Rick, put a glove on and feel what a fully dilated cervix feels like." I examined Lorraine again. Reaching back I couldn't feel her cervix at all. All I could feel was the firm head of the baby. "Can you feel the anterior fontanelle?[3] It's the soft spot near the front of the baby's skull. It's at about 4 o'clock right now. It's LOP, that's left occiput posterior." I tried, but couldn't feel anything there other than more of the baby's head. "That's another thing for you to work on," came Jerry's reply.

With the next contraction, we all began the chorus of "Push!" Lorraine held her breath and pushed as hard as she could lying in bed. I didn't know then that this was the worst possible position for the pushing or second stage of labor. Squatting and other upright positions open the pelvis better and make use of gravity to enhance the effectiveness of the mother's pushing efforts. Here everyone labored and pushed supine, and nobody questioned this dysfunctional practice.

The nurse was now doing the coaching. "Wait for the contraction to build a little. That's it," she was saying. "Now take a deep breath and push for all you're worth. Push! I'll count to 10, then take another breath and push again. One…two…three…"

We all cheered her on, even Rad. Jerry left to tend to other patients. Eventually the nurse left, too. Rad and I stayed with her, exhorting her with every contraction to push as hard as she could. Lorraine was a fighter, and she wanted this baby so much. Even Rad and I could tell she was making progress. An hour and a half later, Jerry came to check her. "She's ready to go to the launching pad," he said. "Rick, you check her, too." My gloved fingers started to reach in, but almost immediately felt the hard head of Lorraine's baby. "That's a +2 station," Jerry explained. "At the rate she's going, she'll be starting to crown in about 10 more contractions."

The "launching pad" meant rolling her bed out of the labor room into the hallway outside the delivery room. There Lorraine continued to labor, with nurses, residents, orderlies, and even members of other patients' families walking past, her flimsy Johnny and active pushing leaving her exposed to anyone who cared to look. A few contractions later Jerry came by to check

---

[3] A baby's skull is made up of multiple bones that have not fused together by the time of birth. This allows the head to mold somewhat to more easily pass through the birth canal. Where the bones come together in the front portion of the scalp is an area not covered by bone, and hence soft, called the anterior fontanelle. Feeling for this allows the examiner to identify which way the baby's head is facing. Some positions of the baby's head are more favorable for labor and delivery than others.

her again. Her vulva was visibly bulging with each push, and Jerry said it was time to take her into the delivery room. A nurse called the anesthesiologist STAT. We rolled Lorraine's bed into the brightly lit delivery room and next to the delivery table. Rad stayed outside, as the hospital had a rule that only partners who had completed childbirth classes could be in the delivery room.

Suddenly, everything was a flurry of activity. As soon as the next contraction ended, the nurse instructed Lorraine to "scoot" from the bed to the table. The nurse quickly set up the instrument table. Jerry and I were outside the room rapidly scrubbing, though it was a more abbreviated scrub than I'd been taught for the operating room. We rushed in, face masks and head coverings on, were gowned and gloved by a nurse, and Jerry began arranging the items on the instrument table. Lorraine's legs were strapped into the stirrups and her bottom now hung at the end of the table. The bucket was moved into position below her. Jerry grabbed a bowl with cotton balls soaked in that orange soap and a ring forceps and began quickly washing her vulva. "That's cooold!" Lorraine exclaimed.

"I'm washing you off," explained Jerry, a little belatedly. The cotton balls went from front to back over her labia, each time ending with a probing wipe of her anus, before being dropped into the bucket.   With the last three cotton balls, Jerry scoured the inside of her vagina as Lorraine tried to pull away. "Now Rick, get in here. You're gonna catch this baby, just like I showed you earlier today." My heart started to pound. Lorraine resumed pushing with the contractions, and the baby's hair appeared through her vaginal opening, at first no more than the size of a quarter, but slowly increasing in size with each push. Her vagina began to stretch, bulging and thinning more and more. When I thought it couldn't possibly stretch any further without tearing, Jerry plunged the needle deep into her and injected lidocaine. He handed me a thick pair of surgical scissors. "Cut right here, between my fingers," he ordered, placing two outstretched fingers between the tissues of her vaginal opening and the baby's head, where he had just injected. I stepped right next to Jerry, placed the scissors between his fingers and cut. The tissue gave way thickly with a gush of blood as I felt the scissors slicing through vaginal tissue, muscle, skin and connecting fibers. I opened a 3-inch gash that poured out blood. With the next push, the baby's head popped out. "Stop pushing," Jerry ordered. He handed me the suction bulb, and awkwardly I suctioned out each of the baby's nostrils and its mouth. "Now we check to make sure the cord isn't wrapped around the baby's neck, a nuchal cord we call it." Feeling along the baby's neck, I touched something round and firm. Jerry confirmed what I had found. "There is a nuchal cord. We have to reduce it, which means unwrapping the cord from the baby's neck. If you deliver the baby without reducing a nuchal cord, the cord tightens around the neck and strangles the baby. Here's what you do," he went on quickly, demonstrating as he spoke. "Get your fingers under the cord and slide it over the baby's head." The slippery cord stretched over the head. "If you ever find a nuchal cord too tight to reduce like this, you clamp and cut the cord with the baby's head right here on the perineum."[4] Then Jerry placed his hands over mine, one on the upper side of the baby's head and the other on the bottom. He pushed my top hand firmly down, and I felt the baby's shoulder pop out from under Lorraine's pubic bone. Suddenly the baby slithered completely out onto my arms, almost sliding into the bucket. Jerry helped catch it against my body, preventing this catastrophe. "Suck out the nose and mouth with the bulb again," he barked. I sucked again. "Now put two clamps on the cord and cut between them." The same scissors cut through the cord, and the baby was free in my arms. I suddenly noticed it hadn't cried yet. Jerry was already calling the nurse over. I placed the baby on a blanket in her arms, and she whisked it away to a warmer on one side of the room. For an interminable moment, there was silence. Then, as she dried the baby off, it let out a lusty cry.

Lorraine started to cry. "My baby! My baby! Oh, what is it? A boy or a girl?" I realized I hadn't even looked.

"You've got a beautiful baby boy," the nurse called out, and Lorraine wept again.

*   *   *

I lay awake in the on call room. After Jerry had sewn up Lorraine's episiotomy, I'd gone out to tell Rad the good news. He had smiled. "I got me a son," he'd said, repeating it several times. "When can I see him?" he finally asked.

"The nurses are bringing him to the nursery. You can see him up there." He started to turn away. Something seemed to be settling inside him. "Lorraine wants to see you. She'll be out to recovery in a few minutes. You can see her there."

"Alright. Alright." And after a pause, "She's a good woman," he'd added.

---

[4] To clamp and cut the umbilical cord on the perineum means to apply two metal clamps to a portion of the umbilical cord while it is wrapped around the baby's neck when the baby's head is protruding out of the vagina but the rest of the baby is still in the birth canal. Then, by cutting the cord between the two clamps, the cord can be unwrapped from around the baby's neck and the rest of the baby can be safely delivered. The cut is made between the two clamps so bleeding won't occur from either the baby's end of the cord or the mother's end.

I caught a glimpse of the baby, sleeping peacefully, as he was wheeled to the nursery in an incubator.

Jerry and I had debriefed after the delivery, reviewing each of the things we'd done and why they had to be done that way—the scrubbing and gowning, the sterile prep, suctioning, reducing the nuchal cord, the episiotomy, the maneuvers for delivering the shoulders, and how to flip the baby over into my arms so it won't drop. He'd then shown me how to write a delivery note in the chart. He included an estimated blood loss of 800 milliliters, almost a quart. It had looked like so much more than that. "Blood always looks like more than it is," he'd said reassuringly. Then he added, "I think you're done for the night. You've had a good first day, and I don't expect there's anything else you need to do. Get some sleep. I've got some more things to do." I looked at my watch. It was 3:45 AM. I hadn't felt tired until that moment.

Lying in the on call room, the words kept running through my head. "I've delivered a baby. I've delivered a baby." My eyes finally closed.

*   *   *

The next morning we checked in on Lorraine during rounds. In spite of getting so little sleep, her eyes were bright, and her face gave off a glow that felt contagious. "You seen my baby?" she asked. "He beautiful, ain' he?" I hadn't seen him, but promised to go up to the nursery later in the day to admire him. "I'm naming him Derrick, Jr. That's Rad's real name, Derrick. So my boy's gonna be Derrick Johnson, Jr.," she pronounced proudly.

We finished up rounds, and I had a lecture that morning. In an effort to be more humane, the obstetrics course director allowed students who had been on call the night before to go home after the morning lecture. I was suddenly very tired. I promised myself I'd see Derrick, Jr., tomorrow.

We arrived at Lorraine's room on rounds the following morning. Her eyes were red and swollen. "You seen him? You seen my baby? I'm so worried."

"What happened?"

"He sick. He so sick. They got 'em up in intensive. They say he might not make it." She started to cry. Suddenly, I couldn't breathe.

"What happened?"

"He got a infection. Tha's all I know." She rocked in the bed.

I couldn't get to the nursery until after the morning lecture. As soon as it was over, I went straight there. "I'm here to see the Johnson baby, Derrick, Jr.," I said to the nurse at the nurse's station.

She looked up. "Who are you?" she asked carefully.

"I'm a med student who helped deliver Derrick. Is he alright?"

"He's not here. He was transferred to the NICU[5] on evenings. That's on the fourth floor."

"Thanks," I barely got out as I hurried for the elevator.

"He's over there, in bassinet 18," the ward clerk said. I walked slowly up to the incubator. Derrick, Jr., was lying on his back, a tube in his throat attached to the respirator that hissed air in and out next to him. Wires connected to white pads that covered his chest ran to another machine that beeped each time his heart beat. Intravenous tubes ran into the incubator, ending in a mound of gauze and tape that tied his tiny hand to the sheet. I stared.

Someone in scrubs and a long white coat came up to me. "I heard you were looking for Derrick," he said.

"What happened?" is all I could get out.

"Derrick has Group B strep. Hi, I'm Hal, the neonatology fellow." He stuck out his hand, which I shook loosely. "He's really lucky we found it so soon," he went on. "The baby started to show some signs of grunting and respiratory distress when he was only 12 hours old. That's what tipped us off. The residents in the nursery did a good job picking it up. He came right to the NICU. We did blood cultures, a spinal tap, chest x-ray, started an IV, intubated him,[6] and got in a first dose of antibiotics, all within an hour of his first symptom," he said proudly.

"What's his prognosis?"

"Actually, there's about a 50-50 chance he'll survive because he got treated so soon. That's better than most who show signs of Group B strep this early."

"Where did this infection come from?" I managed to ask.

"Group B strep is a bacterium that grows normally in the mother's vagina in a certain percent of the population. It usually doesn't make the mother sick, but it can be devastating to newborns."

"Thank you," I muttered, as he turned to respond to an alarm in one of the nearby incubators.

"So I made my baby sick," Lorraine moaned. "It was me made him sick?"

"No, that's not what I said." I tried to explain it to her again.

"So if I don' have this thing in my privates, he don' get sick. Isn't that it?"

---

[5] NICU is an abbreviation for neonatal intensive care unit.

[6] Intubation is the act of placing a plastic tube into the patient's mouth, passing it through the throat and beyond the vocal cords so that it ends in the wind pipe or trachea. This allows air to be forced in and out of the lungs to breathe for the patient who cannot breathe for him or herself for any reason.

"Yes, but it's not your fault."

"I know whose fault it is," she said darkly.

"What do you mean?"

"I ain' told you everything 'bout Rad. I love him and all, but he done some things."

"What kind of things?"

"Bad things."

"Like what?"

"I didn' wannna tell you, but you know that scar you done as' me about? Well I didn' get it as a chil'. Rad gave it to me. He cut me with this knife he's always carrin'." I looked at the scar on her arm. It was 4 inches long and an inch wide.

"It looks like it was never stitched."

"That was Rad. Once he gave it to me, he made me promise not to go to no hospital. So that's how it heal up."

"Has he done other things to hurt you?"

"Yas. Yas, he has, but I done promise him I wouldn't tell nobody."
"Is it something he did, that you can't tell me about, that makes you think he caused you to have that bacteria in your vagina?"

She was rocking again now. "I don' know. I don' know. Maybe." She paused. "When he came to see me in recovery, he tried to have sex with me. I had them stitches and all, but he still wanted to do it, right there in recovery. I told him no. If a nurse hadn't walked in right then, I don' know what he'd a done. I knowed he had his big ol' knife with him then."

"Lorraine, I'm really glad you shared this with me." I thought of the hours I'd sat with Lorraine and Rad during her labor, not knowing all the while Rad had a large knife on him. "We still don't know what's going to happen with Derrick, Jr., but I think it's important that you talk to someone about all of this."

"No. No. Rad said he'd kill me if I told anybody."

I didn't know what to do. She was right. He might kill her if he knew what she'd told me. Who knew what he'd do to me, too. I swallowed hard. "I understand why you're afraid of Rad, but Derrick could be at risk of being hurt by him if you don't talk to somebody about this."

She looked right at me, her eyes suddenly flashing. "Ain' nobody gonna hurt my baby. I'll talk. I'll talk. Long as nobody hurts my little baby."

"I'll set it up for someone to come talk to you."

*　　　　　　　　*　　　　　　　　*

Jerry rolled his eyes. "Not another one," he muttered.

"What should we do? Who can she talk to?"

"Social service takes care of this kind of stuff. I'll write an order to consult social service. I hate getting into this stuff. Why don't you be the one to talk to the social worker?"

"Sure, I'll do it."

Several hours later I found myself speaking with a short, white woman whose stance, legs spread wide apart, arms tightly crossed, told me she wasn't easily intimidated. "We see this all the time," she was saying. "These women come in here to have babies from men who've been abusing them for years."

"But why?" I asked. "Why don't they just leave them?"

"Just leave them? First of all, he'd probably come after her and kill her. He's already shown he's capable of that. Then, where would she go? She's got no money, no place to live, and a newborn baby. If you were her, where would you go?" I had no answer. "We can set it up for her to go from here to a women's shelter. I just spoke to the NICU, and they said the baby's doing better, so it looks like he's gonna make it. She can go to this shelter with her newborn, and we'll arrange for nursing visits for her baby there. Given what's going on, we can probably extend her stay so she can leave when the baby leaves. Where she goes from there, that's up to her."

"I find it hard to believe what Rad tried to do to Lorraine in recovery."

"Don't. A few months ago, another one of these assholes tried the same thing. He got a lot farther. Tore out all that woman's stitches in the process."

"Thanks for taking care of this."

"There'll be some paperwork to fill out. Jerry said you'd be taking care of that stuff."

"I can't do much else right now." I wanted to do more, so much more, but felt frustrated and overwhelmed. Getting her through labor and "catching" her baby seemed so technical and trivial in the face of Lorraine's current struggles. Pulling Derrick, Jr., back from the edge of death from a Group B strep infection was technical, though anything but trivial. It required all the sophisticated diagnostic and treatment means of our most modern medicine. Yet even with this technical feat accomplished, Lorraine and Derrick, Jr., still faced an uncertain and likely tragic future.

As Mulanda before her, Lorraine was teaching me that true healing and health require a mixture of technical, social, and psychological skills and resources. Modern medicine, I was quickly learning, only valued the technical. Perhaps that was because the psychological and social challenges were too hard, as in Lorraine's case. But all too often, it seemed it was because physicians and the healthcare system *chose* not to address them. It was simpler and more rewarding to define the universe of challenges medicine would address as purely technical, and then solve these with ever more effective technology. This allows physicians to simply declare success and move on, even if the patient is not truly well or relapses because we haven't addressed the deeper causes of their difficulties. I was finding it was not within me to think in such a compartmentalized manner.

*   *   *

Several days later, Lorraine and Derrick, Jr. were getting ready for discharge. Lorraine wouldn't meet my eyes. "I'm so embarrassed," she finally said. "I don' know what to say."

"I just want to know you and Derrick, Jr., will be safe."

"I don' know. Rad's a bad man. I'm afraid he's gonna find us wherever we goes."

"I know that's a risk, but that's not a reason not to try to get away from him. He doesn't have a right to hurt you...you or Derrick."

Her eyes flashed again. "That's the only reason I'm gonna try. For little Derrick, Jr. here."

I paused. "Did you ever think of changing his name, given what we've now been talking about?"

"No. No," she said wistfully. "It's important he knows who his daddy is."

"Well, goodbye, Lorraine."

"Goodbye, Dr. Sheff. Thank you for what you done. I know you tryin' to help."

"Please, call me Rick. I'm not a doctor yet."

"Thank you, Rick."

*   *   *

"What do you think of this tracing?" Jerry asked, calling me into the labor room. He reached over to a fetal heart monitor and stretched out the folded pile of paper until it became a continuous, long sheet. A black squiggly line meandered along the full length of it.

"We haven't had our lecture on fetal heart tracing interpretation yet," I responded hesitantly.

"Well let me show you something anyway. See how the tracing has been hovering around 140 beats a minute for the past half hour or so?" I quickly scanned the long piece of paper, taking in that it showed both the baby's heart beat and the mother's contractions. I still couldn't see where he had gotten the number of 140. "Now look at the last few contractions. The heart rate is dipping down toward the end of the contraction and taking about 30 seconds to come back to its baseline." I could vaguely make out the pattern he was describing. "Because the heart rate isn't coming back to baseline as soon as the contraction is over, we call these 'late decelerations.' They are a sign of insufficient oxygen supply to the baby," he ended, with a slight note of alarm in his voice. "Do you know what to do when this happens?" he asked, looking directly at me.

"I…I'm really not sure," came my answer. "I assume we could give her oxygen," I volunteered, noting the nurse was just placing an oxygen mask over the woman's head.

"Yes. Any other ideas of how we could improve blood supply to the placenta?"

I paused. Looking around I noticed the patient was lying on her left side. "You've got her over on her left side, so I assume that's something else we can do."

"Exactly. But why would this help?"

My mind raced. "The inferior vena cava runs behind the uterus. I guess rolling her to her left side gets the uterus off the inferior vena cava, and this allows more blood to return to the heart so more can be pumped out."

"That is precisely right." I beamed. "The other thing we can do is increase her intravascular[7] volume by opening up her IV and giving her more fluids. If her tracing doesn't improve, she's headed for a section."

As we walked out of the room, I realized we'd had this entire conversation with our backs to the woman in labor whose tracing we'd been discussing. Jerry had never even introduced me to her. I didn't know how much she'd heard of what Jerry had explained, nor how much she understood. We kept walking. Something felt wrong. A few steps outside her room I stopped. "Excuse me, Jerry," I said, turning to go back into her room. "Hi. My name is Rick Sheff," I said, introducing myself to a very dark skinned black woman. "I'm a medical student here. I didn't mean to be rude by not introducing myself earlier."

"Oh, thas alright. I's used to it," came the reply, fatigue (or was it depression) in her voice.

"I still don't think it's right. Please accept my apology."

"Ain no need to apologize to me."

"Yes, there is. You deserve more respect than that." I reached out my hand. Hesitating, she finally reached back and took it. "Now we've officially met. Hope your labor goes well and your baby comes out healthy and screaming his or her lungs out."

Finally she smiled. "I'd like that," she said, dropping my hand and averting her eyes.

---

[7] Intravascular is composed of intra which means inside and vascular having to do with the blood vessels. So intravascular means the space inside the patient's blood vessels.

* * *

Several weeks later, toward the end of my month-long obstetrics rotation, a friend came to visit me from out of town. Sarah had been an old girlfriend, actually my first love from high school, but now we were just friends. She said she'd always wanted to see a birth and wondered if I could get her in to see one. I checked with Jerry, and he said we could pull it off, even though it was breaking a few rules.

That night Sarah came to the hospital. She put on a set of scrubs and waited in the on call room for me to get her when a delivery was about to happen. A few hours later we were lucky.

"I'll be scrubbing with Jerry. Since this is a service delivery, I'm going to do it with Jerry supervising. You can stand off to one side. We'll call you in when it's about to happen.

"I'd like to watch the whole thing, if I could," she responded.

"Alright. Jerry's cleared it with the nurse. Just step into the delivery room and stand in the far corner. Put your mask on before you go in." She entered in time to watch the woman's legs being tied onto the stirrups. Jerry and I finished scrubbing, came in with our masks and head gear on, put on the gowns and gloves, and draped and prepped the woman in the usual manner as her perineum bulged larger and larger. At the critical moment, I injected the lidocaine, cut the episiotomy, and delivered the baby. I'd worked out the mechanics, so there was little chance of dropping this one. "It's a girl," I called out. I suctioned the baby, clamped and cut the cord, and handed the baby off to the nurse. It was a great feeling. Jerry still did most of the episiotomy sewing, but I'd been practicing tying knots, so he let me do a few of the stitches. I was happy with the progress I'd made, and Jerry said I'd done pretty well, too.

As we wrapped up, Sarah left the delivery room. I emerged into the hall, beaming with pride for how well I'd handled the delivery. I looked at Sarah. Tears overflowed from her eyes, running down behind her surgical mask. I thought they were tears of joy from having seen her first delivery. But something didn't feel right.

"That was great, wasn't it?" I started.

Slowly, she took off the mask. "That was so sad, so very, very sad." The tears kept coming.

"What was sad?"

"That this is what we've done to giving birth. It should be so special, but instead that woman was tied down on a cold table, everyone was screaming at her, you cut that huge thing in her, and her baby was just whisked away. She didn't get to hold it for so long, not until they did all those things to it. They even gave that baby a shot before the poor thing was held in her mother's arms. It's wrong. It's all wrong." She cried softly.

I suddenly felt embarrassed. My thoughts flashed back to the anesthesia resident. Was it just a few weeks ago that I'd fervently hoped never to be so severely out of touch with the experience of any patient I treated? Yet here I was, ecstatic to have carried out the mechanical process of "catching" a baby, oblivious to this mother's true experience. And what of the newborn baby? I had never even considered what her experience might have been, or that it could have been or should have been any different. Had medical school already taken me so far from simple human sensitivities that I couldn't see what was so plain to Sarah? It didn't matter how sensitive I thought I was. The technology of medicine was exerting its irresistibly seductive draw on me, as it had on countless students of medicine before me and countless ones who have followed. The socializing power of medical training was threatening to prove stronger than I was, already molding me into a prescribed image of what a physician should be, an image I didn't embrace and thought I could overcome. In Sarah's eyes I suddenly caught a glimpse of how close I was to losing this battle. "You're right," was all I could say.

Chapter V

Surgery

"Who's next?"

The words danced on the paper held in my trembling hands, as I tried desperately to organize my thoughts. "John Allen is 78 years old. He has a foot ulcer which has not been doing well and he's now developed an infection," I began.

"No, no, no! Stop him right there," Dr. Lugano shouted. "Didn't anybody ever teach you to present a patient properly?" I forced myself to look into Dr. Lugano's eyes. They were blazing with rage. "Why do you waste my time like this?" he said turning on the chief resident, Mike. Behind Mike stood the fourth-year resident, a second-year resident and two interns. Except for Mike, all of them avoided Dr. Lugano's eyes. "Which one of you admitted this patient besides this excuse for a student here? I want to hear from someone who actually knows something about this patient."

"I admitted him," one of the interns said. Bill, his name was, had been on call with me last night. I knew he'd been up all night, but you couldn't tell from his face which was alert and ready for action. His wrinkled scrubs and white coat with a blood stain on it were the only telltale signs of his night on call.

"Well, get up here and present this patient, damn it. I don't have time to let some incompetent student fumble through this. We've got to finish rounds by 7:15 this morning so we can start in the OR on time, and we've got at least a dozen more patients to go. That's why we start rounds at 6 AM. Haven't I taught you anything?"

Bill stepped forward.

"This-is-the-third-PH-admission-for-this-78-year-old-white-male-with-an-18-year-history-of-type-two-diabetes-and-both-large-and-small-vessel-disease-who-presents-with-a-non-healing-ulcer-and-cellulitis-of-the-right-foot-which-has-failed-outpatient-treatment-now-requiring-admission-for-debridement-local-wound-care-and-IV-antibiotics," he droned.

When would I ever learn to present like that, I wondered, and without any notes at all? Bill went on to complete presenting the patient's history, examination, lab test results, assessment, and plan practically in the time it took me to get the patient's name out. Actually, I don't recall him mentioning the patient's name. But I didn't have time to dwell on this as we were already racing off to the next patient's room. "Rounding" as a team made for quite a sight. Dr. Lugano led the parade, followed in strict status order by the chief resident, fourth-year resident, second-year resident, the two interns, and with myself bringing up the rear. We moved quickly from one room to the next, their long white coats flying out behind as they walked. My short coat, ending at my hips, just didn't have the same effect. Outside each room one of the interns would briefly summarize the case, what had happened over the previous 24 hours, and any plans for that day. "This is the 68-year-old black woman with the non-healing venous stasis ulcer. Her cellulitis is slightly improved. She's due for further debridement[1] today." And we would move on—except that Dr. Lugano was required to at least "see" each patient each day. And so he did. Often "seeing" the patient was a wave from the doorway. When there was a particularly critical wound for him to examine, the intern would quickly cut off the bandages, Dr. Lugano would briefly examine the wound, and speaking as much to the wound as to the rest of us, he would state his assessment, which the intern dutifully noted in the patient's chart. Rarely did he ever speak to the patient.

       *   *   *

"How can you tell whether this ulcer is healing well or not?" I asked Bill. Then I looked up into Mr. Allen's anxious face and said, "Do you mind if Dr. Andrews takes this opportunity to teach me about the problem you have?"

"No. No. Go right ahead. I know you boys have to learn your stuff sometime, and I don't mind if this darn leg of mine can help you. I just hope you can learn something from my 'case', as I heard you guys call it, that will help you treat someone else sometime."

"You're looking for healthy granulation tissue," Bill explained. "This little patch on the edge, right here, that's what you want it all to look like when it's healing well."

"So I got a little part of that hole in my foot that's actually healthy?" Mr. Allen asked, attempting to seem flip, but his voice quivered just enough to tell us he was truly scared.

"Well, yes, at least that part I was showing to Rick," Bill responded cautiously.

"That's good 'cause I got a friend who had sugar like me and it ended up they had to take his leg off. I saw what it did to him, and I don't ever want to have to go through that."

"Well, that's why we're giving you the best antibiotics and checking you out at least a couple of times a day," Bill replied, trying to be reassuring, but I knew what his tone meant. Outside Mr. Allen's door he dropped his voice and said, "Did you see the dark area in the middle of the ulcer? It looks awfully suspicious for osteo."

---

[1] Debridement is the removal of dead or poorly healing tissue to allow the remaining healthy tissue a better chance to grow and fight infection.

"What's osteo?" I asked.

"Osteo stands for osteomyelitis. You know, a bacterial infection that gets into the bone. This guy's foot looks pretty bad. I think he's probably got osteo in the bone at the base of his ulcer. With his diabetes and bad circulation, I'm not sure he's gonna be able to heal this."

"What does that mean for him?"

"He's got a pretty good chance he'll lose that leg."

"You mean amputation?"

"Yeah, that's exactly what I mean." My stomach turned at the thought. "You'd better get used to it. Here on the vascular surgery service we see a lot of that. If the circulation blockage is in one of the big arteries, we can usually perform a bypass procedure and save the limb. But if the patient's got diabetes, that blocks both the big and small arteries, so they can't get enough local circulation to fight the infection and heal the ulcer. Osteo sets in, and sometimes gangrene, and then we have to take the leg off or they'll die. You chose a pretty depressing service for your surgical rotation."

"I didn't actually choose this service. I was just assigned to it. When will I get to see the other kinds of surgeries, like gall bladders, stomach, breast, and stuff like that?"

"Not on this service. You'll see bypasses, triple A's, carotid endarterectomies, and of course amputations. That's about it. Other surgeries will be going on in the ORs around us, but I guarantee you'll be doing a lot of holding retractors for Lugano, so you won't be seeing much other than vascular stuff. And of course you'll have the pleasure of seeing Lugano's antics in the OR."

"What antics?"

"You'll see soon enough. We'll be in the OR tomorrow."

*   *   *

We finished rounds just in time to make it to the OR. It was clear one was not to be late to a Dr. Lugano case. The first case of the day was a fem-pop bypass. This meant, Bill explained, that the patient had a blockage in the femoral artery running through his thigh, the very same artery I'd tried without success to dissect that first day of anatomy. His main symptom was a cramping-like pain in his calf that came on when he walked and went away when he rested. The surgery was to begin with "harvesting" or removing a vein from the patient's leg which would be used to "bypass" the blockage. The vein would be attached to the femoral artery near the groin and reattached to a portion of the artery further down the leg. The place it would be attached was behind the knee, and this portion of the artery was called the popliteal artery. That's what gave the operation its name, femoral-popliteal bypass, or fem-pop for short.

Dr. Lugano was already getting impatient as we entered the room. "Isn't my patient asleep yet?" he badgered the anesthesiologist. "What does it take to get your patient taken care of around here? Am I the only one who knows how to move quickly? I don't know what you think you're doing over there, but it's sure as hell not anesthesiology." Above his mask I could see the anesthesiologist roll his eyes. This obviously wasn't the first time he'd heard such encouragement from Dr. Lugano. He continued prepping the patient in silence.

"Well, look who decided to waltz in here," Dr. Lugano greeted us. We gloved and gowned, and I took my place next to Mike, who stood on the side opposite of Dr. Lugano. By now the patient was asleep and draped so Mike and Dr. Lugano got right to work. Even though Mike was a chief resident, Dr. Lugano began the surgery. He deftly made an incision high in the patient's thigh. He and Mike moved remarkably swiftly, clamping and "frying" each of the little bleeding vessels with the "Bovie," just as I'd seen in gynecology. Soon they had identified and isolated the large, saphenous vein which was exactly where Dr. Lugano had made his incision. This should go quickly, I thought. They tied off one end of the vein. Then to my surprise, they moved down the leg to the lower end of the thigh and repeated the same process, isolating the vein in this area. Then again they made an incision, this time in the upper calf, and isolated the vein, and finally in the lower calf as well. Even though they were moving quickly, I could tell time was passing. I could also tell Dr. Lugano's reputation was well deserved.

"Come on, Mike, act like you've done this before," he chastised. "No, no, no! Not there, for Chris' sake, over here. Can't you get anything right today?" Mike's jaw clenched, but he didn't look up. "Stop it! Stop right there. You'd've killed this patient three times over if I wasn't here to make sure you got it right." As a senior resident, he'd been subjected to this abuse from Dr. Lugano virtually daily for almost five years.

When all four incisions were completed, Dr. Lugano began dissecting out the vein. Again he and Mike moved quickly, carefully identifying, tying off, and cutting each smaller vein connected to it. Finally, after over an hour, they had succeeded in removing, in one piece, the patient's saphenous vein in its full length from his groin to his ankle. Dr. Lugano lifted the wax colored, worm-shaped, precious prize onto a separate, small table. Hunched together over the table, their foreheads virtually touching, they repeatedly injected a clear liquid, into the vein, identified every leak, no matter how small, and carefully placed a tiny suture to tie off each leak.

"This one's looking pretty ratty, but I think we can make it work," he finally pronounced. "Bill, you and the student can start sewing up the harvest sites. We'll start preparing the graft site." They moved to the groin on the opposite leg and began working.

Bill handed me a needle holder with a needle attached to a long black thread and a pair of forceps. I looked at him bewilderedly, not wanting Dr. Lugano to see my panic. Bill nodded and whispered, "Watch what I do. Use the forceps to lift one edge of the wound, like this. Then place the needle into the skin at a right angle." He performed these motions deftly. Out came the needle in the middle of the wound. "Then grasp the needle with the needle holder, like this. Pull it out and place it through the other side of the wound, like this." Out popped the needle from the patient's skin. "Notice how it came out at a right angle. Now you grab it again, pull it through, then pull the suture just tight enough for the edges of the wound to come together with the tiniest hint of a pucker. Don't pull it too tight. That will decrease the circulation, slow healing, and increase the chance of an infection. Also, remember you can expect swelling at the incision site over the next few days, so just approximating the wound edges enough that they barely touch now will mean the swelling will bring them together with the right amount of pressure later." I admired how much Bill already knew about suturing as just an intern.

Now it was my turn. I had learned something about suturing from the little bit of sewing up of episiotomies they had let me do in obstetrics, but this felt very different. Picking up the edge of the skin at one end of the wound in the ankle area, I began. "Place the needle through the skin at a right angle," I said half to myself, my hands trembling slightly. The skin was tougher than I'd expected, and the edge of the skin slipped out of my forceps. "Damn it!" I hissed.

"What was that?" Dr. Lugano looked up at us over his dark black glasses suspended on the end of his nose, a clear edge in his voice.

"Nothing. I'm just giving Rick some pointers on suturing," Bill quickly replied as our eyes met. We each held our breath until Dr. Lugano turned back to his work with Mike. We let out a collective sigh, both knowing we had just dodged a Lugano outburst directed at us.

The work went slowly. I glanced up at the part of the wound Bill was working on, and it looked very different from mine. Each of Bill's sutures was placed an equal distance from the one before it, making a perfect ladder-like effect on the patient's skin. Every one of my sutures looked different. Some were too close together. Some too far apart. Some were placed at a skewed angle. It reminded me all too much of my first grade drawings. Something about my fine motor skills, I once heard a teacher tell my mother. The words of Dr. Sanchez came back to me. "I'm not sure surgery should be your specialty of choice," he had said that first day in anatomy lab. Today I was sure it would not be.

"We're ready here," I heard Dr. Lugano announce. "Alright, Mr. Anesthesiologist, let's see if you can earn your keep and give my patient 5000 units of heparin IV now." The doctor at the head of the table ignored the jab and started to administer the heparin he'd drawn up in anticipation of Dr. Lugano's order.

"Come on, Mike, give me better visualization here," Dr. Lugano barked. He put out his hand for an instrument, but the scrub nurse had just turned away to give instructions to the circulating nurse and didn't see it. "What's a surgeon got to do to get a little help around here?" he barked at her. "Help me. Help me. Somebody help the doctor!" he whined, his voice cracking with frustration.

"Watch out," Bill whispered, "Here it comes."

"Hey, you at the head of the table," Dr. Lugano shouted. "You call yourself an anesthesiologist. How come my patient's still clotting here? I thought I ordered you to give 5000 units of heparin eons ago."

"I gave the heparin when you ordered it," came the reply.

"You may think you gave it. You may have dreamt you gave it. You may have given it to this poor bastard's arm board. You may have given it to his pillow, but by God you didn't give it to my patient."

"I gave 5000 units IV at 9:27," was the tight-lipped answer.

"I know clotting blood when I see it, and my patient is clotting. If you want to take responsibility for the graft clotting off and this guy losing his leg, that'll be on your conscience. But I won't let that happen. You give my patient 5000 units right now while I'm watching so I can make sure you give it right."

"That would be a double dose, and it would be dangerous," the anesthesiologist warned.

"It'll be more dangerous for you if you don't. I'm watching now. I want to see 5000 units of heparin in my patient's arm right now.

The anesthesiologist drew up a syringe with the heparin and injected it slowly into the patient's IV. Dr. Lugano smiled with satisfaction. "That's a good boy. Maybe now I can go on with my surgery, if it's alright with you, Mr. Anesthesiologist."

"Uh, Dr. Lugano," Mike began, the fact that he didn't address him by first name alerting all in the room but me that a real problem was in the making. "There's some excess bleeding in the surgical field here."

"What are you talking about?" Dr. Lugano hissed. He returned to the groin incision where he'd already attached one end of the vein. It was oozing blood rapidly from the site of attachment and the blood was filling the surgical field. "Damn it. Get me some suction here! This shouldn't be happening." He and Mike worked rapidly, but the blood kept coming. "Don't just stand there," he yelled at the anesthesiologist, "give him the protamine. Reverse the damn heparin. Can't you see he's gotten too much heparin?"

Nobody said a word.

*   *   *

"This is the 78-year-old male with type two diabetes, large and small vessel disease, and a nonhealing ulcer over the medial maleolus of the right ankle. I haven't liked the way the wound has been looking the last few days, and we've already established he's not a candidate for bypass," Bill informed the group on rounds. Dr. Lugano then led the parade of white coats into Mr. Allen's room. He looked up, startled at our entrance since we usually stood outside his door for rounds.

"Cut off the bandage," Dr. Lugano grunted without addressing Mr. Allen. Bill cut off the bandage and carefully removed the gauze packing which filled the ulcerous hole in Mr. Allen's ankle. Dr. Lugano turned the foot one way, and then another. He put his nose down close to the skin and sniffed. "Looks like it might be osteo," he said, speaking to the ulcer itself. "How long has he been on IV antibiotics?"

"Twelve days," Bill replied.

"Get a bone scan. If it shows osteo, we've got to move on this."

"Wh-wh-what does 'move on this' mean?" Mr. Allen stammered, tears instantly welling up in his eyes.

"Nothing. Nothing for you to worry about. That's just doctor talk," came Dr. Lugano's reply, and he walked out. In the hall he turned quickly to Bill and said, "I'm pretty sure he's gonna lose that leg, probably a BKA.[2] But don't say anything to him yet. It's best not to worry him prematurely."

"Shouldn't he be given some warning," Bill asked. "He may need some time to get used to the idea."

"Look, Bill," he said in his best effort at a kindly tone. "I know you want to help him. But I've been at this game a long time. Telling the patient too early in treatment he's likely to lose a leg just upsets him unnecessarily. If he's going to get upset, he can do that when the time comes. Who's next?"

*   *   *

The blood had long since drained out of my hands. This was the third triple A of the week, and I had learned early on to dread this particular operation. A triple A meant an abdominal aortic aneurysm, a widening and weakening of the main aortic artery as it ran down through the abdomen.

---

[2] BKA is an abbreviation for below the knee amputation.

Left untreated, it could leak or rupture, and that usually meant certain death. So Dr. Lugano prided himself on how many lives he saved every week by placing an artificial graft in patients with these aneurysms.[3]

Unfortunately, the procedure had to be performed deep in the abdomen, which meant the surgeons were operating in a relatively narrow hole. To be able to properly visualize everything they were doing, they needed the major abdominal organs held up and out of the way. I had learned all about the lot of medical students in holding retractors during gynecology, but those procedures had lasted only an hour and a half. Triple A's could easily take three or four hours of retractor holding. And, unlike during gynecology, I could see almost nothing of what the surgeons were actually doing down in that hole, so there was little to learn. Besides, surgery had already lost its novelty. Gone was my awe and appreciation for the beauty and intricacy of our inner organs. It was replaced by an odd mix of boredom, fear of Dr. Lugano's outbursts, and that incessant desire to help so just once I would hear "Good job" or "That's the right length to cut the suture ends" or simply "Thank you."

I had watched for weeks as Bill or Mike had helped Dr. Lugano out, cutting sutures and suctioning when blood or fluids filled up too much of the surgical site in which they were working. I really did want to help. Or was it that I wanted approval for a job well done. I'd learned enough about myself before entering medical school to know I had a tendency to seek approval from those least likely to give it. (A legacy from my own demanding father, no doubt.) But to seek approval from Dr. Lugano? That was hopeless. Still, I couldn't stop myself.

It was particularly hot that day, and the circulating nurse repeatedly had to carefully dab sweat off Mike's and Dr. Lugano's foreheads so it wouldn't drip into the surgical field and contaminate it. It was critical that triple A procedures maintain the utmost of sterility, because if just a few bacteria got into the graft, it would serve as an infected foreign body inside the patient, and that would result in a life-threatening infection that would be impossible to treat without opening the patient up and removing the graft.

As the operation wore on, I noticed the suction tip sitting idly on the sterile drapes covering the patient's chest. The thick plastic tube, which eventually connected with the wall suction valve, lay coiled tightly under it. I leaned forward and noticed, deep in the abdomen where Mike and Dr. Lugano had just opened the aorta, that blood was pooling around where they

---

[3] At the time of my training, abdominal aortic aneurysms could only be treated through major, open abdominal surgical procedures. Though this procedure is still sometimes required, today these aneurysms are frequently treated by placing a stent, which is a small hollow tube, inside the blood vessel. This can be done from a small incision in the patient's groin and does not require the major surgical procedure described here.

were working, even though the aorta had been clamped off. Bill had his hands full, as did Mike and Dr. Lugano. I realized this was my chance to do something more than just hold retractors. They needed my help. I shifted the retractor to my right hand, and with my left hand, reached out to pick up the suction tip.

All I wanted to do was hold the suction instrument on the edge of the area in which they were working to suck away the excess blood and fluids. They would see that I wanted to help. They would appreciate how much I cared and was sincerely trying. They would know that I had truly helped at this critical time. But it was not to be. I will never know if it was because I used my left hand, or because my hands were stiff and half asleep from holding the retractor for so long, or just my inexperience, but when I lifted up the suction instrument, the tension of the coiled tubing released. The instrument sprung out of my hand, flew up striking Mike on the forehead, and landed squarely in the splayed open aorta in the middle of the surgical field. A stunned silence filled the room.

"Get this menace out of my operating room!" exploded Dr. Lugano. "He's clumsy. I've seen it from the beginning. He's clumsy and doesn't belong in my or anybody else's operating room. He'll never amount to anything. Get him out of here. I don't ever want to see him back in my operating room again!"

Bill came around to my side of the table and took the retractor out of my hand. His eyes, visible just above his surgical mask, let me know he was truly sorry for me at this miserable moment. I skulked away, hearing Dr. Lugano's voice bellowing after me. "If I ever see him in my OR again, I'll have somebody's head. Now get me five liters of saline with two grams of Ancef[4] to irrigate the hell out of this surgical field he's just contaminated, and let's hope he hasn't killed my patient! Somebody help the doctor!"

   *   *   *

Humiliated and ashamed as I felt, it only took a few minutes for me to realize that banishment from Dr. Lugano's operating room was more a gift than a curse. I had at least three hours before my team would be out of the OR. I paused, wondering what to do with this newfound time and freedom. Passing a board where the day's OR schedule stood posted, I noticed one of the surgeons was scheduled to begin a mastectomy shortly. I found him scrubbing, introduced myself, asking if I could observe his surgery without scrubbing in. He seemed flattered by my interest and welcomed me. My awe

---

[4] Ancef is the brand name for cephalexin, an antibiotic used to treat bacteria that contaminate our skin, including streptococcus and staphylococcus.

of surgery returned as I watched him matter-of-factly cut deeply into this woman's breast, peel the entire breast away from her chest wall, and place her breast in a steel surgical pan. While he proceeded to close and reconstruct the chest wall where her breast had just been, I was drawn to the table on which the scrub nurse had just placed the bowl, on its way to the pathology lab for examination. There lay her breast. It appeared noticeably shrunken, as if removal from her living form had already drained any vitality out of this so recently living tissue. Once again the mystery arose— what was the allure of a woman's breast and how did it differ from this mass of flesh, the areola and nipple staring up from it, that already was beginning to putrefy before my eyes.

This was only the first of many new surgical procedures I was able to observe, courtesy of my banishment. I did, finally, watch gall bladder surgeries, as well as surgeries on the colon, small intestines, stomach, spleen, pancreas, and thyroid. The other surgeons appreciated my questions. They were gracious to me. They taught me that Dr. Lugano did not represent all, or even the majority, of surgeons. When the enlightening and educational part of each surgery ended, I was free to leave and find another interesting procedure, rather than spend my valuable learning time holding retractors or cutting sutures. Unbeknownst to Dr. Lugano, he had mightily contributed to my education in surgery, though not in the manner he had intended.

I was also struck by the latitude Dr. Lugano appeared to command in his rude and abusive behavior. Nobody stood up to him, much as nobody stands up to a bully. Clearly he had behaved this way for years. Why, I wondered, was his behavior tolerated? Little did I know it had much to do with the amount of business he brought to the hospital. As long as he kept bringing patients who in turn brought revenue to the hospital, his behavior would be tolerated by fellow physicians and especially hospital administrators.[5] I couldn't help but think every episode of his antics demeaned the profession of medicine, at least the profession I had aspired to join. In this way, Dr. Lugano contributed something else to my education in an unintended manner. He taught me to do the opposite of almost everything he did.

*   *   *

---

[5] Hospitals and physician leaders across the country are finally tackling this problem, which has come to be called disruptive physician behavior. The risk of losing the physician's patients and revenue still makes this difficult for many hospitals. The other challenge is that physicians at hospitals organize themselves into a self-governed medical staff, and it is very hard for physicians to hold their peers accountable for behavior through this self-governed medical staff. By investing in leadership training for medical staff leaders, some are making progress in reducing or eliminating disruptive physician behavior.

Though I was free to explore other operating rooms, I was still tied to our team for rounds and patient care. So I was there when the day came that Mr. Allen had dreaded. Bill had just completed his usual description of Mr. Allen's non-progress in healing his ulcer. The bone scan had confirmed osteomyelitis, but Dr. Lugano had wanted to give a few more days of antibiotics to see if the wound would improve with a change in antibiotics. Dr. Lugano led the team as we swarmed into Mr. Allen's room. He looked around, and was noticeably disturbed that Mr. Allen was nowhere to be found. He paused, as if an idea came to him, and made straight for the bathroom door in the room. He flung it open, and there we all saw Mr. Allen, sitting naked on the toilet.

Dr. Lugano did not pause for a second. "Let me see that foot," he said more to Bill than to Mr. Allen. Bill reached down and dutifully cut off the dressing as we all tried to look away to spare Mr. Allen any further embarrassment. "I don't like it. I don't like it one bit," Dr. Lugano muttered. Mr. Allen's look of shock became one of alarm. Dr. Lugano turned and closed the door to the bathroom. "How many days has he been on the new antibiotics?"

"Seven days, sir," replied Bill.

"That's it. We're done. The foot's got to come off," he said. Turning on his heel he swung open the door. Mr. Allen still sat naked on the toilet. "The treatment's not working. We'll have to amputate your leg. We'll schedule you for surgery tomorrow." Mr. Allen burst into tears. Dr. Lugano closed the bathroom door. We all stood stunned, listening to Mr. Allen's muffled sobs.

Dr. Lugano strode out of the room. "Who's next?" he asked.

*   *   *

"Great party," Marsha cooed as we tried to wend our way off the crowded dance floor.

"Two secret ingredients make this a great party," I replied, my lips close to her ear so she could hear over the throbbing music. The smell of her hair made me want to linger. "First, we planned it for Saturday night of a transition weekend." She looked at me puzzled. As a student in the social work program, she needed an explanation. "Everyone in our medical school class had a final exam yesterday. Mine was surgery. So all the medical students here are blowing off steam. We've just busted our butts, and now the pressure is off for one weekend before we start our next clinical rotation. Monday I start internal medicine, the toughest of them all."

"What's the other secret ingredient?"

"Lab alcohol," I said, pouring her a glass of punch. Again she looked puzzled. "My housemate smuggled some alcohol out of the lab he's working in. It's 190 proof and virtually tasteless. Makes for a great punch and an even greater party."

"Thanks for the warning," she said, sipping her glass more cautiously.

Marsha and I didn't need any alcohol to make the evening exciting. Her eyes, her smile, her gentle touch on my arm as I had introduced myself— these had already achieved an intoxicating effect. When we awoke in each other's arms the next morning, I knew I'd found the woman I'd been searching for. All the next day, we couldn't let the other out of our sight. Finally Sunday night came.

"Marsha…"

"Here comes the speech."

"What speech."

"The one when you tell me not to expect you to call."

"No. Well…yes. But not for the reason you think. I'm starting internal medicine tomorrow. I've been told it's more work than any other rotation. But the last 24 hours have convinced me we're supposed to be together. So if I don't call or see you much for the next two months, please don't think I'm not interested."

Marsha looked at me uncertainly. She, too, had felt the immediate pull between us. Now she wondered if I was letting her down gently. "Look, just be straight with me. I've been around enough to know when a guy's giving me a line. But I can't read you. You're confusing me."

"Don't be confused. This isn't a line. If I don't call you for the next two months, that's because of internal medicine, not you. Please be patient. I don't want to lose you."

"Something tells me I might regret getting involved with a doctor." She paused as our eyes locked. "But I'll be patient…for now."

"Two months. Just two months and then I'll call you."

Chapter VI

Internal Medicine

"So datz vot psychology iz!"

"Hi, Marsha, it's Rick."

"How was your first day of medicine?"

"OK. What are you doing tonight?"

"No plans. Why?"

"Can we get together?"

"Sure. It's five o'clock now. When will you get off?"

"I'm leaving now. I'll be there in half an hour. Let's do dinner."

So began my internal medicine rotation. No patient workup, no textbook reading, no on call requirement in the hospital ever seemed important enough to keep me away from Marsha. I didn't study much during that first month of medicine. But I did fall in love.

            *   *   *

"I can't breathe! I can't breathe!"

This was not the way I expected to be greeted by Mrs. Jackson as I entered her room. After all, I was only trying to get my patient workup done for tomorrow's teaching session with Dr. Davis. My intern, Barry, had said this patient had an exacerbation of COPD, which stood for chronic obstructive pulmonary disease or emphysema. We both thought she would be a good case for me to workup since I hadn't done a history and physical on a patient with this problem before, and it would be a good way for me to learn more about COPD.[1] But as I casually entered the room, I hadn't expected to find this terribly thin woman laboring with every breath, the muscles of her neck standing out as she gasped for air, imploring me with panic in her eyes. "I…I'll get some help," I stammered.

"Help? Help? Why can't you help me? Aren't you a doctor?"

"No. Actually I'm a medical student."

"Isn't that a doctor?" She could barely get the words out. "Can't you help me?" she begged.

"I wish I could, but I'll get someone who can." I ran for the phone at the nurses' desk. "Barry, you've got to get here immediately. Mrs. Jackson can't breathe. She looks like she's going to code any minute."

---

[1] COPD (pronounced by saying the name of each letter, C-O-P-D) stands for chronic obstructive pulmonary disease. This refers to the spectrum of illnesses, including emphysema and chronic bronchitis that usually result from a lifetime of cigarette smoking.

"I'm working up the guy in 217 who's just started complaining of chest pain, so I can't come right now. I left Mrs. Jackson a half hour ago. She was pretty anxious, but I told her the nebulizer treatment could get moved up. The respiratory therapist should be there any minute. Besides, the aminophylline will be kicking in pretty soon. The steroids we gave her won't really be helping until tomorrow morning. We're doing everything we can for her short of intubation.[2] Do you think she needs to be tubed[3] right now?"

"I don't know."

"Does she look like she's tiring? Can you tell me how much air she's moving with each breath?"

"I'm not sure I can answer either of those questions. This is my first patient I've ever seen with emphysema, at least with emphysema this bad."

"They're all scared when they get an exacerbation. I guess it's kind of understandable since they feel like they can't get enough air, but she'll get through it. Just call me if she looks like she's getting worse."

"How will I know?"

"Listen to her chest. Her expiration will be longer than her inspiration, the opposite of healthy lungs. That's just the way these chronic lungers are. When I examined her she was about three to one, meaning her expiration was three times longer than her inspiration. That's one of the things you can follow. Also, if her breath sounds are getting louder, and you can hear wheezes and rhonchi and stuff, then she's moving air better, which is another good sign. But if her chest starts getting quieter and her expiration gets even longer than three to one, then you'll know she's going in the wrong direction. A really ominous sign is if she gets somnolent, meaning if she gets sleepy. This can be a sign she's fatiguing from breathing so hard and starting to retain CO_2.[4] It's called CO_2 narcosis.[5] Page me immediately if that happens. Gotta go."

---

[2] Intubation is the procedure of placing a plastic tube into a patient's mouth or nose, through their vocal cords and extending into their trachea, the upper airway leading to the lungs.

[3] Tubed is a short hand term for intubation.

[4] CO_2 is the chemical formula for carbon dioxide, the waste product of our body's metabolism that is eliminated through respiration. When physicians use this term, they usually say it as C-O-2, using the letters and numbers of the chemical formula.

[5] Healthy people feel an urge to take a breath when their CO_2 levels rise. Patients with COPD become accustomed to living with higher CO_2 levels than normal. Their urge to take a breath comes from a drop in their oxygen level, a phenomenon called hypoxic drive. If we provide oxygen to COPD patients, their oxygen level rises and they lose this hypoxic drive to breathe, which causes their CO_2 to rise to dangerous levels. The earliest symptom of this rise in CO_2 is the patient feeling sleepy, the phenomenon called CO_2 narcosis. If CO_2 narcosis develops, the amount of oxygen

The phone went dead. With a pit in my stomach, I returned to Mrs. Jackson's room. Would I know whether she was getting better or not? Would I be able to know with any certainty if she was moving into a life threatening state in which emergency intubation would be needed? I certainly didn't feel up to the task. As I entered her room I found to my great relief that the respiratory therapist had already started giving her a treatment with a cold air vaporizer called a nebulizer that mixed cool steam with a medication called albuterol. This same time last year I'd been sitting in a pharmacology lecture hearing about beta receptors on the surface of cells in the airways and the group of drugs like albuterol that could stimulate these receptors and relax the smooth muscles of the bronchial tubes. Tonight I watched as this magical elixir, delivered straight into the lungs through an odd looking plastic pipe and mouthpiece, miraculously opened up Mrs. Jackson's airways. From across the room I could see she was breathing better, the panic gone from her eyes as she clutched the mouthpiece in her hands and eagerly sucked in the healing mixture. She paused just long enough to take the mouthpiece out and say, "Thank you so much for all your help. I don't know what I would have done without you." With that her lips grabbed hold of the mouthpiece as she resumed her hungry, near desperate drags.

*   *   *

"This is the third PH admission for this…uh…uh…" I glanced down momentarily at the notes in my hand. "…uh…64-year-old woman with a history of COPD and…uh…" I glanced down again.

"Alright. Why don't you try it again?" The voice was so warm, so fatherly, like he really cared. "Take a moment to compose yourself and start it over one more time. Before you do, think to yourself which of her medical problems are relevant for me to hear in the beginning so I can understand what's going on with her present illness. Also, we usually say male or female rather than man or woman. It's just common usage. Oh, and by the way, I don't think you mentioned her name, did you?"

I glanced around at the group of 20 students who were all participating in this internal medicine rotation with me. Though I was a little nervous, unlike doing rounds with an impatient attending and team of residents, under Dr. Davis' leadership this session made us all feel like we were learning together. I took a deep breath, closed my eyes for the briefest of moments, but not so long that anyone would notice, and started again. "This is the

---

the patient is receiving must be reduced or the patient can become dangerously ill or die.

third PH admission for Barbara Jackson, a 64-year-old female with a history of COPD and diabetes who presents with an exacerbation of her COPD and chest pain that began just a few days ago."

"That was better. I need to know about her diabetes up front because when we treat her with steroids it will make her diabetes harder to control. You could shorten that last part by saying 'new onset chest pain'. Just saying that phrase tells the listener the category of conditions to start thinking about in a differential diagnosis of the patient's illness. And given the racial and ethnically mixed population we treat here, it is customary to mention the patient's race and sometimes ethnicity in this opening sentence of your presentation."

It was finally beginning to make sense. Here I was, taking internal medicine as one of my last basic rotations, but it seemed like the best foundation for all the others. At least that's the way Dr. Davis taught it. I wished I'd started with medicine. But from my classmates who had started that way, I'd heard they had felt lost for much of their medicine rotation. It seems no one rotation is the best way to start. (Of course, the order of my rotations had been determined by the scheduling computer without the least concern for the order that might be best for me or any other particular student.) During my psychiatry rotation I had learned how to do a mental status examination and to be comfortable with emotionally disturbed and disturbing patients. During gynecology, I'd learned to do a pelvic exam, to handle myself in an operating room, and to write preoperative and postoperative orders. I'd learned the basics of how to deliver a baby during my obstetrics rotation and to handle sleep deprivation for the first time, along with starting an IV, a little bit about tying knots for suturing, and how to use medications to manage pain. I wasn't quite sure what I'd learned during my surgery rotation, but somehow I'd eventually been exposed to a wide range of surgical conditions and procedures and had truly begun to understand the wound healing process. Along the way, I'd slowly absorbed information about the countless illnesses to which the human body can succumb as one patient after another became my teacher, showing me how each of those conditions, which had previously only been words in a textbook, appeared in a live, feeling person. The same was true for the mind-numbing number of medications to which I'd been exposed during my basic science course in pharmacology as their names, which had run together so much in my mind, became linked one by one to my experiences with specific patients and their very personal stories. There were so many more illnesses and medications to learn about, in fact an overwhelming number. But somehow, by the time I'd arrived at my internal medicine rotation, I'd developed enough of a framework of understanding and knowledge that new pieces of information found a ready home. The half-life of this information was still usually a matter of weeks, but I was beginning to see the truly important and clinically relevant information again and again, and it was definitely sticking with me better. I was beginning to grasp the "whole" of medicine, and this before the end of my second year of medical school.

Was there any way to avoid the sense I'd had of feeling so uncertain and overwhelmed during my early medical training? I had chosen this medical school because of its progressive program of covering all the basic sciences, such as anatomy, biochemistry, physiology, pathology, and others, in the first year. This allowed students who so chose to go directly into clinical rotations at the start of the second year, as I had done. I loved this approach. As anxiety provoking as it had been, immersing myself in the clinical world of caring for patients had brought all of the dry book learning from the basic sciences to life. Most other medical schools covered the basic sciences in two full years of classroom work before students were allowed into clinical rotations. My friends who had followed this traditional curriculum had spent much of their first two years of medical school feeling increasingly bored and frustrated, though of course overwhelmed with the volume of material to be mastered, while I had spent that same time being anxious but stimulated, and equally overwhelmed. My encounters with patients, both their stories and their challenging clinical problems, motivated me to learn for their sake, and for the sake of the countless patients I now knew would come to me for my knowledge and skills as their doctor.

With this motivation, I'd thrown myself into the unique training Dr. Davis offered during the first month of my internal medicine rotation. He was known for being the best teacher in the school for learning how to perform a really first rate history, physical exam, and write up, though his requirement for details and comprehensiveness was legendary. "I read your write up on this patient quite carefully," he began, "and I want to commend you for an excellent history of present illness. However, your review of systems certainly left something to be desired. You only listed three conditions under each system."

"How many should I have listed?" I asked. After all, the review of systems had seemed like an endless list of symptoms the patient might have. I emphasize *might*, because that was the whole point of the review of systems. Once I'd completed asking about the history of her present illness that had caused Mrs. Jackson to seek medical care, as well as her previous personal medical history, current medications, and family medical history, I had to go through that list of symptoms for every single system. Oh yes, I'd also had to ask about her social history, like who she lived with, what her job was, and whether or not she smoked, drank, or did any illegal drugs, as well as learn that her husband had died four years ago, before getting to the review of systems. But once I got to this list, it did seem endless. "Do you have headaches?" I had started.

"No," she'd answered.

"How about double vision?"

"Nope."

"Loss of hearing?"

"No."

"Ringing in your ears?"

"Not that I recall."

"Difficulty chewing or swallowing?"

"No."

"Do you wear dentures?"

"Yes." Finally, something positive on the review of systems.

"Do they fit well?"

"Yes, of course."

I had gotten no helpful information and my questions seemed to irritate her. She was tired, just beginning to be able to catch her breath, and still quite anxious. It had seemed like a purposeless exercise, especially since I hadn't even completed asking questions about the systems in her head, and I still had the rest of her body's systems to go through. Hence my question to Dr. Davis: "How many symptoms should I have listed for each system?"

"As many as it takes," came the reply. "At this stage of your training, you need to list all of the negatives. Later on you'll learn which negatives are pertinent to the specific clinical problems with which the patient is presenting. At that stage you'll be able to present just the pertinent negatives. For now, I don't expect you to know which are pertinent and which aren't. Besides, by learning all of the possible symptoms for a truly comprehensive review of systems, you are creating a great foundation for your future clinical practice." He paused. "And one more thing. If you learn nothing else from me this month, I want you to come away knowing that a good history will tell you far more about your patient's diagnosis than any exam finding or laboratory test."

This did make sense, but it was a definite pain in the butt. In fact, my completed history and physical write ups for Dr. Davis averaged 12 pages. Not only did I have to include all of the content I'd gathered from interviewing and examining the patient, and all the negatives of what the patient didn't have, I had to learn to build a comprehensive list of the patient's problems and to include an assessment and plan for each of the problems. This meant spending time in the library reading about the diseases the patient might have, understanding the different ways they might present, and developing a differential diagnosis. This differential diagnosis was a list of possible diseases which might be causing the clinical picture with which the patient was presenting. My analysis needed to include specific steps I'd recommend to either rule in or rule out each of these possible diagnoses.

For my previous clinical rotations, I had done lots of write ups, but frankly, I'd given the assessment and plan portions short shrift. One reason was that I didn't have enough clinical knowledge of each of the specialties to do this well, so I had provided the minimum necessary to get by. My fellow students seemed to spend more time reading about the diseases we saw than I did. Was it my imagination, or were they spending a lot more time both in the hospital and in the library than I was? Though this thought lingered in the back of my mind, I consistently felt there were better ways to spend my time. After all, I had assumed I'd be practicing psychiatry, so I wouldn't need to know much about these other specialties. But over this past year, I was surprised to find myself more intrigued by learning about the clinical conditions treated by other specialties than I had expected. Now, during my internal medicine rotation, I finally felt motivated to deepen my knowledge of the medical diseases I was seeing. And who better to do this with than Dr. Davis?

"So what is your differential diagnosis for Mrs. Jackson's chest pain?" he asked.

"It seems a little unusual for angina," I'd begun.

"That's true. By the way, the best way to communicate this in your introductory sentence of your presentation is to refer to it as 'atypical chest pain'. Again, this cues your listener to be thinking about a particular list of potential diagnoses and hints at where your assessment and plan is going early on. However, you didn't answer my question. When I ask for your differential diagnosis, I expect to hear a succinct list of diagnoses you are considering for this patient."

"The differential diagnosis for Mrs. Jackson's chest pain includes musculoskeletal pain, coronary insufficiency, and pleuritis."[6] I actually sounded like a doctor.

"That's much better. Did you consider the diagnosis of pulmonary embolus?[7] This is a frequently missed and potentially lethal condition."

---

[6] Pleura is the name for the thin tissue layer that lines the outside of the lung. Itis is a term that means "inflammation of." So pleuritis is a term for inflammation of the outermost lining of the lung. It can occur with lung infections as well as with other lung conditions.

[7] A pulmonary embolus is a blood clot that originates in the venous circulation, usually in the legs, and breaks off, traveling through the right side of the heart and out to the lungs where the clot becomes lodged. It is a potentially serious, even life threatening, condition and can mimic many other clinical conditions that cause chest pain.

*   *   *

The next morning on rounds, Mrs. Jackson looked remarkably better. The rate of her breathing had slowed, and she no longer appeared to be straining her neck muscles with every breath. She smiled at me as our rounding team entered the room. "There's the doctor who helped me so much last night," she'd announced to the team, pointing in my direction.

"It's good to see you feeling better this morning," I said somewhat sheepishly.

Outside the room Barry said sarcastically, "How does it feel to be a hero?"

"But I didn't do anything," I began protesting.

"You know that, and I know that, but Mrs. Jackson doesn't know that. Besides, you going back in there when I couldn't actually let her know our team was keeping an eye on her. Why don't you take over rounding on her each day? I'll just look over your shoulder."

"I'd be glad to." I was beaming.

*   *   *

"Haloo, Dr. Sheff," Mrs. Jackson called out as I entered her room three days later. I had explained to her my status as a medical student multiple times, but she insisted on calling me Dr. Sheff. Despite my protests, I enjoyed it. "I was hoping you'd come in to see me early this morning," she went on. "I think a stranger came into my room last night. Who do you think that was?"

"I'm not sure," I started. "There are lots of people who might have to come into your room at night, like the nurse or the custodial staff." Something about her manner had changed. There was a slight pressure to her speech.

"No. It wasn't them. I'm sure of it. Someone was in here, and he meant to do me harm," she said darkly. "I'm really quite worried."

"I'm sure there is nothing to worry about, but I'll ask the nurses to look into it."

I mentioned her concern to the head nurse, and she said the night nurse had reported that Mrs. Jackson had become fairly agitated. They had to give her a sedative to get her back to sleep. This morning her nurse had found her pacing in her room. I discussed this with Barry who thought she might be "sundowning." This was a term for elderly patients who became disoriented, particularly at night, and showed signs of delirium. I'd learned about this phenomenon during my psychiatry rotation, though Mrs. Jackson was a little young for this condition. The best treatment was lots of contact with staff and family, and discharge from the hospital back to a familiar environment as quickly as possible. Barry thought she'd be ready for discharge tomorrow and that an extra dose of a sedative this evening would get her through until she could be discharged.

The next morning I came to see her before rounds. She was wide eyed, pacing frantically, and speaking nonsense.

"I don't know where it is?"

"Where what is?" I asked.

"My dishpan. I had it a moment ago. I was just cooking supper for my husband." I knew her husband had died four years ago. "Where is it? I must get supper ready." Then she looked at me. "Who are you?"

"I'm Rick Sheff, the medical student who's been taking care of you the last few days."

"Really. I don't recall. What are you doing in my home? Are you coming to dinner tonight? Where is Harry? He's never this late."

"Will you excuse me, please? I have to check on something." At the nurses' desk, the head nurse confirmed that she had been terribly confused all last night, and it hadn't gotten better this morning. "Barry, something's really wrong with Mrs. Jackson. She is confused and hallucinating. She thinks her dead husband is alive and that she's cooking dinner for me right now."

"She's a COPDer, right? How much has she been getting for a steroid dose?"

"It looks like her solumedrol has been kept at her admission dose for the last four days."

"This could still be sundowning, but the differential diagnosis also has to include steroid-induced psychosis."

"What's that?"

"Sometimes patients placed on high dose steroids develop a syndrome that looks exactly like psychosis, along with visual and auditory hallucinations. You'd better order a psych consult. Hey, didn't you tell me you wanted to be a shrink? This should be right up your alley."

We had to delay Mrs. Jackson's discharge until the psychiatry resident could see her. He confirmed the diagnosis of steroid-induced psychosis. Though her COPD had improved enough for discharge, she was clearly in a full blown psychosis and could not go home. The psychiatry resident recommended transfer to the inpatient psychiatric unit. This was in another building two blocks down the street from the hospital unit she was in now. By the time the transfer order was written by the psychiatric resident, it was change of shift time for the nurses. The head nurse, who had taken a liking to me because of my personal interest in Mrs. Jackson, asked if I wouldn't mind accompanying her for the two block walk to the psychiatry building since her staff was tied up in change of shift report. Given the comfort I'd developed in working with psychotic patients during my psychiatric rotation, I agreed to help out.

"Mrs. Jackson, we have to go for a walk." She was now dressed and her IV had been removed. This left her free to pace even more wildly around the room.

"Oh my. Oh my. What will I do?" she was asking nobody in particular. "I have to go now. I have to get home to my husband."

"You can't go home now. I have to take you to another building where you can get more treatment."

"Oh, that's so nice of you. Are you the one who will be taking me home? I really must get home. Harry will be wondering what's become of me. I must make him dinner right away."

"As I said, you won't be going home just yet." We had started walking toward the elevator.

"Oh, you are a nice man. Where are we going shopping?"

"We aren't going shopping. I am taking you to another part of the hospital so you can get more treatment."

"Yes. Yes. Of course. And what will you do while I am shopping?" The elevator had reached the ground floor, and we went out the front door of the hospital.

"This way. The other hospital building is this way."

"I'm in such a hurry." She began striding off in the opposite direction. I ran to catch up with her. Her COPD was starting to get a little worse with all her exertion.

Taking her elbow firmly, I said, "This is the wrong way. Come with me."

"No. No. No. I'm on my way…"

"Mrs. Jackson. I'm afraid you're still quite confused. I need you to come with me." I began to panic. We were out on the open street, in the middle of the city, and getting farther from the psychiatric building. What if I couldn't get her to come with me? I'm just a medical student, entrusted with this flagrantly psychotic patient. What will I do if she bolts? She can't get very far with her COPD, but she could get sick or hurt in the process. "Mrs. Jackson…" She paused and looked into my eyes. "Mrs. Jackson, I am trying to help you. Will you let me help you?"

"You…You…Do I know you?"

"Yes. I am Dr. Sheff. Your doctor. I've been taking care of you for the last few days."

"A doctor? Yes, a doctor. Yes."

"That's right. We are going in this direction." She allowed me to steer her back toward the psychiatric building.

"A doctor, did you say?"

"Yes, a doctor." Suddenly an image of Dr. Wilson, the attending from my psychiatry rotation, flashed before me. He was leaning forward taking the pulse of that hysterical woman. "Playing the medical card" he'd called it. I reached for Mrs. Jackson's wrist and started taking her pulse. She paused, apparently willing to allow me to do so. "I will be checking your pulse as we walk," I suggested, and with this, she started accompanying me without resistance. "You've got a strong pulse," I volunteered.

"Yes. My pulse..." She seemed content to just walk with me. The closer we came to the psychiatry building, the more reassured I felt that I would not be responsible for "losing" an ill, confused patient on the streets of the city. Finally, we walked through the doors of the psychiatry building. Once she was safely in the hands of the intake staff, I let out a long sigh of relief.

"Goodbye, Mrs. Jackson," I said, finally releasing her wrist. She seemed distracted by something or someone only she could see. She didn't say goodbye.

*   *   *

The second month of my internal medicine rotation was spent at the University Hospital. After our orientation meeting, I was sent to find the intern to whom I'd been assigned. To my pleasant surprise, it was Jerry, the OB/Gyn resident I'd worked with during my obstetrics rotation. He was now doing his six months of medicine, and we were assigned to work together for the next four weeks.

"You wanted to be a shrink, didn't you?" he began.

"Yeah, but to my surprise I'm enjoying this other medical stuff more than I'd expected. In fact, this month I really want to get more involved in what goes on in the hospital at night than I've done in some of my other rotations. I also want to take this month to learn more about how to do scut work."

Though I'd done a little of this scut work in my other rotations—starting IVs, drawing blood, looking at urine specimens under the microscope, and checking lab results—I had managed to avoid much of it. I had simply told each of the interns I was working with about my interest in psychiatry, and they seemed to be very understanding. What I hadn't known was that among interns and residents a common saying was, "Show me a medical student who only doubles my work, and I'll kiss his feet." In other words, teaching medical students slowed them down, which meant keeping them from some of the precious few moments of sleep they would be able to grab. So my "willingness" to skip much of the scut work was actually a blessing to them.

Jerry and I joked about my previous foray into starting IVs and drawing blood during obstetrics, and he agreed it would be good for me to get some more experience. I didn't have to wait long. "Mr. Jones in room 347 just spiked a temperature. He needs blood cultures and a CBC.[8] Why don't you get his bloods?"

"What's he in for?"

"Lung cancer. He's been getting chemo and has been having some side effects. He's all yours."

As I headed off to find room 347, my steps quickened and my shoulders pushed back from their usual hunched over posture. I began to feel important. I wasn't just doing workups for my own education. I was actually a member of the team and was about to make a contribution. I was going to help Mr. Jones and Jerry.

First, I had to stop in the supply room to get the equipment I'd need. When a patient spiked a temperature, I had learned, it meant they might have a new infection. Determining whether or not this was the case meant getting cultures of their blood and urine along with a CBC. If they were very sick, it sometimes meant getting a spinal tap, also called a lumbar puncture or LP, to make sure they weren't developing an infection in or around the brain, like meningitis or encephalitis. For now, Mr. Jones didn't need an LP, but to get his blood tests done I was going to need to draw at least 5 cc's of blood for each of two blood cultures and an additional cc for his CBC.

Five cc's was a lot of blood. Most blood tests could be run on 0.5 cc or so, and this was drawn using one or sometimes a few vacuum tubes that would suck up the blood into the tube once the needle was properly placed in a vein. Drawing a blood culture meant using a large syringe and manually having to pull back on the syringe plunger to draw up the blood. It also meant cleaning off the skin very carefully with Betadine, that yellow-orange iodine based soap whose smell I'd come to associate with the wards and the OR. I therefore loaded myself down with a pile of sterile gauze packs, a bottle of Betadine, two 10-cc syringes, two blood culture bottles, and the CBC tube. Now it was time to select a needle size for drawing the blood. The usual needle size for most routine blood drawing was 21-gauge. I thought to myself that it would take forever to draw up that much blood through a 21-gauge needle. Looking through the needles I selected two 18-gauge needles. These are significantly larger in diameter than the 21-gauge needles, and I reasoned they would allow me to draw up the blood much faster. This would increase my chances of getting all the blood I needed, and I assumed that faster would mean less pain for Mr. Jones.

Awkwardly clutching my treasures, I made my way to room 347. There I found a bald-headed black man looking much older than his 56 years. In

---

[8] CBC stands for complete blood count. This test measures the number of red blood cells, white blood cells, and platelets per milliliter of blood, as well as some of the characteristics of the cells. Platelets are the particles in the blood that initiate a blood clot whenever bleeding occurs.

fact, he looked both tired and ill. "Hi. My name is Rick Sheff. I'm a medical student here to draw your blood tests," I said cheerfully.

He looked at me with long suffering eyes and put out his right arm. "That's usually where they can get blood from me this week," he said, motioning to a tiny vein on the back side of his right hand. I stared at the thin, dark blue line I could barely make out through his black skin.

"I'm sure that vein will plump right up once I get this tourniquet on," I said, trying to be reassuring as I applied an elastic band around his arm. The thin blue line barely changed. "Let me look around before deciding which spot to use," I added, trying to maintain my upbeat tone. I moved the tourniquet from one arm to the other, but to no avail. I was having my first lesson in what happens to the veins of patients on chemotherapy. They experience so many blood tests and IVs that all the usual veins quickly become scarred and no longer usable. Mr. Jones was a veteran of this process and showed no surprise when I returned to his right hand.

"I done told you so," was his only reply.

Slowly I prepared the skin covering the back of his hand. The elastic tourniquet was back in place, tied as tightly as I could so the vein would swell as much as possible. Three separate times I soaked a gauze with the Betadine and swabbed it in the circular motion I'd been taught, slowly spreading outward in ever enlarging circles. I was aware of how long this was taking and imagined that Mr. Jones' hand was likely throbbing with the tightness of the tourniquet. "Now a little bee sting," I said, using that phrase I'd heard many times from both residents and nurses as they prepared to draw blood or start an IV. I paused with the large, 18-gauge needle poised just above the tiny vein. The needle was easily three times the diameter of the vein. Something about this didn't make sense, and I thought of abandoning the effort to ask for help. But if I could just get the tip of the needle into the vein, I thought, I'd have the best chance of getting out the most blood. Holding the skin down to stabilize the vein as Jerry had taught me, I pushed the needle tip under the thin skin towards my target. The syringe remained empty without a trace of blood in it.

"Yeow!" came the response from Mr. Jones. After this initial outburst, he seemed to settle back into his long-suffering mode. I glanced up at his eyes briefly only to find them glazed over, as if he was seeing something far away. When the initial thrust of the needle failed to produce the "flashback" of blood I had hoped, I was forced to reposition the needle under the skin. I could sense Mr. Jones' muscles tightening as I probed first in one direction and then another, all the while pushing this large bore needle further and further under the delicate skin on the back of his hand. A tiny trickle of blood appeared in the syringe, and I hoped I had finally entered the vein. I pulled back on the plunger of the syringe, which produced a harsh in breath from Mr. Jones, but no more blood in the syringe. Something felt very wrong. After what seemed like an eternity, I withdrew the needle. A dark stream of blood oozed from his hand which I covered with yet another gauze. "Did ya get it?" he finally asked.

"No. Unfortunately, I was not able to get it. I'm sorry. I'll ask Dr. Kosky to try." He said nothing, but his eyes took me in for a moment, and then looked away. He rolled over in his bed, leaving only his back facing me.

I didn't tell Jerry what I had done, meaning using the absurdly large needle and causing Mr. Jones so much unnecessary pain. I had already figured out my error and felt shamed by it. In fact, I never told anyone.

*   *   *

During the ensuing weeks, I tried my hand again at blood drawing, as well as starting IVs, performing urinalyses, and other routine elements of scut work. Though it took some effort to put my shame about Mr. Jones aside enough to try again, I carried with me two lessons he had taught me— never be afraid to ask for help, and don't proceed if something inside is telling me to stop.

One day Jerry said, "Here's your chance to learn how to do an LP. Follow me." We entered a room in which an elderly, demented woman laid moaning and tossing in her bed. "The first thing you need is a good, strong nurse or orderly holding the patient so she won't move." He motioned to a strapping black man on the other side of the patient's bed who grasped her simultaneously by the head and by her knees and rolled her onto her side in a tight fetal position. She struggled briefly in her confusion and then became still.

All the while speaking to me and not saying one word to the patient, Jerry first explained and then demonstrated the technique for performing an LP. He began by opening up the equipment tray in a sterile fashion and washing off the woman's back with Betadine, much as I had learned for blood cultures. This was followed by injecting a local anesthetic into the skin and a little into the track the LP needle would be traveling. He showed me how to identify the boney prominences on each vertebra, the spinous processes I'd learned about in anatomy, in order to determine the location and angle for placing the needle. He demonstrated how to set up the four sterile tubes for collecting the spinal fluid. Then he raised the spinal needle and demonstrated how it had an outer casing and an inner, removable core. I shuddered at its five inch length, imagining the needle embedded in this woman's back. Angling the needle parallel with the spinous processes, first he pushed it rapidly into her skin about an inch, and then slowly advanced it. Deeper and deeper into her back it went. Periodically he would remove the inner core to see if anything came out, but nothing did. She was quite thin, so there didn't seem to be much further to go. Suddenly Jerry informed me he felt a subtle pop. This was the signal he had been looking for that told him the tip of the needle had just penetrated the tissue layer that covered the spinal cord called the meninges. Sure enough, this time when he removed the core, a drop of crystal clear fluid appeared at the outer end of the needle. Moving quickly, he captured a few drops of this precious fluid in each of the four tubes he'd set up. Next he removed the needle and instructed me to hold a sterile gauze over the tiny pinhole he'd left in her back. With that, it was all over.

Moments later Jerry was explaining to me which test to order from the fluid in each of the four tubes. I was still amazed at the simplicity and elegance of the procedure I'd just witnessed. Something in me wanted to try it myself.

I didn't have long to wait. Several days later, Jerry informed me it was time to fulfill the long-standing dictum of medical training: watch one, do one, teach one. I had already watched one LP, so it was time to do one. This time it was on a very obese man who was unfortunately quite awake and alert. Again, we had the assistance of a strong and imposing male nurse who held the patient tightly in a fetal position. However, since he was quite with it, I spoke to him throughout the procedure, explaining each step just before I did it. Jerry was right there with me, but somehow I seemed to know exactly what to do. I palpated each of the vertebral spines, counting upward from his pelvic bone, and identified the space between his third and four lumbar vertebrae. I set up the tubes, washed off his back with Betadine, and administered the local anesthetic. Then lining up the LP needle, I poked it through the skin. In my mind's eye I began to visualize the anatomy of the spine and the muscles and tissue layers the needle was passing through. I stopped to check for the clear fluid often, but none came out. Slowly I advanced the needle. Deeper and deeper it went, far deeper than the first woman's had. Jerry and I discussed this and agreed the man's obesity meant the needle would have significantly farther to travel. Jerry stared at the needle, now embedded a full four inches, and worried out loud that it should have entered the spinal canal already. For a moment we discussed what other anatomy was in the area, identifying that a needle embedded too deeply might miss the spinal canal completely and hit the aorta, a thought that made me shutter. But both of us agreed this was very unlikely given his size and obesity. I had also not bumped into any bone, a sign the needle would have been off course. I paused, looking for that sense I'd so recently learned to identify that something was wrong which would tell me to stop. It wasn't there. Instead the needle seemed to be appropriately placed. I slowly guided it even deeper. Just as the hub of the needle was about to reach the skin, meaning the needle could not be advanced any further than its full five inches, I sensed an ever so subtle pop. Removing the inner core, I was thrilled and relieved to see a drop of that crystal clear liquid form on the end of the needle. I collected a few drops of the liquid in each of the four tubes and removed the needle.

As I held the gauze in place over the puncture hole from which the needle had just been removed, Jerry murmured, "Holy cow. That is probably the best LP I've ever seen. You're a natural at this." Finally I had found a procedure in medicine that came easily.

> *   *   *

"The ED[9] just called. There's another admission coming up for us," Jerry sighed. It was close to midnight, and this was the fourth admission of the evening. "They say three more are coming up behind this one." It was going to be a very long night.

I had decided this second month of medicine I would stay all night in the hospital, learning everything I could about not only scut work, but what it took to care for patients in the hospital through the night. In my other basic clinical rotations, my interns had been all too glad to let me go home by 10 PM (except for obstetrics, of course). They knew I was mainly interested in psychiatry so they didn't want to waste their time at night teaching me, especially since this would just slow them down and steal a precious hour or two of sleep from them. I knew many of my fellow students were staying all night. I could tell by how exhausted they looked the following mornings. Staying all night was encouraged but not required. I put such "encouragement" in the category of the bottomless pit demands of medicine, yet I knew I was not earning any "Brownie points" by cutting out at 10 PM. All too soon, I knew, it would not be my choice, and I would have to stay overnight in the hospital. I dreaded working through the night, forced to stay awake for 36 hour shifts straight, grabbing an hour or two of sleep if I was lucky. I'd had my first taste of this during obstetrics, and I was in no hurry for more. My physical health and sanity were more important than trying to impress my intern or attending physician.

Perhaps this attitude, outwardly so different from many of my classmates, came from having spent several years doing other things between college and medical school. During my senior year in college, I'd already been accepted into medical school when a scholarship to study at Oxford University in England literally found me. I had not even applied for this scholarship, but suddenly found myself faced with a choice of going straight to medical school or taking a two-year all expenses paid detour to

---

[9] ED stands for the emergency department, also referred to as the emergency room with the abbreviation ER.

study any subject of my choice at Oxford. I'd chosen the scholarship. The first year in Oxford had been intellectually exhilarating, as I chose to study politics and philosophy, and pursued tutorials and lectures with some of the most stimulating minds in the world. During that year my heart had been shattered by the breakup of a relationship with the woman I'd thought I would marry. As I became increasingly depressed and lonely, I had felt very much the expatriate in a foreign country. Yet I continued to pursue the study of politics and philosophy, passionately searching for unshakable foundations for truth and morality made all the more urgent by my unhappiness. I hadn't found such foundations, but rather more and more questions. Fascinating though they were, the pursuit of these questions did not meet the longing I felt, a longing for something that had no name. As I took my seat on the plane to return to the United States for the summer after my first Oxford year, a thought formed within me that I did not have to go back. I had never stepped off the lock-step order of one school year followed by another. I suddenly realized I could choose not to go back. I could turn my back on a full scholarship to Oxford and an Oxford degree to pursue something else. But what would that be? I knew I wanted to go on to medical school, but now had a full year ahead of me while I reapplied. Given a chance to do anything for a year, what would I do?

Piece by piece, a plan for the coming year fell into place. It began with arranging for a training experience as a group therapy counselor with a wise and wonderful man who became my cherished mentor as I took my first, faltering steps toward becoming a healer. I pursued long-standing interests in dance and theatre, eventually dancing full time in two performing companies. It was a year of exposure to many experiences that stretched and expanded my understanding of myself, of others, and of my world view. I began to find the kernel of answers in work and in relationships instead of more academic questions.

With all of these experiences, I'd developed a sense of myself and what is truly important to me that served as a resource as I confronted the demands and pressures of medical training. It was this resource I drew upon in making decisions for myself at every turn when the demands of that training pressed in on all sides and threatened to overwhelm me.

And now, I found myself more interested in medicine than I'd expected. Now I *wanted* to stay late in the hospital and learn more. Plus I had learned enough in my previous basic rotations to be of at least some assistance to Jerry through the night. It was now my choice to stay in the hospital overnight, and that made all the difference—or so I thought. Yes, choosing to stay in the hospital, choosing to work through sleep deprivation, was better than blindly and obediently doing so. But this capacity to make choices for myself in the face of the pressures of medical training, valuable though it was, would not be enough. I knew the relentless work pressure and sleep deprivation had the potential to warp me. I'd seen their insidious results in the older students and residents. I thought I could prevent it from happening to me. Tragically, even this knowledge and that sense of self I felt so strongly now would prove inadequate to protect me. I just didn't know it yet.

"How 'bout if we both check out this new admission together," Jerry suggested. "You can finish her workup once we get her orders going. She's a COPDer."

"I actually know something about COPD," I chimed in, as we walked briskly to the stairs. Sprinting up the three flights two steps at a time even though we'd both already been working for 16 hours straight, we arrived at the new admission's room quickly. There we found June Warner, a 58-year-old woman in some respiratory distress, and I instantly recognized the wide-eyed look of panic I'd seen on Mrs. Jackson's face. She was hungry for air and struggling to breathe.

"Hi, Mrs. Warner. My name is Dr. Kosky. This is Rick Sheff, a medical student working with me. I understand your emphysema has taken a turn for the worse," he said, pulling the curtain between our patient and the patient sleeping in the other half of the room.

"Yes. Yes. I can't breathe." She pushed out the words through pursed lips.

"Do you have any other medical problems?"

"I can't hear you," she panted. "You'll have to speak up. I'm afraid I'm quite hard of hearing."

"DO YOU HAVE ANY OTHER MEDICAL PROBLEMS?" Jerry said loudly.

No."

"EVER BEEN IN THE HOSPITAL BEFORE?"

"No. I'm so scared."

"Let's just take a quick listen to your lungs."

"What?"

"LET'S JUST TAKE A QUICK LISTEN TO YOUR LUNGS." With that Jerry sat down beside her on the bed and signaled for me to move to the other side. He opened up her johnny from the back and we both listened to her lungs. The breath sounds were faint, with a few wheezes. I noted her expiration was about three times as long as her inspiration. Jerry took a quick listen to her heart. "Alright," he went on. I noticed you got a bolus[10] of aminophylline in the ED to get you started. I'll write some more orders that should help you. I'm sure you'll do fine. But if you start to get sleepy, make sure you have the nurses call me. It means you're retaining CO_2, you're

---

[10] A bolus means a large amount given together. It can be of any size.

going into CO_2 narcosis. That's an ominous sign." With that, he flew out of the room.

"What did he say? I couldn't really hear him. But he said 'ominous', didn't he? What did he mean 'ominous'? He said something about getting sleepy. I'm so tired. I haven't slept at all the last two nights…sitting up all night trying to catch my breath. I am sleepy. I'm already sleepy! Is that ominous?" Panic rose in her voice.

In a move that was becoming almost instinctive, I took her wrist to check her pulse. "YOU HAVE A STRONG AND STEADY PULSE," I said, trying to be reassuring by playing the medical card at the same time I was forced to speak loudly. I knew Jerry had been trying to garner her help in identifying if CO_2 narcosis was setting in, but in his hurry he had inadvertently only exacerbated her fear.

"I do? Is that good?"

"YES. IT'S A VERY GOOD SIGN. NOW TELL ME MORE ABOUT YOUR PAST MEDICAL HISTORY. WHAT MEDICINES ARE YOU TAKING?" She seemed to settle down a little and began to tell me the information I needed. I learned she was widowed, lived alone, had smoked two packs of cigarettes a day for over 30 years, and had worked as a secretary until her emphysema had caused her to retire prematurely the previous year. As I went through the exhaustive review of systems, the questions seemed to take her mind off her breathing a little. As we talked, the nurse came in several times to start an IV drip of aminophylline and to hang her first IV dose of steroids, both of which Jerry had ordered.

I moved into examining her. Looking into her ears I said, "THESE ARE THE CLEANEST EARS I'VE SEEN IN A LONG TIME. YOU HAVE NO IDEA HOW DIRTY THE EARS ARE I'VE GOT TO LOOK INTO ALL DAY LONG. WHAT BEAUTY SECRET ARE YOU KEEPING FROM THE REST OF US?"

For the first time she laughed. "I…I don't know. I don't suppose I do anything special."

As I moved to sit beside her to listen to her lungs, I placed one hand on her shoulder, deliberately trying to apply as reassuring and warm a touch as possible. For the first time, I felt like I was self-consciously applying the "laying on of hands." I listened intently to her lungs. They had opened up some, so I could hear significantly more wheezes than I had earlier. Her expiration to inspiration ratio was now closer to two and a half to one. "YOUR LUNGS ARE ALREADY STARTING TO OPEN UP. I CAN HEAR YOU MOVING A LOT MORE AIR THAN WHEN I LISTENED JUST A LITTLE WHILE AGO WITH DR. KOSKY. THE MEDICINES HAVE STARTED TO WORK. YOU'RE ALREADY GETTING BETTER."

"I am? Why, yes, I do believe I feel a little better."

"I THINK YOU'RE WELL ENOUGH TO LIE DOWN ON YOUR BACK SO I CAN GET A PROPER LISTEN TO YOUR HEART." She lay down, visibly relaxing into the bed. "YOUR HEART SOUNDS VERY HEALTHY." She didn't say anything, but appeared to beam ever so slightly. I lay a reassuring hand on her abdomen, allowing it to linger there a little longer than usual so she could relax under my hand before I began the process of examining her abdomen. By the time I had finished examining her, she was clearly far more relaxed.

"Thank you. You have helped me so much. I feel much better now." Then, panic suddenly rising in her voice, she sat straight up. "But I do feel sleepy. Is that a problem? Is that 'ominous'?"

"NO. NO. BY EXAMINING YOU I CAN TELL YOU'RE ALREADY STARTING TO GET BETTER. DR. KOSKY WAS DESCRIBING SOMETHING THAT HAPPENS TO ONLY A FEW PATIENTS WITH EMPHYSEMA IF THEY ARE NOT IMPROVING. YOUR FEELING TIRED IS PERFECLTY NATURAL AND NOT AT ALL RELATED TO WHAT DR. KOSKY WAS TALKING ABOUT." With that, she lay back into the bed, exhausted from her long ordeal but feeling comforted. She was ready to go to sleep.

"Good night, Dr. Sheff" she said.

"GOOD NIGHT, MRS. WARNER," I said, backing toward the door.

Just then a voice came from the other side of the curtain. "So datz vot psychology iz!"

Peering around the curtain into the darkened half of the room, I could barely make out a figure lying in the bed. She seemed to have a cloth wrapped around her head. I had completely forgotten this was a semi-private room and that Mrs. Warner had a roommate who had been forced to listen to the entire exchange I had carried on near the top of my voice. I suddenly felt very self-conscious. "Excuse me," was all I could get out.

"Vot you did for her. Datz vot psychology iz. I vish someone had done zat for me."

I was now intrigued. "Why is that?"

"I'm dying." Suddenly my breath went out. "Ven I came into ze hospital a few veeks ago, I vas very confused. I vish someone had done for me vat you did for her."

I sat down in the dark. Just by my sitting down she seemed to know it was OK to go on.

"I hov cancer, and I am dying." Again my breath went out. I had never sat with anybody who was dying. I had never even had a conversation with a patient with cancer that so directly addressed death. I had read Elizabeth Kubler-Ross's first book on death and dying. I had even attended a lecture on death and dying by a national expert during my first year. The most important message I'd taken away from the lecture was that dying patients need just to talk about their experience. It doesn't mean the doctor needs to do anything about it, but it is important to listen and to let the patient know you care. Although it was past one in the morning, and this woman was not my patient or a patient on our service, I wanted to stay with her.

"Tell me about what happened."

"I hov cancer. It has spread everywhere."

"Are you in pain?"

"Yez. Yez. Of course I am in pain."

"I'm so sorry. Is there anything I can do?"

"It iz not pain zat bothers me. But ven I came into ze hospital and vas very confused, I vas so scared zat zis iz vat dying vould be like. It vas a terrible feeling."

"It must have been."

"I do not vant to die so confused."

"I know there are many things doctors can do to help you during such a time. I don't have the expertise to know what they are, because I am just a medical student, but would you be interested in talking to someone about what you are going through? There are doctors who specialize in working with people who are dying."

"Yez. I think dat vould be good." Her voice trailed off. She gazed toward the wall. We sat together in silence.

After a while, something felt complete between us. "I have to go now," I finally ventured.

"Of course. Sank you so much for helping me."

"It's been my pleasure."

Emerging out of that dark room into the brightly lit hallway, I knew something important had happened for both of us. For her, she had finally been able to talk honestly with another about her impending death. I took it upon myself to write a note, attaching it to her chart so that her attending physician would know she was interested in talking to someone about dying. For me I knew that, for the first time tonight, I had been a physician as I had always wanted to be. It felt extraordinary.

*   *   *

For the next several days, I took care of Mrs. Warner. I presented her on rounds and visited her during the day. She slowly improved. As for her roommate, each day I would look in on her as well. We would talk briefly, though never again about her dying. I did inquire each day if anyone else had come to talk to her about her dying, but each day the answer was, "No. Not yet." I became increasingly frustrated and left a second and a third note for her attending physician. One day I found her dressed and ready to go home. I again asked if anyone had come to talk to her about dying. "No. No. Zey hov not," she replied wistfully. "But talking vis you vas enough for now. Sank you."

Chapter VII

Pediatrics

Something Important Fell Out

"Now just let me look in your ear," I said in my sweetest, most sing-song voice, even though I fully realized this nine-month-old baby had no idea what I was saying. Just as I was about to get the ear speculum[1] into his ear, he jerked his head away and let out a lusty cry. "Shit!" came my reply, just under my breath. A nurse tending to another baby on the other side of the room snickered. I looked up, but by then she was busy with whatever she was doing. "Come on. How hard can this be?" I muttered to myself. Trying again, I took firm hold of this poor baby's ear, pulled it up and back as I'd been taught, only this time I didn't even have time to get my otoscope[2] anywhere near his ear before his head began flinging violently from side to side as he cried.

"At least he's giving me a good look at his throat," I thought, grabbing the moment of his latest scream to stick a tongue depressor into his wide open mouth and catch a fleeting glimpse of his tonsils before his head jerked away again. "What's next?" I thought to myself. "Oh yes, time to listen to his heart and lungs." I had bought a smaller, pediatric head for my stethoscope with a special diaphragm to pick up the more subtle sounds of a young child's heart, and with great anticipation placed the head of my stethoscope over his heart. Instantly, I tore the stethoscope off my head as a piercing cry assaulted my ears. "How the hell can you listen to his heart when he's wailing like that?" I muttered. I was able to listen to his lungs, if that is what I could call it, because he had to pause in his screaming just long enough to take in a breath, which at that moment allowed me to catch a vague impression of his breath sounds. They sounded OK to me.

After that came his abdominal exam. I placed my fingertips on his stomach trying to visualize the liver, small and large intestines, and other organs within his tiny belly. But all I could feel was his taught abdominal muscles as he let out another scream. I had no idea what was really going on in his abdomen. Next, I had to examine his genitals. Having never had experience with young siblings or cousins, or even with babysitting, I stared

---

[1] Just as a speculum is the instrument inserted into the vagina to allow a clear examination, an ear speculum is the plastic piece inserted into the ear to allow a clear examination of the ear canal and middle ear.

[2] An otoscope is the instrument to which the ear speculum is attached that physicians use to look into ears. It provides a light source and magnification.

at the diaper. Were those actually pins on either side holding it in place?[3] I
carefully unsnapped the first one, trying to be sure not to let the sharp point
touch the baby's skin. "Maybe I can examine him leaving one pin in place,"
I thought, convinced that if I ever took off both pins I would have no chance
of getting the diaper back on. I pushed the now free end of the diaper back,
exposing his little penis and scrotum. The instructor had said something
about checking to make sure both testicles were descended into the scrotum.
I leaned over and held the loose skin of his scrotum between my fingers,
concentrating on trying to feel for his testicles. Everything seemed to mush
together under my fingers. Was that one that just slipped between my finger
tips? Suddenly, a yellow stream erupted from his penis, rising in an arc and
landing directly on top of my head. Another snicker came from across the
room as I jumped back, striking my head on the crib railing.

"You'll learn about that pretty quick, I bet," the nurse called out.

Assuming I'd found what I had been looking for in his scrotum, I folded
the diaper back over and picked up the safety pin. It seemed huge for his
small body. I managed to pop it open with one hand and, holding my fingers
between his skin and the diaper, tried to slide the pin through both ends of
the diaper. I got through one, but the second seemed much tougher. I pushed
harder and harder. Suddenly it popped through right into my thumb. "Ow!" I
looked up sheepishly. The nurse had stopped what she was doing and was
watching me with a broad smile. Squeezing the safety pin back together, I
realized the diaper was skewed to one side and now barely covered his
bottom. That was as good as it was going to get.

I quickly checked his reflexes, which were plenty brisk in his agitated
state, and turned to leave, feeling quite dejected with my first attempt at
examining a pediatric patient.

"Yo, you want to get my baby here hurt?'" the nurse shot at me. She
scurried over. I paused, bewildered with what I had done wrong. "Didn't
they tell you never to leave the side of the crib down? I thought that was the
first thing they told all you medical students," she said, lifting up the crib
side until it snapped into place. She was right. At this morning's orientation
session to my pediatric rotation they had emphasized the importance of
putting the crib sides up before leaving any pediatric patient.

"I'm sorry. I guess I forgot," was my lame reply as I started to walk out
of the room.

"Looks like you done forgot something else," she said sternly, glowering
at me. I paused, knowing she was undoubtedly right but having no idea to
what she was referring. "Your hands? You forgot to wash your hands. You
want to get my babies sick up here?"

---

[3] This experience occurred just prior to hospitals switching over from cloth to
disposable diapers.

"Oh, right," I said sheepishly, rushing over to the sink to wash my hands. More than in any other rotation, on pediatrics they emphasized rigid hand hygiene with the requirement that we wash our hands before and after touching every child. They made clear this was to prevent the spread of infections among the vulnerable, very sick children who populated this pediatric hospital. But they also made clear it was in our best interest in order to minimize the chance we would get the pediatric crud. That's what they called the continual series of respiratory and intestinal infections every student who rotated through the pediatric hospital was almost guaranteed to come down with. Give up, the upper classmen had said. Nobody escapes it.

I slunk out of the room. Hadn't I just started to feel comfortable in the hospital, especially on my medicine rotation? Now this, a renewed sense of humiliation just by trying to do my first physical exam on a pediatric patient. This was going to be harder than I'd thought.

*   *   *

"I have a problem," the speaker had begun this morning. He was a tall man, in his early fifties, with a mostly bald head, wearing a tweed sport coat and bow tie. "They tell me that everything I do can be done just as well or better by a computer. The computer can ask a patient all the questions, including the ones I might forget. The computer can analyze the laboratory results and not miss one of them. The computer can remember every medical fact ever entered into it. Can we say the same for each of you?" The two week half-life rule popped into my head as I nodded in silent agreement. "You see, this is my problem. I don't want to be replaced by a computer. So what can I do—what can each of you do—that a computer cannot?" He paused. I liked him already. His manner was warm, his smile genuine. He seemed like the kind of doctor kids and their moms would trust quickly. I wanted to be like him, though without the mostly bald head.

"So what can you do that a computer cannot? To begin with, we don't yet have a computer that can properly examine a child. That day may come, but it is not here yet. But a computer can't sense the love in the room that reassures you this mother is doing all she can for her child. A computer can't read the body language as a three year melts into his mother's lap, letting you know there is deep trust there. A computer can't hear the frustration in the voice of the burned out, frazzled parents of a child who is still awakening four or five times a night at the age of 15 months. And a computer can't hear the shame and humiliation I see in the eyes of both the child and his mother when they have to tell me he is still wetting the bed at the age of eight and getting teased by his friends. You see, I'm not ready to move over and be replaced by a computer." Now I really wanted to be like him.

He had gone on, exhorting us to bring our full clinical skills, listening skills, and observing skills to working with children. He explained how it was different examining a pediatric patient. The history usually came from someone else, so you had to filter the answers through your understanding of the source. Some of the examination techniques would be different, though I hadn't appreciated how important those differences would be until I'd tried to examine my first pediatric patient. "And," he had said, "you need to know how to ask questions so a child will answer them honestly. For example," he'd gone on, "you never ask a child if he picks his nose." There was a titter in the room. "Of course he will answer 'no'. Instead, you simply assume he picks his nose and ask him which finger he uses. He'll readily answer, 'This one!'" he exclaimed, holding up the index finger of his left hand. We all laughed, but I knew he was sharing an important wisdom.

*   *   *

The pediatric services were divided up by patient age. I had been assigned the zero to two service (hence my frustrating experience examining a nine-month-old child). The service for children ages two to 12 was considered the best service for students, since you could do more with these kids. They could talk, you could examine them more like you could adults, and they had problems, like asthma, that weren't so unique to infants. The adolescent service was depressing because the majority of those children had cancer. I walked past the adolescent ward and caught a glimpse of three teenagers, each completely bald, lying listlessly in their beds. My stomach turned. The sadness had been almost too much to bear. With a morbid curiosity I wanted to keep staring at them, but, both out of courtesy and my own discomfort, I had to look away.

Even then, they were teaching me an important lesson. A sick child is much more upsetting than a sick adult. By "sick" I didn't mean just having the flu. Instead I saw patients with profound inborn errors of metabolism, congenital[4] diseases that would prevent them from ever living as normal children and often from achieving normal mental development. I saw patients with life-threatening infections like meningitis that moved so quickly, with devastating neurological damage if the doctors and nurses didn't do exactly the right things at the right time. I saw patients whose small size made them extraordinarily vulnerable to rapid changes that could be life threatening, such as dehydration from a simple case of diarrhea. In

---

[4] Congenital means something with which a patient is born. In this case, Jennifer had a structural problem with her heart from birth.

fact, part of what I began to learn was how to distinguish a child with a self-limiting illness, such as a cold or other viral infection, from a child who was truly "sick," one who needed medical intervention to prevent deterioration or death.

I was shocked to learn how complicated simply calculating proper IV fluid doses for a child could be. We spent hours on rounds discussing how to calculate the exact amount of sodium, potassium, bicarbonate, and water needed and how quickly it should be infused. It all depended on how much weight the child had lost through dehydration. We were constantly being asked to estimate whether the child was mildly, moderately, or severely dehydrated.

What was at stake? A child whose sodium was too low could seize. A child whose potassium was too low could suffer a fatal heart arrhythmia and die. A child whose fluid was replaced too quickly with too much "free water" without the proper amount of sodium might suffer brain swelling and ensuing brain damage. A severely dehydrated child who didn't receive enough fluids quickly could die before your eyes. Sick children, I was learning, were extraordinarily fragile. One wrong decision and the child could suffer a lifetime of harm or even die. I wasn't sure I wanted this much responsibility. And yet I was intrigued.

*   *   *

As I hunched over the pediatric textbook in the library, the words on the page swam before my eyes. I was trying to read up on the condition of the patient I'd examined that morning, but my head felt so congested it throbbed. The cough that had racked my body for the past week wouldn't let up. Now my stomach was giving out on me, and I'd come down with a case of the runs. The pediatric crud was everything they told me it would be.

I had done what they'd asked. The nurses had hovered over me and the other medical students, seeming to take great pleasure in reprimanding us for not putting up crib sides and not washing our hands. They'd done more than that, of course. In fact, I was deeply impressed by the knowledge and skill of these pediatric nurses. They handled the infants with ease and grace, not a single wasted movement. They were able to start IVs in the tiniest of veins. They seemed to sense even subtle changes in the conditions of their charges, alerting the physicians in time to make life protecting changes to their care. For the first time I came to understand how much these nurses were a crucial part of the treating team.

Thanks to their vigilance, washing my hands before and after touching each patient had become engrained in me, a habit I hoped never to change. But still it had not protected me from the pediatric crud. So here I sat in the library, trying to focus on page after page of the textbook, but to no avail. I needed a break.

I stood up and wandered out to the pay phones.[5] Time to call Marsha. She'd moved in with me by the end of my internal medicine rotation. Before she did, she'd made it clear she wasn't interested in a relationship that placed her second to anything, especially my work. I loved Marsha. I'd made a promise to myself never to allow my work as a doctor to keep me from creating a full life outside of medicine, including sustaining a healthy relationship. I'd recited for Marsha the words of that exhausted medical student who'd warned me that medicine was a bottomless pit I would never be able to fill up, and that I would always have to be the one to decide when enough was enough and walk away. We'd made an implicit promise to each other. I would never allow medicine to prevent me from maintaining that life outside of medicine, including my relationship with her. She would serve as my link to this life, especially when medicine's demands became overwhelming.

So now I wanted to call Marsha to let her know when I would be home. Receiver in hand, I stared at the numbers on the phone. What was my number? That's right, 643-1746. I dialed, but got a wrong number. Was it 1764? I dialed that, and again got a wrong number. Bewildered, I searched my mind. Then, slowly, it dawned on me—I'd forgotten my own phone number. It had finally happened. I'd crammed so many facts into my head that something important had fallen out. With disbelief, I realized I had only one option. "Hello…information…in Philadelphia…could you please give me the phone number for Richard Sheff?"

---

[5] Obviously, given the years I was in training, cell phones were not in use yet.

Chapter VIII

Choosing a Specialty

"What's love got to do with psychiatry?"

"If you have the heart of a lion," the speaker began, "the mind of an Einstein, and the hands of an artist, then surgery is the specialty for you." Some of us in the audience began to laugh, but he was quite serious and went on, "I say the mind of an Einstein because surgeons are ever so much more than technicians. They must always be thinking. Whether making the correct diagnosis, selecting the procedure of choice, making split-second decisions during an operation, or providing optimal post-operative care, a surgeon must maintain a mind as sharp as the scalpel he wields. In short, surgery is the best of all the specialties."

"Move over, Einstein," came the retort from the next speaker. "The real thinkers of medicine are the internists. If you love to learn, love to solve the most challenging of problems, love to constantly keep your mind open to new possibilities, then internal medicine is the specialty for you." He went on, of course, finishing with, "So that makes internal medicine the best of all the specialties."

The speakers each took their turn, exhorting all of us, now at the beginning of our fourth year of medical school, to choose their specialty. "That's why pediatrics is the best of all the specialties." "That's why ophthalmology is the best of all the specialties." "That's why radiology is the best of all the specialties." I'm not sure dermatology was represented, but virtually all the others had impassioned representatives exhorting us to choose what was obviously the best specialty—theirs.

The representative for family practice was a little odd in this group because he wasn't a family physician. He was the "family practice advisor" for medical students, but he was actually a pediatrician who had never studied or practiced family medicine. That was the best this medical school could do. After all, family practice wasn't a real specialty in the eyes of this elite, academically renowned east coast medical school. They saw it rather as an aberrant bump on the curve in the history of medicine that would soon fade away, replaced once again by "real" specialties, just as the old general practitioner had been in the decades before.

Though not quite sure why I had come to this special meeting called by the medical school for all fourth year students to "help" us choose a specialty, I sat through it with a bemused smile. After all, hadn't I already chosen the specialty that felt so right for me?

*   *   *

There was little question in my mind as, in the midst of my year of dancing and training as a group therapist, I had reaffirmed my decision to go to medical school. I knew more than ever that I wanted to be a psychiatrist. I'd spent this year training to lead therapy groups under the tutelage of Dr. Saul Silverman,[1] a pioneer in the field of marriage and family counseling. A voracious reader and student in the fields of psychology, theology, and ethics, Saul, as everyone called him, had actually begun his career as an attorney. After graduating from Harvard University and Harvard Law School, he had practiced law for 10 years yet found he had been a terrible lawyer. He was always writing contracts taking into account the feelings of the other party, which made him, by definition, a very poor attorney. Eventually he concluded that to be a truly excellent attorney, he had to spend his time tearing people down. Instead, he wanted to help build people back up after they'd suffered. So he had quit his job as an attorney, pursued a doctorate of education in marriage and family counseling (which when he did so in the 1950's, was just a fledgling field) and then opened a small private practice, offering counseling to individuals and couples.

He continued treating individuals and couples, but began experimenting with treating his patients in groups. Saul, along with others during this time, came to find that people reproduce in a therapy group the same behaviors that are getting them into trouble or causing pain in their lives. As a leader of the group, he created a safe, loving environment in which participants could get the "Aha!" that can only come from seeing your behavior and its impact on others reflected in the eyes of the group participants. But why could his patients take in these often difficult and disturbing realizations through the experience of his groups? It didn't take long for me to learn his secret. In fact, it was no secret at all. He went out of his way to establish within each group a loving, caring environment. With clear boundaries and strong leadership, Saul saw to it that every group participant felt deeply loved by every other group participant.

This was not romantic love at all, not the "falling in love" type of love. Instead it was what Saul called "existential love." The existentialists believed that everything we do in life is a choice, and that each individual is personally responsible for every choice he or she makes. Saul would say, "We can choose whether or not to love another just because that other person is a fellow human being. We are accountable for this choice just as much as we are accountable for choosing whether or not to act morally." In

---

[1] Dr. Silverman is one of only two individuals in this book whose names have not been changed to protect their identity. I've kept Saul's name here as a way of thanking him, in a manner that can never fully express the depth of my gratitude for his life-long brilliant and courageous work, the generosity of his time and his heart he shared with me, and the profound impact he has had on my life and my work. Sadly, Saul passed away shortly before this book went to print.

fact, for Saul, love was the essence of morality, and morality was not merely an academic issue. For his generation, the horrors of World War II and the explosive rise of science had made it imperative to seek a non-theological foundation for ethics. Rooted in the Judeo-Christian doctrine to love another as oneself, Saul, as were many during that time, was passionately seeking a foundation for love and ethical behavior that didn't depend upon a willful, omnipotent God to provide the fundamental commandments of an ethical life.

Into this environment I had stepped, fresh from an undergraduate major in intellectual history, fresh from studying politics and philosophy at Oxford, a restless and devouring intellect, seeking answers to the very same questions. After each therapy group, we would discuss the events of the group. I learned to understand the "why" of the members' behaviors and the "how" of Saul's extraordinary skill in leading the group. Then our talk would turn to the great questions, and late into the night Saul and I would debate. I did feel committed to love others, but why? With urgency and passion, I sought the foundation of this commitment. It was the basis of my calling to be a physician, to be a healer. Eventually I came to embrace Saul's concept of "existential love." It gave voice to a truth that had driven me, though I had never before named it.

So this was my image of being a psychiatrist—to love my patients with an existential love, to help them to love themselves and others, and to use the environment of the therapeutic relationship or group to help them gain insights about themselves that would allow them to live happier, more fulfilling lives. Saul ran his groups with humor, caring, and skill that led me and the participants to the most authentic and satisfying interactions with others we'd ever had. Saul would often say he preferred a good therapy group to a good party any time, and I had to agree with him. I could see myself spending a lifetime in this work.

*   *   *

During my first year of medical school I took a chance—a chance to bring together my life before medical school, my passionate explorations with Saul, and my choice to become a psychiatrist whose practice would be rooted in existential love.

Clutching the precious manuscript in my hands, I looked up, surveying my fellow medical students in the room. There was Jim, seemingly always with a warm smile, but those black, heavy rimmed glasses with the "Coke bottle" lenses made his face appear a little distorted. Roger was very earnest, but couldn't seem to ever buy pants quite long enough for his gangly frame. And Sam, with his flaming red hair and beard and penchant for radical

politics, always had an opinion that set others on the defensive. There was
no doubt; every member of this Psychiatry Interest Group was just a little
odd.

"Love: An Existential Commitment and Process. A paper by Saul
Silverman, D.Ed., and Richard Sheff," I opened.

> She loves you, yeah, yeah, yeah,
> She loves you, yeah, yeah, yeah,
> She loves you, yeah, yeah, yeah,
> With a love like that, you know you should be glad.
>   The Beatles

But you're not glad.

And so I began reading our paper, this precious paper that summed up
the endless hours of impassioned debate, exploration, and, yes, love, I had
shared with Saul. Why were we, one of the most free societies in history—
free to follow our hearts in love—surrounded by divorce, broken
relationships, wounded hearts, and countless other waste products of failed
love? What were the essential differences between romantic love, the love
we seem to endlessly pursue, and existential love, the love we so deeply
need? With a rising, at times strident voice, I poured out my heart to my
fellow medical students. Surely they would understand. Surely, given their
interest in psychiatry, in the emotions of their fellow human beings, they
would grasp these insights. "Why should I care about another?" began one
passage. "I care because I recognize in the other my own humanity, my own
frailty, my own failings. I must turn to myself, to my own pain, with
compassion. For it is only through compassion for one's own inner struggles
that genuine empathy for another flows."

"And what is healing about this?" began another passage. "Why should
we love another, even when that person is acting in a cruel manner, a
manner destructive of others, or suffering from a failure to love themselves?
The answer lies in the timeless wisdom of Jesus, of Gandhi, and of Dr.
Martin Luther King—that love in the face of unloving behavior is the
essence of the healing element."

As I concluded with the implications of this insight for psychiatry and for
medicine in general, I felt a deep sense of satisfaction. Here, at the midpoint
of my first year of medical school, I had bridged the gap between my life
before medicine and the life that lay before me as a physician. Eagerly, I
awaited their response.

An awkward silence followed. "Ahem." Jim cleared his throat. His smile
was suddenly gone. "I'm not sure I agree that romantic love is so bad," he
began. "In fact, romantic love expresses a deep drive in all of us. This drive

to procreate, 'Eros' as Freud called it, is one of the two great human drives. It cannot and should not be denied."

"I wasn't saying that romantic love itself is bad. Romantic love, without existential love, is the problem. That's when love can turn so destructive," I retorted.

"But how have you defined this existential love?" he replied. "According to you it is the love that flows when I look at another and see myself. Isn't that the essence of narcissism? Aren't you creating an ethics of narcissism?"

"Well, I suppose you could put that spin on it. We do have so much in common with others, including our shared humanity. If recognizing that is narcissistic in your view, then there is no arguing with your interpretation. For me, I find that kind of interpretation stifling. How else can we talk about our shared humanity as the foundation of ethics, of caring for others, without reducing it to psychological jargon like "narcissism"? Can't we get away from this jargon?"

"Yes, but what research evidence supports your position. This sounds more like some warmed-over philosophy you are spouting rather than a scientific paper," came Roger's contribution.

"Yes. Yes. This is rooted in philosophy, and ethics, and history too," I acknowledged. "These serve as guides to our behavior today. We can't find our direction, our moral compass, without these. Science, particularly when applied to human behavior as in psychiatry, must be informed by these non-scientific insights."

"So you agree your paper has no scientific footing," Roger exclaimed, barely able to stay in his seat, a light in his eye I'd never seen before. He was deriving some kind of pleasure out of tearing down my heart-felt offering. I glanced around the room again. They were all on the edge of their seats, eager to come forward with their own insights, each of which was critical of some aspect of the paper. The battle raged on as one by one every element in the paper was dissected, criticized, and, in the minds of most of those in the room, ultimately discarded as wrong-headed. This was the academic process in action. This was how science "progresses." One person putting forward a new insight, approach, or finding is followed by everyone in his or her field descending to tear it to shreds, preferably as publicly as possible to further their own reputation. What survived this attack was considered valid enough to be incorporated into the compendium of acknowledged scientific "truths," at least until some new "truth" supplanted it in the future. I valued science, though I was learning how thick skinned one needed to be to make science one's vocation. The scientific method is certainly how medications and surgical procedures should be tested for effectiveness. But my intention was different. My goal was to inform our application of science to the ever-so-human activity of psychiatric treatment. It was to address the "why" of psychiatry, not the "how."

Actually, upon reflection I realized I was also attempting to address the "how" by arguing that compassion was a necessary element for healing to occur in the therapeutic encounter. Why wasn't this clear to my fellow students? Why didn't they value empathy enough to acknowledge its critical value in healing, especially in psychiatry? How could one scientifically test the impact of empathy? Attempting such a test would be an example of the Heisenberg Uncertainty Principle[2] at work, because almost any attempt to control and measure empathy would change the clinical interaction so much the results would be hopelessly invalid. No wonder psychiatry had struggled for decades to prove the efficacy of therapy.

Thankfully, the meeting was winding down. I wanted nothing more than to slink away from this room and this Psychiatry Interest Group. I had opened my heart to my fellow students, and they had seen fit to tear into it. At least this was my experience of the past hour and a half. I had learned many lessons, but none so clear as when the final comment was made.

"You know, Rick, in the end I just don't see what love has to do with psychiatry anyway."

                *   *   *

The sign seemed almost lost among the myriad posters plastered on the bulletin board at the med school entrance, but it caught my eye. It read:

Treating the Whole Family: The Art and Science of Family Practice

A lecture by Hiram Curry, MD,[3] chair, Department of Family Medicine, Medical College of South Carolina

---

[2]The Heisenberg Uncertainty Principle states that any attempt to measure the location of an object affects that object so that its location can never be known with certainty. For large objects, the degree of uncertainty is small compared to the size of the object, so this phenomenon is irrelevant. For subatomic particles, the degree of uncertainty is on the same order of magnitude as the size of the particle being measured, so we can never accurately determine the exact location of a subatomic particle at a moment in time. The same may be said of attempts to measure the presence, absence, or degree of empathy. The very act of attempting to measure empathy will change a therapeutic interaction so much that it won't be clear what is actually being measured and to what degree.

[3] Dr. Hiram Curry, founder and chair of the department of family medicine at Medical College of South Carolina, is the second figure in this book whose name has not been changed. Through this one-hour lecture, and his lifetime of work it embodied, he changed my life. For this, I am forever grateful.

Thursday, June 6
4 PM
Lecture Hall A

Sponsored by the Family Medicine Interest Group

It was a miracle I had even come to school this day to have a chance to see the poster because, in the waning days of my first year of medical school, I had burned out. The pressure to learn all the basic sciences in one year had been grueling. As we ended the year with pharmacology, I skipped most of the lectures and just studied the textbook. I did attend the first lecture since it covered the history of pharmacology, a subject that interested me because of my fascination with the history of science. Two memorable comments by our world famous professor of pharmacology stayed with me. The first was when he had quoted one of the great physicians of the early twentieth century (whose name I had already forgotten) who had said, "If all of mankind's pharmacopeia throughout the ages were to be thrown into the sea, it would be all the better for mankind and all the worse for the fishes." He was referring, of course, to the use of such toxic substances as mercury and arsenic. I wondered at the time how future generations of physicians would view our current chemotherapy drugs which are so toxic that we must bring patients to the edge of death to gain their benefits. The image of bloodletting came to mind.

The second memorable comment came from the professor's brief discussion of homeopathy in his cursory romp through the history of pharmacology. I had known very little about homeopathy, so it was enlightening to hear that Dr. Hahnemann, the founder of homeopathy, had been rebelling against the toxic medical regimens of his early nineteenth-century compatriots, seeking a less poisonous therapeutic approach. He believed the body had great self-healing powers, and that giving just the right medication at the right time could trigger a patient's optimal healing reaction. He went wrong, according to our professor, when he diluted his remedies to 1 part in 10 to the 10th concentrations, or 1 part in 10 to the 20th, or even 1 part in 10 to the 30th. Then he said something that caused me to pause. He had said, with a grand gesture and mockery in his voice, "Why we all know that, based on Avogadro's number,[4] there's not a snowball's

---

[4] Avogadro's number is the number of molecules in a mole of a substance, which is a measure chemists use to calculate chemical reactions. Avogadro's number is 6.02×10^{23}. So the professor was right that a substance diluted 10^{30} times would be very unlikely to have even one molecule of the original substance left in it.

chance in hell of even one molecule of the substance being in his dilution. So of course homeopathy couldn't work. It was no more than another example of the countless quackeries modern, scientific medicine has had to overcome." With this peremptory dismissal of homeopathy, he went on.

Now staring at that unassuming, letter-sized announcement of the family practice lecture, I found myself intrigued. My interest in psychiatry had sensitized me to the critical impact families had on each of us. My limited patient encounters throughout the first year, including that very first patient interview with the man with COPD who had so poignantly described his last moments with his father, had cemented my belief that understanding families is important in treating patients. So at the last moment, I decided to pull myself away from the library and go to this lecture on family practice.

The lecture hall was only about a quarter full. One of the students introduced Dr. Curry, and a balding man with a round face approached the lectern. "It may seem a little odd to you to have a neurologist deliver a lecture on family practice. Some of you may not have even known that I am a neurologist. But for the first 20 years of my clinical career, I practiced neurology."

What was a neurologist doing delivering a lecture on family practice, I wondered.

"You see, during those 20 years, I treated countless patients with headache, fatigue, weakness, and other such complaints. That is the nature of who comes to see neurologists. Of course, if you have a stroke, the neurologist comes to see you, but that's not what I'm here to talk about. I'm here to talk about what my patients taught me. They came with problems that didn't seem to be difficult to diagnose, but were quite often difficult to treat.

"Why were they so difficult to treat? After all, it didn't take a rocket scientist to diagnose a tension headache. Nor did it take brilliance to diagnose dizziness that stemmed from anxiety. And it didn't take an academic appointment to diagnose fatigue that came from a depression. Yet no matter what treatment I prescribed, few of the patients I saw with these problems ever really improved.

"Now some may say this is nothing new for a neurologist. We are used to treating patients who don't get better. After all, neurologists are famous for simply answering the question, 'Where is this patient's lesion?' without being able to offer any treatment once the lesion is diagnosed. But I was looking for professional satisfaction that went beyond simply identifying the 'site of the lesion.'

"Our research literature says that tension headaches respond well to biofeedback and medication. Why didn't these work in my hands? To answer this question, after years of frustration, I finally asked my patients. They were oh so willing to come forward with the answer. The rest of their lives—the source of their tension and stress—wasn't changing. A mother, whose son had attention deficit hyperactivity disorder, told me she found it impossible to follow any regimen for herself due to the chaos her son's behavior was causing for her entire family. But I hadn't known about her son, because I was only treating her. A man I'd been seeing for fatigue told me, once I finally asked, about his wife's alcoholism. It was everything he could do just to keep their family together, and keep his wife's drinking from being discovered by the neighbors and her boss. I realized that what happens in my patients' families had everything to do with what made them become ill, and with what strategies would and would not work to help them get better. I began realizing I was treating the wrong patient. My patient wasn't the individual in front of me. Often my patient was the entire family."

It felt as if Dr Curry was speaking directly to me. The rest of the room faded away.

"It was this realization that led me to the field of family practice.[5] You see, family practice arose as a discipline in response to the explosion in specialization that occurred throughout the 1950's and 60's. Yes, we need specialists, with their in depth knowledge of a single field and their skill in procedures it takes a lifetime to master. But physicians, and indeed the entire health care system, stopped looking at patients as a whole person, a person with a family and social context. This was what the old GPs of the previous generation excelled at, but their downfall was that they did not have enough knowledge of the rapid advances in medicine to treat their patients with the most up-to-date approaches. Would it be possible for a single practitioner both to possess enough technical knowledge to treat patients in an up-to-date fashion and also to understand and treat their personal and family context?

"From this dilemma arose the new specialty of family medicine. This is a specialty in breadth. This means a family physician should be competent to treat 90% of all the problems presented to him or her as a primary care physician. A family physician should be able to go toe to toe with a pediatrician on immunizations, toe to toe with an internist on the diagnosis

---

[5] The original name of this specialty was family practice, which was meant to distinguish it from general practice. General practitioners were often referred to as GPs, so the practitioners of family practice were then referred to as FPs to distinguish them from GPs. In recent years the name of the specialty has been changed to family medicine in part to fit with the way we name other specialties, such as internal medicine, and in part to reflect that there is a growing science to the "practice" of family medicine. Throughout this book the terms family practice and family medicine will be used interchangeably.

and treatment of community acquired pneumonia, and toe to toe with an orthopedist in the diagnosis and treatment of non-surgical back pain. At the same time, a family physician should be expert in assessing and treating the family context of any presenting problem. This is the field I left neurology to join. In my generation, there were no training programs for family practice. GPs completed one year of training after medical school, called an internship, that rotated them through all the specialty areas. They then hung out a shingle to practice medicine as a general practitioner. But to train today's physicians to care for the array of problems they'll see in family practice with enough scientific and practical knowledge to be competent requires a three year residency program. Today, you have hundreds of family practice residencies from which to choose. I can tell you with complete honesty this is the most satisfying and effective way I can imagine to practice medicine."

As he closed and everyone around me stood up to leave, I remained in my chair, an electrical feeling running through me. I could see myself practicing medicine like that for a lifetime. For the first time, I opened to considering the specialty of family practice along with psychiatry.

        *   *   *

Dr. Schmidt turned his wheelchair to face me as I entered. "Sit down, Rick," came his slightly hoarse greeting. I had chosen Dr. Schmidt as my adviser first year because he was a psychiatrist with an excellent reputation. He had been warm and thoughtful in our first meeting. Perhaps having a chronic, progressive neurological disorder himself had made him a more sensitive clinician and advisor. Now he sat across from me, giving me his full attention as we met near the end of my second year to plan my third year schedule.

"So how has your thinking progressed on selecting a specialty?" he began.

"I started med school so sure I wanted to be a psychiatrist, as you know. At that time, when I pictured myself entering a room full of books by the great writers in psychiatry, I felt I could spend a lifetime studying their works. I've kept open the option of family medicine during this year of my initial clinical rotations in obstetrics, gynecology, surgery, medicine and pediatrics, but mostly assumed I would still go on to psychiatry. To my surprise, I found myself enjoying learning about various medical conditions and taking care of patients this year more than I'd expected. I'm not exactly thrilled with entering a room full of books on diabetes, asthma, cancer, and pneumonia, but I'm at least intrigued enough to consider this possibility."

"If you are considering both psychiatry and family medicine," he responded, "then you should choose courses in your third year to spend some time doing what a psychiatrist does day in and day out and some time doing what a family physician does day in and day out. Then you can see which specialty is best for you."

*   *   *

Walking into the clinic to begin my third year experience in outpatient psychiatry, I recalled a conversation with Dr. Wilson from a year earlier when he had been the attending physician on the psychiatry unit. I'd asked him about how happy he'd been with his choice of psychiatry and how to know if it was right for me. He'd paused, in his thoughtful way, and responded, "I think you'd enjoy psychiatry. But you have to be comfortable being alone."

"What do you mean being alone?" I had asked, thinking of how much my experiences with Saul and the therapy group had been about anything but being alone.

"When I sit with a patient, one on one, I really can't share that experience with anybody else. It's between me and the patient. I do that eight hours a day. You have to be content keeping all that to yourself to be happy as a psychiatrist." He had paused again, looking increasingly wistful as the silence deepened.

"What about therapy groups?" I'd asked.

"Groups? They are generally run by the social worker, not the psychiatrist," had come the reply.

I had looked at him, sensing his underlying sadness. This moment had been all the more sad because with Saul I had experienced an approach to therapy that had been both collegial and fun. Dr. Wilson's life was neither.

Arriving at the outpatient psychiatry office pulled me out of this memory. A deep black woman's voice called out, "Come on in." She looked me over up and down. "You must be this month's student. No mistakin' that. Dr. McPhee, he's this month's attending on the teaching service; he'll be down in a few minutes. We've already got a new intake for you this morning so you can jump in with both feet." Was that a muffled snicker I caught her trying to cover up? "By the way, my name's Nancy. I run this department."

"You look like the kind of person that if I know what's good for me I'll do whatever you say."

"Y'all got that right. You do right by me and I'll do right by you. Ya know what I mean?"

"I sure do. Just want to make sure you'll give me a signal if I'm making a wrong turn in your book?"

"Be more than happy to." Now a big, white toothy smile covered her face. We'd worked out our pecking order, and she liked that I knew my place.

Dr. McPhee arrived, his white coat hanging loosely off his gaunt frame. As we introduced ourselves, I stuck out my hand. He looked down, hesitated for just a moment, and then reached out his hand to shake mine with a grip that was more limp than firm. Why did psychiatrists here at the University Hospital wear white coats? That wasn't how it had been during my psychiatry rotation at the Institute. It was a good thing I'd brought my short white coat this first day of my outpatient psychiatry rotation.

"You'll begin by doing intakes for the clinic. This means that each new patient gets a diagnostic interview. You'll be doing these interviews, and I'll observe you through the one way mirror in the intake room. Then we can talk about what you found and what the patient's treatment should be. Once you've done the intake, if they'll be coming for therapy you can start the therapy for them."

"But I'm only going to be on this rotation for a month. It doesn't seem right for me to start therapy if I can't be here for the long haul."

"Don't worry about that. These are clinic patients. They are lucky to have anyone to see. Once you're gone, they'll be transferred to one of the social workers in the clinic." With a wave of his boney hand, he dismissed my concern.

＊   ＊   ＊

"Hi. I'm Rick Sheff, a medical student here. I'll be doing your intake interview." The young man looked up slowly from his seat in the waiting room but didn't say anything. "I can see by the form you filled out your name is Brad Josephson. Is it alright if I call you Brad?"

"Actually, I prefer Mr. Josephson," came the stilted reply. I was somewhat taken aback as we were about the same age, but accepted his preference.

"Alright, Mr. Josephson. Please come with me." He didn't move. "Are you OK, Mr. Josephson?"

"Yes. I'm fine. It just takes me a while to get up."

"Is this psychomotor retardation from depression?" I wondered. Nothing seemed to be happening. "Is there a problem, Mr. Josephson?"

"Yes, there is a problem. You seem to be terribly impatient, and it is quite rude."

"I'm sorry," I replied, feeling scolded and puzzled and not sure how to proceed. After what must have been more than a minute, he finally stirred, slowly lifting himself up awkwardly. "Would you like some help?" I offered.

"No," came the curt reply. "I can manage just fine on my own."

We began walking down the hall toward the interview room, making very slow progress because of his awkward gait. "You seem to have some trouble walking," I observed. There was no response. "Is there anything I can do to help?" I again offered.

"Yes. You can carry this," he replied, holding out a flat briefcase with a strap.

I took it, eager to have something to do to help as we moved painfully slowly down the hallway. Finally, we entered the interview room. He looked around very carefully, taking in the large mirror on one wall.

"What is that for?" he asked in a challenging tone.

"That is a one-way mirror. Dr. McPhee, a senior psychiatrist is on the other side watching both of us. He is my supervisor. After our interview I will share my findings with him and together we will come up with a recommended plan of care which we will share with you." He gazed at the mirror for a long time. Finally, he looked around the room again and walked directly, though still awkwardly, toward the chair in which I had planned to sit. It was clearly the larger of the two chairs in the room. He turned the back of the chair to face the mirror. I took a seat in the smaller chair. "So, Mr. Josephson, what can I do to help you today?"

"For starters, you can lower that shade over there," he said, pointing to the window. "It's awfully bright in here."

"OK," I replied, jumping up to lower the shade. I wanted to do anything to make him more comfortable. "Now, what brings you here?" There was a long pause as he looked around the room again. "I said, 'What can I do to help you today?'"

"I don't know that there is anything you can do to help," he said flatly.

"How do you know if you don't give me a chance?"

"It is still too bright in here. Could you turn down the lights some?"

"Yes, I can do that," I replied, wanting to gain his trust by proving I can respond to his needs. At the same time, I was becoming progressively more angry at him. I went to the light switch. "How is that?" I asked with a slight edge in my voice, turning off half the overhead lights.

"That's better, but still a little bright."

"How is this?" I turned on the lamp on the desk in the room and turned off the overhead lights completely. He responded with silence. I wanted to hit him. "You still haven't told me why you are here."

"I've been to so many doctors, and nobody has been able to help me."

"What are you asking help for?"

"You saw how I was walking, didn't you?"

"Well, yes…"

"Then you don't seem to know much."

"I still don't understand." My jaw clenched as I strained to maintain my composure.

"I can't move my body right."

"How long has this been going on?"

"I don't really know. A long time."

"Was there a time when you used to be able to move more normally?"

"I'm not sure."

"If you are having trouble moving, have you been evaluated by a neurologist?"

"That's why I'm here. The neurologist sent me here."

"What is the name of your neurologist?"

"Which one?"

"Which one?" I repeated, becoming a little suspicious.

"Well, I've been to a few of them."

"And what do they say?"

"They say they can't help me. That's why they sent me here." The picture was finally becoming clear. He had a gait disorder, and perhaps a more generalized motor disorder, but none of the neurologists could "find the lesion." This meant his symptoms didn't fit any known anatomical structure or pathway in the brain or nervous system. To a neurologist this could mean only one thing—it's all in his head. He's crazy. That's why they sent him to the psych clinic.

The rest of his history of present illness produced no additional helpful facts. He was infuriatingly vague about his past medical history and family medical history. He was not on any medications, but did admit to fairly regular use of marijuana. He lived alone. His mental status exam was perplexing as he didn't seem to fit any of the major categories I'd learned about in psychiatry such as depression, schizophrenia, or anxiety disorder. It wasn't even clear to me if he was truly psychotic or not.

At the end of the interview I stood up. Again I had to wait an interminable time for him to get up. He began to leave without his bag which I offered to him. "That's for you to carry," came the dismissive reply, as if I was his servant. We took forever to return to the waiting room. He left without saying goodbye. Nancy just rolled her eyes.

Exhausted and frustrated, I slumped into the chair next to Dr. McPhee. "So what is your assessment?" he asked, not responding to my obvious frustration.

"Multiple times during the interview I wanted to haul off and smack him upside the head."

"I'll give you a hint," said Dr. McPhee. "The fact that you feel that way is a sign."

"A sign of what?"

"A sign that he was angry. If you find yourself becoming angry in the presence of a patient who is not overtly angry, then that patient has repressed rage. You are feeling it for them. Besides, he was clearly manipulating you. I was waiting to see what he'd come up with next."

"Why would he want to do that?"

"That's what I'm not sure of, but you fell into every trap he set. He was clearly enjoying the power he had over you. You were too eager to please and appease him. You'll have to learn to recognize this kind of manipulation when it is happening and claim your own power in the moment. Now, how would you sum up his differential diagnosis?"

"I'm not really sure. His mental status exam was infuriatingly confusing. He doesn't fit any of the major diagnostic categories I've learned."

"That's right. He doesn't. If his presentation is atypical, perhaps his diagnosis is. Are you familiar with schizotypal and schizo-affective personality disorders?"

"I've heard of them, but I've never seen anybody with those conditions."

"I believe these are the two most likely conditions in his differential diagnosis. He appears to also have a somatoform disorder, meaning his gait disorder that doesn't have an anatomical or physiological basis, as his primary diagnosis, in addition to one of those two personality disorders. Why don't you read up on these conditions tonight and we'll discuss this gentleman and a psychopharmacology plan for him further tomorrow when I receive your write up."

So this was psychiatry in today's medicine.[6] You get the patients that other specialties give up on because their problem is "all in their head." You get to deal with truly manipulative, difficult patients. It's the social workers who do the group and individual therapy. A psychiatrist's job is to be the diagnostician and prescriber of medications. Perhaps the Psychiatry Interest Group was right when they questioned what love had to do with psychiatry.

*   *   *

"You can start with the child in exam room 3. He has a fever and a cough." So began my month-long experience in family medicine. Actually, it had begun the day before when I had driven into this small town in the

---

[6] In assessing this experience, I was well aware there were other aspects to psychiatry, such as Dr. Wilson's practice and other options for spending more time providing one-on-one or group therapy. However, the role of the psychiatrist was beginning to change during my training, affected in large part by insurance payments. Psychiatrists were being increasingly shifted to a role that emphasized diagnosis and psychopharmacology (medication) more than providing therapy.

mountains of upstate New York. To get there I drove for miles past
farmland and pastures, past a few small towns with church spires and
clusters of houses. Finally, arriving in town, I drove through the
"downtown" area, which meant a few blocks of stores, the Grand Union
super market, one movie theater, and several gas stations.

What would it be like to live here, I wondered half out loud. The area
boasted a junior college just outside of town, but that was it for culture.
Would I be happy here?   That's what I had come to find out.

I stood outside the child's exam room. The aide's note said he was a
seven-year-old boy presenting with a three-day history of a runny nose,
fever, and a cough. This should be pretty straight forward, I thought. After
all, isn't family practice about handling a lot of colds, sore throats, and
backaches? I opened the door. On the exam table sat a scrawny, wide eyed
boy who began speaking the moment I stepped into the room.

"Areyouthenewdoctor?Idon'trememberseeingyouherebefore.Mynameis
Michael,butmyfriendscallmeMickey.Iguessthat'sbecausemydadalwaysliked
MickeyMantel,butInevermethim.MickeyMantelthatis.Ididseemydad,atleastI
usedtoseehim,buthedoesn'tcometoseeusmuchanymore.What'syourname?"
He finally paused for a breath.

"My name is Rick Sheff. I'm a medical student who will be spending a
month with Dr. Christian and the other two doctors here in the practice. I
reached out to shake his hand. He grabbed mine and shook it violently.

"Pleasedtomeetyou.Soyou'renotfromaroundhere?Whereareyoufrom?"

"I'm in medical school in Philadelphia." He paused long enough for me
to glance around the room. His mother, an exhausted looking woman, who
couldn't possibly be as old as she looked, had an infant in one arm and was
grabbing with her free hand for an agile three-year-old boy who kept staying
just out of her reach. A five-year-old girl sat on the floor playing with a
toddler who couldn't have been more than a year old because he kept trying
to pull himself up to walk. He'd take one or two steps and fall over. His
sister kept trying to keep him from banging his head on the corner of the
exam table, a chair, or any one of countless sharp edges that seemed to
surround him. I turned to Mickey's mother and introduced myself. "Hi. As I
said to Mickey, my name is Rick Sheff. I'm a medical school student here in
town for a month. What can I do for you today?"

She looked at me with bloodshot eyes. "Mickey's been sick. He's been
coughing and coughing. He's got that green goop comin' outta his nose all
the time. I hadda keep him out of school the last two days. He done been up
at night, and so's the baby. As you can see, Mickey's a handful. I just gotta
get him back to school. I can't take it much longer. What can you give
him?"

"Well, first let's see what's wrong with him." I looked at the chart for his
vital signs and realized the aide who'd brought him into the exam room

hadn't taken his temperature. I glanced around the room for a thermometer and found it in a steel basin soaking in alcohol. I had never had to take a patient's temperature before. This had always been done for me by the nurses in the hospital. I picked up the old fashioned mercury and glass thermometer. The last person I'd had to take a temperature on was me, and that had been back in high school. I glanced at the thermometer which read 99. I vaguely remembered you had to shake it down, so I shook the thermometer a couple of times and rechecked the temperature. It now read 97. "I guess I have to shake it harder," I thought and gave it a couple of more violent shakes. Suddenly the wet thermometer flew out of my hands, smashing against the wall, sending fragments of glass and mercury all across the floor. Stunned, I stood for a moment, not knowing what to do. Suddenly, in a panic I looked down at the toddler. "Pick him up right away," I called out to the five-year-old girl. There's glass and mercury on the floor, and we don't want him getting into them." She picked the child up. "Don't let any of the kids back onto the floor," I called out as I ran out of the room.

I bumped into the aide in the hall. "I broke the thermometer," I blurted out. "There's glass and mercury all over the floor. That mercury's really toxic. What should we do?"

She rolled her eyes and said, "I guess we got to get those kids out of there. Lucky this is Jim's day off, so we got an extra exam room not bein' used right now. Let's get them in there. We'll just close that room off until it can be properly cleaned."

"Thank you. Thank you," I said, exploding in a sigh of relief. She ushered the entire family into another exam room and sealed off the scene of my humiliation with a sign that read "Closed for Cleaning."

Just then Dr. Christianson, who was known to all his patients by his first name, Dave, came up. I sheepishly explained what had happened. He looked me up and down for a moment. "That's a helluva way to start your preceptorship with us. Hope it's not a sign of things to come. But, you know, I had a mishap or two like that during my training, so don't let it get you down. Now what did you find out about Mickey?"

That was it? No stern lecture? No harsh glare? I liked Dave immediately.

"I didn't have a chance to find out much since I'd just started with taking his temperature. I did find out he's had a three-day history of URI symptoms,[7] a persistent cough, and purulent[8] nasal discharge."

"Any sore throat or ear pain?"

---

[7] URI is an abbreviation for upper respiratory infection. The upper respiratory tract includes the nose, throat, and sinuses, so this is what most people call having a cold.
[8] Purulent means showing signs of infection, such as dark yellow or green mucous or pus.

"I didn't get a chance to ask or look. But his mom seemed awfully frazzled. She's got the whole family in there, and that sure looks like more than I'd be able to handle."

"Yeah. Mickey's got a pretty severe case of ADD, so he's usually bouncing off walls. Nancy, his mom, isn't real good at making sure he takes his Ritalin. Her life would be at least a little more manageable if she did. Their dad ran off some time back. Nancy's taken up with a new guy a few years ago and had her two youngest kids with him. She's pretty stressed on a good day. Let's go see them." I was impressed with how much he knew about Mickey and his family. We went in.

"Hi, Nancy."

"Hi, Dave." Nancy's face seemed to brighten when Dave walked in.

"Hey, Mickey," he smiled at the child as he turned to him. "Give me a high five." Their hands met in the air. "Rick here tells me you've had a cold for a few days. Any pain in your throat or ears?" Dave had already picked up the otoscope and was looking in his ears. Mickey was visibly fidgeting, but let Dave examine him. "These ears look clear, but I must say, it looks like it's been a while since they've seen any soap or water."

"Awe, come on Dr. Dave, you know how much I hate that."

"Open your mouth and stick out your tongue." He looked in his throat. "I know you hate to take showers, but it's important that you mind your mom. Lift up your shirt." He listened to his lungs. "So how about it? Will you mind your mom about taking showers?"

"OK. I will," he said sheepishly.

"Now, Nancy, you been giving him his Ritalin every day?"

"Well, sometimes I might miss a day here and there." Her eyes looked down at the floor.

"I know how hard it is, but this is important. You got a calendar at home?"

"Yeah. What's that got to do with Mickey's medicine?"

"How about marking the calendar with a check each day he takes his medicine? Can you do that?

"I'll try."

"Not just try. Now I need you to give me your word you'll do this calendar thing."

"Alright. You know I can't stand to get you mad at me."

"Good. I won't get mad, but you can bet I'll ask you about it when you're in with the next kid for something or you're in for your next pap smear. Now, here is a prescription for an antibiotic for Mickey. He has to take it three times a day until the prescription is finished. He'll be feeling better within a few days, but it's important you finish the whole bottle. OK?"

"OK. Thank you so much, Dave."

"You take care, Nancy, and get some sleep. I haven't seen your eyes this bloodshot in a long time." Then more softly, "I know it's hard. Hang in there."

"I will. Thanks for everything."

Out in the hall Dave said, "You write up the note in the chart. I'll see the next patient. When you're done writing it up, we can talk about it." He turned and dashed into the next exam room.

Dave's entire visit with that family had taken no more than five minutes. In that time he had evaluated Mickey and diagnosed a sinus infection that needed antibiotics. He had addressed Nancy's noncompliance in giving Mickey his Ritalin. I knew enough about family therapy already to recognize that he had done an intervention to strengthen Nancy's parental authority in the face of an acting-out child like Mickey. And he had shown genuine empathy for Nancy's difficult situation so she knew at least one person cared, and that clearly had meant a lot to her. He had been masterful.

Later, as we discussed the case, Dave said, "Antibiotics are often a tough call. We don't like to prescribe antibiotics for a viral infection. It can be expensive and lead to resistant bacteria. The literature keeps changing as to whether or not purulent nasal discharge really means it's a bacterial sinus infection that needs antibiotics. In this case, when it was a judgment call, I made the decision to err on the side of giving antibiotics because Nancy seemed pretty desperate, and she needed to go out feeling like somebody had thrown her a life preserver. I don't like treating viral infections with antibiotics, but sometimes it's not just the infection I'm treating, if you know what I mean." It made sense to me.

*   *   *

"There's a four-year-old child in room 2 with a fever and a cough. Why don't you take that one?"

"Alright," I answered. Another URI, I thought to myself. Just what I'd expected from family practice.

As I entered the room, I found a listless child, lying limply in her mother's lap. Something about her eyes looked odd, but I couldn't put my finger on it. She was breathing fast. "Hi. My name is Rick Sheff. I'm a medical student spending this month with the doctors here." I stuck out my hand, and the mother shook it limply.

"My name's Mary. This here is Suzie."

"What seems to be the problem?"

"Suzie just ain' right. She just ain' right."

"In what way isn't she right?"

"She just lies here like this. She won't do anything but cling to me. I can' get her to eat or drink anything."

I began to be concerned she may be dehydrated, as I'd learned in my pediatrics rotation. "Is she complaining of anything, like a sore throat or ear ache?"

"No. She just keeps coughing."

"Let me look in her ears." Both ears seemed normal. "And now her throat." It, too, seemed normal, but her lips were dry. "Could you lift up her shirt so I can take a listen to her lungs?" I thought I might have heard some rales[9] in the lung base on the left side. I also took the opportunity to count her respirations and found she was breathing at 24 breaths a minute.

"Dave, I'm worried about the girl in room 2. She has a three-day history of fever and a non-productive cough. Her respiratory rate is 24. I think she has a few rales at the left base, but most of all she just looks sick."

"Let's go see her." Dave took one look and his eyes told me he, too, was concerned. He listened to her lungs, moved her head up and down, and quickly checked her heart and abdomen. He took a breath in and paused before speaking. "Suzie is looking pretty ill. I'm concerned about her. I think we need to get her up to the hospital right now. She looks like she might be a little dehydrated. I also want to get some blood work on her to see how serious her infection is. We'll also get a chest x-ray. Once I get these results, I'll let you know better what may be causing her to be this sick. Are you OK taking her up the hill to the hospital? Do you want someone to drive you?"

"No. I got her here. I guess I could get her a couple hundred yards up the hill to the hospital. Dave, do you think she's really sick?"

"Mary, I do think she's sick. I think there's a good chance she has pneumonia and that we'll have to keep her overnight in the hospital." Mary's eyes welled up with tears. "But I think what she has is something we can treat. She should respond really well to some IV fluids and antibiotics. In a day or two, I'll bet she'll be back to giving you hell like she usually does. I don't think you need to worry."

"Thank you, Dave." As we began to leave, she asked, "Dave, will you be seeing her later?"

"Yes, I'll come up and see her once we get some test results. Brian is on call, and I'll fill him in on everything so he can keep an eye on Suzie tonight."

---

[9] Rales are the sound heard through a stethoscope when there is fluid in the lung's air sacs. This happens in pneumonia from an infection. It also happens in congestive heart failure when the heart doesn't pump blood forward well enough, so it backs up in the lungs, causing fluid to ooze out into the air sacs. Rales sound like crackling, much like what you hear if you rub a tuft of hair between your fingers just outside your ear.

"Thank you again, Dave. Brian's OK, but I really trust you."

Out in the hall again, Dave explained, "Remember the way Suzie looked when you walked in. That is a sick child, a toxic child. You can make that diagnosis from across the room. Notice how she just clung limply to her mom, not wanting to move. Her eyes weren't focusing very well. She is usually a bundle of energy, but she sat there like I could have lit a fire under her and she wouldn't have moved. I agree with you she may have pneumonia, but I'm concerned she may have a bacteremia.[10] Her neck isn't stiff, so I doubt she has meningitis, but that's what a kid with meningitis could look like, too. I'll tell the hospital to run a CBC with diff,[11] Chem 6,[12] and blood cultures. We'll start her on IV fluids. She is sick enough I'd start her on amp and gent[13] and see how she does.

Later that day, her CBC came back showing a high white blood cell count with a left shift[14], confirming Dave's suspicion of a serious infection. He called up to the hospital and heard from the nurses she had become a little more alert with a bolus of IV fluids. That was a good sign. Her chest X-ray had a suggestion of a pneumonia in the left lung base, but wasn't definitive. At the end of the day, we both went up to visit Suzie. Mary was at her bedside, looking tired but relieved. It was obvious Suzie was already doing better.

---

[10] Bacteremia means an infection with bacteria that has spread throughout the bloodstream. It is a serious and potentially life threatening diagnosis if not treated in a timely fashion.

[11] Diff is an abbreviation for a differential blood count which evaluates the percentages of different types of white blood cells. These percentages are very helpful in determining how severe an infection is.

[12] Chem 6 is an abbreviation for a collection of six blood chemistry tests: sodium, potassium, chloride, bicarbonate, glucose, and blood urea nitrogen (BUN). The BUN goes up in dehydration. If the bicarbonate is low, it is often a sign of more severe illness—dehydration, an infection, or both. The sodium can be either high or low in dehydration, but whichever it is helps guide the right kind of IV fluid replacement. The potassium must be maintained within a relatively narrow window or else serious and potentially life threatening heart arrhythmias or abnormal rhythms may result.

[13] Amp is an abbreviation for ampicillin. Gent is an abbreviation for gentamycin. Both of these are antibiotics. At the time I was in training, combining these two antibiotics together was the treatment for some of the most potentially serious infections. Due to changes in antibiotic resistance and the development of newer antibiotics, today different antibiotics would be chosen for this clinical situation.

[14] A left shift means her white blood cells showed a higher than usual percentage of young white blood cells. These are cells churned out by the bone marrow in large numbers when the body fights a serious infection.

After we left Suzie's room, Dave turned to me. "Her chest X-ray was pretty unimpressive today. Once she gets rehydrated, I expect tomorrow's chest X-ray to show an extensive pneumonia. You need enough fluids running through your veins to make an infiltrate show up on X-ray. After all, it's fluid in the lungs you're seeing when you see the pneumonia." He paused. "So what do you think about seeing a bunch of URI's in family practice now? You think it's boring?"

"I guess not. We did see six patients with URIs today, but one allowed you to intervene in Mickey's Ritalin non-compliance, and this one seems to have even been about saving Suzie's life."

"Not bad for one day's work as a family doc, huh?"

"Not bad at all."

          *   *   *

"Julie's contractions are two and a half minutes apart. She's had a few decelerations on the fetal heart monitor, but they look more like variables than lates.[15] Two hours ago she was 4 centimeters and -1 station. I haven't checked her since then," the nurse reported as Brian and I arrived on labor and delivery. Actually, labor and delivery was a small wing of the one-floor hospital. It held four labor beds, but it was rare to have more than one woman in labor at a time. Yet I was impressed with how competent this nurse appeared to be.

Brian looked at the heart tracings. "I agree with your assessment, Jean, but I'm concerned these last few variables may be developing a late component. Rick, see here where the baby's heart rate is recovering just a little more slowly in the last four contractions than it had in the previous ones. Let's get her over on her left side and increase her IV rate. Let's also start her on oxygen at 2 liters per minute. Rick, do you know why we are doing these things?"

"Getting her over on her left side gets the uterus off her inferior vena cava and increases blood return to her heart and the resulting blood flow to the placenta. Increasing the IV improves her intravascular blood volume so

---

[15] During labor, fetal heart monitors can keep track of the baby's heart rate. Certain patterns in the heart rate are considered signs of the baby not getting enough blood supply. If they become severe enough, they are an indication for a Cesarean section. A variable deceleration is considered a sign of the baby's head being compressed during a contraction and is not considered a danger sign. A late deceleration is one in which the heart rate slows down toward the end of a contraction and doesn't speed up again until the contraction is completed. Late decelerations, if severe and persistent, are an indication for a Cesarean section unless a vaginal delivery can be accomplished quickly.

she can get better circulation to the placenta, and giving her oxygen increases the oxygen she is delivering to the placenta."

"Very good. I guess they actually taught you something down at the university." I silently thanked Jerry for his patience and teachings in those early morning hours. Over the next few contractions, the heart tracing improved. "Alright, while we're waiting for Julie to deliver, let's finish rounding on the other patients we've got in the hospital." Our first stop was the intensive care unit, which in this small, rural hospital was composed of three beds. This morning one was occupied. "Mr. Harkins is a 67-year-old male now three days post anterior myocardial infarction.[16] He threw some PVCs[17] last night, so we started him on a lidocaine drip.[18] The PVCs seem to have settled down." Brian obviously liked taking care of sick patients. Of the three physicians in the group, he kept most current with the latest research literature and prided himself on practicing medicine every bit as up to date as was practiced in most urban hospitals. He was less interested in counseling and interpersonal patient issues than Dave. Jim, with whom I spent the least time, was the best in the group at procedures. He was excellent at suturing, had great skill in applying forceps during deliveries, and did the bulk of minor surgeries in the office. Each had his strengths and knew when to call on a partner for help.

"We're about to visit Jim Kendrick," Brian explained as we headed down the hallway. "Jim's a real nice 86-year-old codger I've cared for in the nursing home for years. This summer we found he had colon cancer. We tried to operate on him, but when we opened him up, the cancer had spread everywhere in his abdomen, so we just closed him back up. He did pretty well until last week when he started vomiting. That was a sign the cancer was starting to obstruct his intestines. We've brought him into the hospital just to keep him comfortable because he was becoming too difficult to handle at the nursing home. We put an NG tube[19] down which worked for a

---

[16] Myocardial infarction is the medical term for a heart attack. Myocardial means the heart muscle, and infarction means that some of the muscle tissue died because of not receiving enough blood supply.

[17] PVC is an abbreviation for premature ventricular contraction. This is a spontaneous electrical impulse in the heart that arises separately from the heart's internal pacemaker. Having rare PVCs can be normal. After a heart attack, PVCs can be a warning sign of a life threatening heart arrhythmia, which is a rhythm problem in the heart.

[18] Lidocaine is a medication, usually used as a local anesthetic, which also helps prevent dangerous heart arrhythmias.

[19] NG stands for nasogastric. An NG tube is a tube that goes through the nose (hence naso) and down into the stomach (hence gastric). It is usually attached to gentle suction to draw out air and secretions from the stomach.

while, but they called me last night to say it was getting clogged. Let's see what's going on."

A foul odor assaulted us as we entered the room. Jim had just thrown up and the nurse was cleaning him. "Let me smell that," Brian said, leaning over the towels the nurse was using to clean him. "What do you think that is," Brian asked me. I sniffed, and immediately recoiled.

"It smells like feces."

"That's exactly what it is," he said quietly to me. Then to the patient he said, "Good morning, Jim." Jim didn't reply. His head waved slowly back and forth, but his eyes didn't seem to recognize Brian or anybody else. We listened with our stethoscopes briefly to Jim's abdomen, but heard nothing. It was firm to the touch. "You take care, Jim," Brian said, briefly grasping Jim by the forearm. "Alice," he said, turning with a kindly manner to the nurse, "please do your best to keep Jim comfortable."

Out in the hall Brian said, "It won't be long now for Jim. Why do you think his vomitus smelled like feces?"

"I really don't know. Could the cancer have obstructed the colon so much that he has stool backing up into his stomach?" The thought repulsed me.

"It's possible, but I think it's more likely he's developed a fistula[20] from his large intestines to his stomach. In either case, it's a time to be thankful Jim is as out of it as he is so he doesn't know what's happening to him. We'll keep him sedated, but I expect he's aspirated some of this foul stuff and he'll have a rip roaring pneumonia shortly. It's another example of when pneumonia is an old person's best friend." I understood what he meant.

On the way back to labor and delivery Brian looked over at me and said, "You bored yet?"

"Bored? What are you talking about?"

"Trust me. If you end up choosing family practice, the 'experts' down at the medical center are going to try to talk you out of it. They'll tell you family practice is boring, that you'll spend all day taking care of stuffy noses, back pain, and healthy people in for physicals. They'll say you'll just be a traffic cop, so that whenever anybody presents with something the least bit interesting, you'll have to refer them out to specialists. Do you think they're right?"

---

[20] Fistula is an open passage way between two structures that is not usually present in a healthy person. In this case, we are discussing the possibility of a fistula or opening between this patient's colon and his stomach.

"Not from what I've seen. You guys never know what's waiting for you in the next exam room. Or like today, we've gone from labor and delivery to the ICU[21] to Jim. There's nothing boring about that."

"Exactly."

Back on labor and delivery, Brian didn't like what he saw in Julie's tracing. "I think she's going to need a section. Jean, please call the surgeon."

Surgeon? I thought to myself. "Aren't you going to call an obstetrician?" I asked Brian.

"In this town, there are no obstetricians. As family physicians, we do the obstetrics. When we need a C-section, we call one of the two general surgeons in town. We do the first assist for the surgery and then break scrub to resuscitate the baby if that's needed. It works out rather well."

Over the next hour, that is exactly what happened. The baby was delivered by a Cesarean section performed by a jovial general surgeon who seemed completely at ease and competent throughout the procedure, joking with the scrub nurse, Brian, myself, the anesthesiologist, and even the patient. He made everybody smile. When the baby came out, Brian asked for suction equipment and suctioned out the baby's mouth and throat just before it took its first cry. He then stepped away from the operating table and briefly intubated the newborn to make sure no meconium[22] had gone down into its lungs. As with the surgeon, I was amazed at how comfortably and competently Brian performed these procedures. Would I ever get there someday?

I paused at the thought. Did that mean this is what I wanted to do? Had I inadvertently shown myself that I had made a decision about a specialty? Perhaps, but something in me still wasn't sure.

        *   *   *

"I just don't feel right."

"In what way do you not feel right, Harriet?"

The woman paused, looking up at me with beseeching eyes. "I don't know. Something's changed. I'm always nervous. I can't sit still. And my heart feels like it's pounding out of my chest."

---

[21] ICU stands for intensive care unit.

[22] Meconium is the term for fecal material passed by the infant inside the uterus. If meconium is present, it is usually considered a sign the infant underwent a period of inadequate blood supply at some time (though not necessarily during the labor, itself). It is important to suction out the airway of an infant born with meconium to make sure none of this material is aspirated into the lungs.

This sounded like classic anxiety to me. But when I took her hands to begin examining her something felt strange. Her hands had a silky, smooth texture, as if they were coated with a fine layer of sweat. They trembled as I held them in my hands, with a fine, rapid tremor. "Could this be hyperthyroidism?" came the question in my mind. I had never seen this condition before but had only read about it. I ran over the symptoms of hyperthyroidism in my mind: fine tremor, excessive sweating, rapid heartbeat, heat intolerance, weight loss. "Do you find yourself hot when others are cool?"

"Yes. My husband has been complaining that I always want the window open in our bedroom, even though it's almost winter. It wasn't always like that. Just for the last three months or so."

"Have you had a change in your weight recently?" I asked, becoming increasingly excited.

"Yes. Yes. I've been losing weight, which I haven't minded. It doesn't seem to matter what I eat, which I also haven't minded. Does my losing weight have something to do with what's wrong with me?" she asked with a note of alarm.

"I'm not sure. We'll have to do some more tests to be sure." Her face became even more alarmed.

"Tests? Tests? What kind of tests?"

She became increasingly agitated. I reached out to take her pulse to play the medical card. Her pulse raced at 120 beats per minute. She didn't calm down. This was definitely different from the cases of anxiety I'd previously seen. I finished her exam, including a close examination of her thyroid. One of my fellow medical students during my physical diagnosis course had had a somewhat enlarged thyroid which we had all felt, and this one felt similar.

"Harriet Jones is a 31-year-old female with a three-month history of sweats, agitation, tremors, weight loss, and heat intolerance. Her exam reveals smooth, moist skin, a fine resting tremor, and a resting heart rate of 120. Her thyroid exam shows a symmetrically, mildly enlarged thyroid. I couldn't feel any thyroid masses. The differential diagnosis is hyperthyroidism vs. anxiety. I think she actually has hyperthyroidism." I could barely contain my excitement.

Dave smiled. "You may have just diagnosed a virgin case of hyperthyroidism. You don't see much of that down at the university, do you? How about her eyes? Any proptosis?"

"What's that?"

"Proptosis is a finding in Graves disease, one of the most common causes of hyperthyroidism. When it's fully developed, patients look like their eyes are bulging out. Frankly, I don't remember what causes it or the best way to elicit early signs of proptosis, so let's look it up." He paused. "Remember Rick, never be ashamed to say you don't know something. You can always look it up. I had to do that a lot for the first bunch of years in practice." Together we pulled out a textbook. It described the placement of the upper lid in a normal eye as just covering the top of the iris, but not the pupil. As the eye moved from up to down, if the white of the eye showed, this suggested lid lag, an early sign of proptosis. If the proptosis was more progressed, the white of the eye showed above the iris even at rest. These eye changes were caused by deposition of material at the back of the orbit which pushes the eye forward.

With this information we went back into the exam room. Sure enough, she had subtle lid lag. "Harriet," Dave began, "it's clear something has changed. Rick here thinks he knows why. Why don't you tell her, Rick?"

"Well," I began hesitantly, "your symptoms and signs are quite typical of a condition called hyperthyroidism."

"What is that?" she asked, fear in her voice.

"It is a condition caused by an overactive thyroid gland. The thyroid gland is here, in your throat." I touched the lower part of her neck gently. "This gland secretes a hormone that controls the rate of many other reactions in the rest of your body. Your tremor, sweats, racing heart, and weight loss are all caused by too much of this hormone. The good news is that this is treatable. You will be able to go back to feeling completely normal."

"Thank you! Thank you! I thought I was going out of my mind. I'm so grateful this is treatable. Just tell me what I have to do next."

Out in the hall Dave smiled his warm smile. "Feels pretty good, doesn't it?"

"Yes it does."

*   *   *

"I'm here for my post partum check. The baby was born six weeks ago." The woman's voice sounded flat.

"How have things been going, Charlotte?"

She paused. "OK, I guess."

"Just OK?"

"Yeah, kinda."

Something didn't feel right. I couldn't put my finger on it. "Are you still nursing?"

"No. That didn't work out."

"Why not? What happened?"

"The baby didn't seem to wanna do it. I guess she liked the bottles better than me."

Did I detect sadness in that answer? "You seem kind of sad." She didn't answer, but tears welled up in her eyes. "What are you sad about?"

She was silent for a little while, and I decided to wait. Finally she said, "You seem like a nice man. Can I tell you something, something I haven't told anybody else?"

"Yes. Of course you can."

She hesitated again. Then the words came out in a torrent. "I can't stand it. I can't stand it. Being a mother is supposed to be so wonderful, so special, but I don't feel any of it. I don't want to do this. I don't think I can keep doing this. I love my baby…at least I think I do. But right now all I want to do is run away and leave everybody. Leave my husband. Leave my baby. Just run away and start all over again. I can't keep doing this anymore." She was crying, but through the tears her eyes blazed fiercely as she gave voice for the first time to the truth she'd kept pent up inside. "There. I've said it. Do you think I'm a bad person for saying it?"

"No, I do not think you're a bad person. I think you're actually courageous for being willing to tell me that. Obviously you've felt you've had to keep your true feelings hidden from everyone in your life."

Her tears spilled over. "That's exactly what I've had to do. I'm afraid they will have me locked up. Worse yet, I feel so bad for feeling this way. I'm the one who thinks I need to be locked up. I'm afraid of what I'll do to the baby."

"Do you really think you'd hurt your baby?"

"No, I know I wouldn't hurt her. But right now I feel trapped. I feel trapped by her, by my husband, by my whole life." She cried softly.

"If it is OK with you, I'd like to share what we talked about with Dave. I think he'll understand."

She looked up, frightened. "I don't know. I'm afraid of what he'll think."

"I've been working with Dave for almost a month now, and I've seen how he reacts. You can trust him."

"Alright. You can tell him." She resumed crying.

I stepped out of the exam room. I had to find Dave. "She's really overwhelmed," I told him. "She said she has a serious urge to run away from her family. She said she doesn't want to take care of the baby or her husband. She just wants out. At first she didn't even want to tell me."

"She's probably got postpartum depression, plus her hormones are all over the place. Her marriage hasn't been the best either, so there's a lot going on. It's a good sign she was willing to tell you. I've known Charlotte a long time, and she doesn't open up much to anybody. You must have said something or done something to really make her trust you. You've got a real knack for this. For now, I think it's best I go in with Charlotte alone."

He left me standing in the hallway. At that moment, something rose up within me—something warm and powerful. I'd made a difference. I'd made a difference for Charlotte. I'd made a difference for Harriet. And along the way I'd discovered something very important. I enjoyed practicing medicine. I enjoyed understanding the human body, how it can fail, and how to help it heal. I now deeply wanted the skills to be a truly excellent physician. And I wanted to be a family physician, not just a psychiatrist. That was because I had learned something else very important: I enjoyed the psychiatry I could practice as a family physician better than the psychiatry I could practice as a psychiatrist. My decision was made.

Chapter IX

Electives

"You can use this power for the good of patients or to fulfill your own needs. The choice will be yours every day."

The couple on the left side of the screen scampered through an open field, shedding layers of clothing as they went. Collapsing together, naked in each other's embrace, their arms and legs sinuously wrapped around each other. His erect penis filled the screen. She reached down, stroked its engorged, throbbing head with her hand, and then, in a close up, took it in her open mouth.

Meanwhile, the couple on the right side of the screen slowly undressed each other, one piece of clothing at a time. As her bra came off, the camera zoomed in, revealing an erect nipple on a dark areola. His tongue came down to give it a sensuous lick. She moaned.

I looked around the classroom, shifting nervously in my seat. Was this really a medical school course, or had I wandered into a triple X theater? We'd spent most of the first day of this course watching movies like this, but always two at a time. First we'd watched a movie of a man masturbating. Of course on the other side of the screen was a woman masturbating. Then came the heterosexual couples having sex in every imaginable position and combination. One movie, entitled *Joy in My Lady's Pleasure,* showed a man performing cunnilingus on a woman, causing her to come again and again. As one couple after another had sex on the left side of our screen, the couple on the right side kept going and going. "What's with the 'marathon couple'?" one student called out, and the name stuck. We began to cheer them on, giving us an outlet for our nervous laughter. When, after what seemed like days of sex, he finally ejaculated, in full close up, his semen shooting out in slow motion onto her lips, cheeks, forehead, and eyelids, we burst into applause. Next came men with men and women with women. We watched an elderly couple having sex, his bald head bobbing up and down as he, too, performed oral sex on his aged partner. We watched as a wheelchair-bound paraplegic launched himself into his lover's arms. She had found a few isolated areas of sensation below his waist he could enjoy, and she lovingly pleasured him in these intimate areas.

After the first few movies, we had broken up into discussion groups that included both men and women, requiring each of us to verbalize reactions to what we'd seen. Some were offended, others surprised. Many were aroused. All of us were uncomfortable to experience these feelings with our fellow students. In each case, our group facilitator encouraged us to express our reactions and to clarify our personal values about the sexual activities we'd just watched. Then we returned for more movies, followed by more discussions. Our course instructors called this process flooding, and flooded we were for two entire days. They knew that anyone bombarded by viewing countless sexual acts initially becomes increasingly anxious, and we certainly did. But, as they also knew would happen, toward the end of the second day something shifted. We were so saturated with images of the most intimate acts that they lost their impact. The instructors had continuously shown two movies at a time so the audience couldn't take either one too seriously. Eventually the anxiety in the room evaporated. We still laughed, but the anxiety gave way to comfort with each other and the acts we were watching.

At this point, in the eyes of our instructors, we were now educable. We could be taught to take a detailed sexual history without being judgmental or blushing. We were also taught to treat sexual complaints by applying the PLISSIT model, a step-wise model for providing help of increasing intensity depending upon a patient's needs. "P" stood for permission, because many sexual concerns, we learned, could be treated by simply giving permission to an individual or couple to experience the feelings they felt or to try a healthy but new behavior. "LI" stood for limited information, since other problems could be treated by providing basic education about the different arousal, climax, and recovery cycles for men and women; the wide range of normal human sexual experiences; and physiological changes that accompany aging. "SS" stood for simple suggestions, referring to specific treatment techniques, such as the "squeeze technique" for premature ejaculation or the "non-sexual intimacy" approach to impotence. Each of these three types of interventions or steps, we learned, could be provided by physicians in the course of their routine office practice. "IT" stood for intensive therapy, because we were taught to recognize when a sexual problem was too complex or manifested a deeper pathology, warranting more intensive treatment.

So what was I doing in this course entitled "Relationship Counseling for Healthcare Professionals" after deciding upon a career in family medicine rather than psychiatry? The answer was simple but challenging. Once I'd made the decision to go into family medicine, almost every course in each specialty became relevant, including psychiatry. I became focused on developing into a truly competent family physician, but then felt overwhelmed by the vast information to be mastered. I was thankful once again for having chosen to attend a school that completed the basic sciences in the first year and the initial clinical rotations in the second year. This left two full years of electives to fill in the pieces. And now that I'd chosen family practice, I felt the urgency of so little time to fill in so many pieces.

                    *   *   *

"Listen carefully to this patient's murmur," the resident, Harry, directed. I strained to pay close attention to the subtle sounds in my ears. Lub-shshshsh-dub. Lub-shshshsh-dub.[1] "Notice how it starts in early systole,[2] initially rising in pitch and intensity, then decreasing just before the end of systole." I could barely hear the murmur, and what I heard sounded just the same throughout. Her heart beat so quickly, it was all I could do just to identify the murmur was there. "This is a grade III crescendo-decrescendo systolic murmur," Harry proclaimed.

"I don't really hear any change in the murmur," I admitted. "It's happening too fast."

"What year med student are you?"

"Third year."

"That's a typical response for a third year. This cardiology rotation will be good for you. We do consults on all the cardiac surgery cases, and a lot of them are here for valve replacement. That means their heart valves have stretched or become so blocked that they are barely functioning. Those patients have great murmurs. Nothing is subtle. The valve surgery patients usually have at least a grade V, meaning you can hear it loudly with your stethoscope barely touching the chest. If you're lucky, you'll hear a grade VI, which means you can hear it before your stethoscope even touches the patient. Those are pretty cool. Why don't you start by working up the guy in 342B? He's in for a triple bypass and an aortic valve replacement. I'll meet you up there in a few minutes, and we'll do his cardiac exam together."

As I headed off to find room 342, I realized with some satisfaction that I was getting exactly what I had hoped for when I'd signed up for this cardiology elective. I had needed to learn a really good heart exam, beyond the basics I'd already been taught, and to get better at reading EKGs. Both were already starting to happen. It had been the same when I'd taken neurology to learn how to do a good neurological exam and to see lots of patients with neurological problems. Radiology had been great because we'd seen films of so many different kinds of cases. Not only did I learn the basics of how to read X-rays, ultrasounds, and other studies, but I'd been

---

[1] When listening to the heart with a stethoscope, there are two clear sounds, called the first and second heart sounds. Through a stethoscope they sound like "lub-dub…lub-dub." Subtle changes in these sounds, as well as murmurs that occur before or after either of the sounds, can be heard and tell the examiner important information about the functioning of the heart.

[2] The heart's cycle is divided into systole and diastole. Systole is the contraction phase during which blood is pumped out of the heart to the rest of the body. Diastole is the relaxation phase during which blood fills up the heart so it is ready to pump it out again with the next cycle.

exposed to a wide range of clinical problems I hadn't even known existed. My ectopic brain, that little black book, wasn't so little any more, as I filled it every day with new information, some of it basic and some of it subtle clinical pearls that might one day be helpful. Ahead of me were orthopedics, ENT,[3] anesthesiology, ophthalmology, emergency medicine, and as many other electives as I could fit in. Suddenly everything seemed important. Though I sometimes felt I was bouncing from one subject to another, slowly all the pieces were coalescing, creating a coherent web of understanding of human illness and its treatment.

Arriving on the third floor, I reviewed the chart for the gentleman in 342B, a Lawrence Williams. Only 58, he had not only blockages in three of his coronary arteries, but also severe aortic stenosis. This meant his aortic heart valve had become narrowed. Since all the blood that goes to the entire body must pass through this valve, if it is severely narrowed, it puts a great strain on the heart and deprives the rest of the body of adequate blood supply. According to the chart, his cardiac ultrasound showed calcium deposits on his aortic valve which made it impossible for the valve to close tightly. So not only did his heart have trouble pumping blood forward, but once the blood got out a large portion of it leaked back into his heart, a condition known as aortic insufficiency. He was becoming increasingly short of breath from congestive heart failure[4] as well as developing angina.[5] He was definitely heading downhill if he didn't have his valve replaced and the blockages in his coronary arteries bypassed.

Just then Harry arrived. "Let's go see Mr. Williams," he said briskly. I raced after him to room 342B. "Good afternoon, Mr. Williams," Harry said cheerily as we entered the room.

A white haired, overweight gentleman with a ruddy complexion and heavy jowls looked up at us. "Hi, Dr. Berkowitz," he smiled. "I'm ready for tomorrow. Sounds like a piece of cake."

"Pretty routine around here. We do lots of these operations all the time, and Dr. Carlson, your cardiac surgeon, is as good as they come," Harry

---

[3] ENT is an abbreviation for the surgical specialty of otorhinolaryngology, a Latin term for ear, nose and throat.
[4] Congestive heart failure, abbreviated CHF, is a very common problem in which the heart muscle becomes fatigued and stretched out. The heart enlarges and doesn't pump well.
[5] Angina, short for angina pectoris, is a condition in which a partial blockage develops in a coronary artery, one of the arteries bringing blood to the heart muscle. The blockage occurs from atherosclerosis, also known as hardening of the arteries. When the heart is called upon to pump harder, it needs more oxygen. But due to the partial blockage in the coronary artery, not enough oxygen gets to the heart muscle, and the experience of not receiving enough blood supply to the heart muscle causes a pain or a cramp, which is the symptom of angina.

assured him. "Rick here is a medical student. Mind if we take a listen to that great murmur of yours? After tomorrow, you won't be in such demand when you have a valve that works again." Mr. Williams' smile grew wider. The celebrity status that came from having a classic murmur everyone wanted to hear agreed with him.

We moved to the right side of his bed. "Let's start by looking at his neck." Harry pushed the button on Mr. William's hospital bed, and the head of the bed reclined backwards until Harry's upper body was propped up at a 30-degree angle. "Please look over at the left wall," he instructed Mr. Williams. "Now look at his neck, here, what do you see?"

I peered at the right side of his bull-sized neck, but didn't see anything.

"You see this," he said, placing his forefinger at the mid portion of Mr. Williams' neck. "Do you see a subtle pulsing of the skin right here?" Suddenly I did. I had completely missed it, and it seemed so clear once Harry had pointed it out to me. "That's jugular venous distension.[6] Now watch this." He pushed slowly and firmly on the right upper portion of Mr. Williams' abdomen. The spot that was pulsing moved two inches up his neck. Harry released his hand and the pulsating point returned to its original location.

I looked at Harry, my eyes wide. "I've never seen hepato-jugular reflux so clearly."[7]

"Mr. Williams, would you please role over on your left side, like you did for me this morning? That's great. Now Rick, get in there, your body right up against Mr. Williams." I remembered first being shown this approach to examining a patient's heart in my physical diagnosis course during the first year when they had taught us how to examine a patient. I had thought to

---

[6] Jugular venous distension, as its name implies, means the jugular vein in the neck is more visible and at a higher level of the neck than is seen in a healthy patient. This occurs when blood flow to the right side of the heart, the side to which blood returns from the rest of the body, backs up. If the right side of the heart is failing, meaning it isn't contracting well, then blood returning to the heart isn't pumped out as well so it backs up, and this backing up of the blood can be seen in the jugular vein.

[7] Hepato means liver and reflux beans backing up into something. So hepato-jugular reflux means blood backing up from the liver into the jugular vein. This happens in a patient with congestive heart failure when the examiner places firm pressure over the liver that's congested with excess blood due to the back up of blood flow from the right side of the heart. The examiner's pressure over the liver pushes blood up the inferior vena cava, the large vein that returns blood from the lower portion of the body to the right side of the heart. This pressure transmits up and through the superior vena cava that brings blood back to the heart from the upper portion of the body and raises the pressure in the jugular vein where we could see it.

listen to a patient's heart, you simply put the stethoscope on the chest. My anatomy course had taught me enough to know the heart sat a little to the left of the midline, so I'd thought that's where the stethoscope went. I was surprised to find that by rolling the patient over on his left side and listening way out on the edge of the chest, I could best hear some of the heart sounds. As I had first strained to hear those faint noises from the heart that would someday make sense to me, I had been impressed that the simple act of listening to a patient's heart was so challenging. Little did I know how challenging, and how important, listening to a patient's heart would prove to be.

I sat on the side of Mr. Williams' bed, leaning into the right side of his more than ample abdomen. "Place your right hand on the left anterior axillary line[8] over on the left side of his chest. What do you feel?"

I placed my hand firmly on his chest, just below and to the outside of his nipple. I smiled, as my hand was pushed up with each heartbeat by the enlarged heart as it contracted. "Is that what a left ventricular heave feels like?" I asked.

"That's right."

"I've read about this and even thought I might have felt a subtle example of it once or twice during my medicine rotation, but it never felt this clear."

"Can you feel anything else? Concentrate now."

I closed my eyes, concentrating all my focus on the sensation in my right palm as it pressed firmly upon Mr. Williams' chest. Was that a fine vibration I was feeling? It was subtle but definitely there. A smile broke out on my face as I looked up at Harry. "Does he have a thrill[9] here?" I asked.

"Exactly!"

"So that's what a thrill feels like."

"Now listen with your stethoscope over the same spot where you felt the thrill." I placed my stethoscope on that spot and immediately heard lub-shshshsh-dub. Lub-shshshsh-dub. There was nothing subtle about it. "Can you hear the crescendo-decrescendo quality to the murmur that you couldn't hear on our patient earlier this morning?" I closed my eyes again and concentrated, tuning out all other sounds. There it was. Lub-shshaaashsh-dub. Lub-shshaaashsh-dub. It was definitely getting louder in the middle of the murmur than at the beginning or the end. Harry could tell I heard it. "Remember what that sounds like. When you listen to a softer murmur see if

---

[8] Axilla is the medical term for the armpit. So the anterior axillary line is an imaginary line drawn down the side of the body beginning at the farthest forward portion of the armpit.

[9] A thrill is the sensation of vibration that can be felt in the area of the heart when a murmur is so powerful that it can not only be heard but felt as well. The sound and vibration both come from turbulence in the flow of blood through the heart, in this case resulting from Mr. Williams' diseased aortic valve.

you can identify that quality of getting louder and softer when it is more subtle. Now, speaking of subtle, listen in diastole and see what you hear. Tune everything else out and just listen to the space in diastole."

This was hard because the murmur was so loud now it was all I could hear. Again with my eyes closed, I tuned out the murmur and the first and second heart sounds, trying to listen to the space just after the "dub" when the heart relaxed. It was supposed to be silent, but I could hear something there. At first all I could tell was that it wasn't completely silent. Slowly, my ears began to distinguish a very soft blowing sound. Lub-shshaaashsh-dub-ssss. Lub-shshaaashsh-dub-ssss. It was all happening so fast since his heart was pumping 90 times a minute, at least. But I heard it. "That's a grade III aortic insufficiency murmur. He has both aortic stenosis and aortic insufficiency. Needless to say, he really needs a new valve." I suddenly looked up at Mr. Williams' face wondering what he was making of this, of our talking about all the ways his heart wasn't working. He returned a kindly look, as if to say, "Go ahead, kid, I'm happy to help you out."

"There's one more thing I'd like to show you, if it is alright with Mr. Williams." Mr. Williams nodded. "Place your fingers gently on his carotid pulse here, in his neck." I placed three fingers on his neck. After pushing aside the large "strap" muscle that runs along the side of the neck, I could clearly feel his pulse. "Now, with your other hand feel your own carotid pulse. Do you feel a difference?"

"Not really."

"Notice how fast your pulse pushes against your fingers. Now notice how fast his pulse pushes against your fingers. Do you feel a difference?" Once again, now that he pointed it out, I could feel a difference. Mr. Williams' carotid pulse rose up against my fingers just a little more slowly than mine, though this was so subtle I had missed it before. "That's known as a delayed carotid upstroke. You don't often feel it, but when you do, it means his aortic valve is so narrowed the heart can't pump the blood out as fast as yours does."

Finally he turned back to Mr. Williams. "Thank you so much for allowing Rick here to examine you. You may not know it, but you've taught him some things he'll use for the rest of his career. They may even save somebody's life one day."

Mr. Williams beamed. "Glad to be of service." I reached out my hand to thank him. He grasped it with surprising firmness. Without prompting he added, still holding onto my hand, "Like I said, I'm ready for tomorrow. I've just got a feeling nothing can go wrong." Harry and I exchanged a glance as we silently acknowledged a difficult lesson I'd already learned: nothing in medicine is certain.

"Remember, this is routine surgery for us around here," Harry replied. "You'll be in the OR early tomorrow, so Rick and I will look forward to shaking your hand up in the ICU tomorrow afternoon." Mr. Williams finally released my hand.

The next morning on rounds, bed 342B was empty. We finished rounding on all the cardiology patients and headed to the cardiology clinic. The sensation of Mr. Williams grasping my hand stayed with me all morning. After lunch we headed up to the ICU to round on the post op patients. Mr. Williams had just arrived, still intubated and on a ventilator. Jim, a PA[10] specializing in cardiac surgery, stood at Mr. Williams' bedside reviewing his vital signs and status with his nurse. Jim had been working with Dr. Carlson for a number of years. In fact, he seemed to run the post op care of all the cardiac surgery patients. He watched them closely, adjusting IV rates, medications and ventilator settings. I was amazed at his skill and expertise, which was definitely not what I'd expected from a PA.

Harry and I approached the bedside. Mr. Williams lay still, eyes closed, the ruddiness drained out of his cheeks, replaced by a pasty white pallor. A large white dressing covered the center of his chest from the base of his neck down to his upper abdomen. Two clear plastic tubes emerged from under the dressing, attached to reservoirs collecting blood that drained from somewhere inside his chest. "MR. WILLIAMS," Harry said loudly enough to be heard over the hiss of the ventilator. His eyes opened half way. In his post-anesthetic fog, a light of recognition came over the otherwise glazed eyes. He started to smile, but the ventilator tube pulled his mouth to one side. He tried to speak, but again the tube stopped him, just as the ventilator forced another breath into his lungs. Finally, he tried to reach out his hand for the promised handshake, but his hands were tied to the bed rails so he wouldn't be able to pull out his ventilator tube. I reached over and grasped his hand where it lay tied at his side. "GOOD TO SEE YOU," I half shouted. Mr. Williams could only nod in acknowledgement. His eyes slowly rolled up in his head.

Just then an alarm went off. The nurse hurried over, followed closely by the PA. Mr. Williams' blood pressure had dropped below a critical level. Jim pushed me out of the way. "Something's not right," he called out. "What are his vital signs?"

"BP 90. Heart rate 115. Pulse faint. He's had a sudden mental status change."

Jim listened to his heart and lungs. He noticed an increase in the blood coming out of one of the drains in the center of his chest. "Get me a stat

---

[10] PA stands for physician assistant. PAs have been trained in many medical and surgical specialties. They practice under the supervision of a physician and often develop great competence in their area of expertise.

blood gas and run his IV wide open," Jim barked. "Get a dopamine drip started at 10 mics per minute.[11] I need a stat portable chest x-ray. And get Dr. Carlson on the phone, STAT!"

For the next 15 minutes, chaos swirled around Mr. Williams' bed. Harry helped, but I could only stand to the side, horrified as the urgency in everyone's voices escalated. For the fourth time Jim spoke with Dr. Carlson. As he hung up the phone, he barked out, "Dr. Carlson believes he's leaking from an anastamosis.[12] We're taking him to the OR stat. Call the OR and tell them to expect us. When we get there I'll crack his chest so by the time Dr. Carlson arrives, we can have the bleeding site exposed. Now let's move it!"

Three nurses frantically disconnected tubes and wires from the ICU wall. "One, two, three, lift!" With loud grunts, the entire team lifted Mr. Williams onto a stretcher, and Harry joined the nurses in pushing him as quickly as they could out of the ICU. I ran ahead to hold an elevator, the one thing I could do to help. They burst through the OR doors. Jim broke away to scub as quickly as possible. The anesthesiologist joined the race at the head of the stretcher, getting a report from one of the ICU nurses as they ran.

"I'll need to do a crash induction,"[13] the anesthesiologist said as much to himself as to other members of the team as they disappeared into the OR.

"Throw on a pair of scrubs and go in," Harry said to me, panting and winded. "You're not likely to see something like this very often." With a nod, I ran for the OR locker room. In moments, I was in the OR, surrounded by swirling activity everywhere. The anesthesiologist was frantically drawing up several syringes of medications, placing wires on Mr. Williams' chest, adjusting his IVs and the dopamine drip, and frequently listening to his lungs. The circulating nurse raced in and out of the room, collecting the instruments Jim needed to get started. The scrub nurse set up her instruments, creating clang after clang as the metal instruments crashed together in her rush to lay them out for Jim. The other nurses had removed Mr. Williams' dressing and were washing his fresh wound with Betadine,

---

[11] Dopamine is a medication related to adrenaline that causes the heart rate to increase and blood vessels to constrict. It is effective in raising blood pressure but is usually reserved for critically ill patients. Mics is an abbreviation for micrograms and is pronounced "mikes."

[12] Anastamosis means a place of joining. In this case, it refers to the location where a vein used to bypass his heart blockage is sewn to the coronary artery.

[13] Induction here means to induce an anesthetized state. This is the medical term for putting a patient to sleep with anesthetic drugs.

the steel sutures catching on the orange soaked gauze pads as the nurses worked quickly.

"Are we ready to go, people?" Jim called out, throwing on his gown and gloves.

"Anesthesia's ready!"

"Srcub's ready!"

"Circulating's ready!"

"Alright. Let's get this chest cracked." With that, I heard one loud snap after another as Jim quickly cut the metal sutures that held Mr. Williams' sternum together using a large wire cutter. Next, the skin sutures, which had been placed so carefully, were literally torn open as Jim attacked them with a scissors. In one stroke, he drove a scalpel deeply through the chest incision, and pried open the two halves of Mr. Williams' chest from the neck to the abdomen.

Just then Dr. Carlson burst into the OR. Before he could speak Jim called out, "We've got the chest open. I can see the bleeder. It's the proximal anastamosis site on the LAD.[14] I'm getting you as much exposure as we can for the repair." Without a word, Dr. Carlson stepped up to the table, accepted a hemostat from the scrub nurse, and deftly placed it across the bleeding site.

"We've got the bleeder controlled," he called out. "What's his hemodynamic status?"

"BP 80. Pulse 130," barked the anesthesiologist. I've got the first two units of packed red blood cells transfusing now under compression. Four more are on the way. If you've got control of that bleeder, there's a chance he's going to make it."

The room burst into spontaneous applause.

Later that night, back in the ICU, I approached Mr. Williams' bedside. He was still deeply sedated. Though I didn't want to disturb him, impulsively I took his hand. Mr. Williams turned half-opened eyes in my direction. We held that moment in silence, knowing this second handshake was one neither of us had planned.

*   *   *

"Welcome to this class on stress and disease," Dr. Herman began. Crouching on the seat of his chair, leaning forward, supported on his haunches, his eyes darted rapidly about the room. "Stress is not what you think it is." His sharp movements matched his staccato words. "What you go through as medical students isn't stress." Was he serious? "You are subject

---

[14] LAD is an abbreviation for the left anterior descending artery, one of the largest arteries supplying blood to the heart.

to strain." I was puzzled. "Strain is pressure exerted upon a system. Every one of you is a system. You are a system in homeostasis. Homeostasis, a term coined by a Dr. Hans Sele, describes a system in dynamic tension but at rest at the same time. It is at rest because all the forces and counter forces are in balance." In other words you feel like you're being pulled in all directions but get to stay in one place. "Stress is the subjective experience individuals feel when they believe they do not have enough resources to cope with the strain to which they are subjected." That's a good description of being a medical student. "Stress impacts health and disease, not strain." He finally paused.

Returning from the bedside to the classroom had initially felt like a letdown. How could classroom study of the basic sciences compete with the drama and satisfaction of working with real patients? Yet, once I'd completed the introductory clinical rotations in medicine, surgery, and the others, I had been required to return to the classroom for at least four more months of basic science electives some time in my last two years of medical school. Virtually all my fellow students had done so by taking microbiology and the second half of pathophysiology at the start of the second year. This was because the basic sciences faculty had been slowly undermining the curriculum designed to squeeze all the basic sciences into the first year. Under pressure to reduce the length of their courses to fit all of them into the first year, each basic science professor had remained convinced that his or her subject was critically important, and was not getting as much classroom time as it deserved. They believed the school's medical students were not getting the necessary grounding in basic sciences by the time they graduated to be truly excellent physicians. So, instead of condensing the necessary material into the first year, they had simply left out key basic science content, like half of the pathophysiology course, and had placed this information into basic science "electives." That's why most of my fellow classmates had chosen to take microbiology and pathophysiology at the start of their second year, delaying entering the clinics for half a semester.

Always the rebel, I had gone straight into psychiatry, gynecology, and the other clinical rotations, planning to return to the classroom for those basic science "electives" when I would have had more clinical experience, and they'd be more relevant and valuable. This had worked well for me, since my clinical experience had already led me to choose a basic science elective I hadn't initially anticipated: nutrition. I'd begun to realize how important nutrition was for everything physicians did in medicine and how little nutrition training I was receiving on the wards. So I opted for an elective course in nutrition to help fill in this important gap.

I also attended many of the interesting lectures that cropped up on the medical school bulletin board, even though they weren't offered for credit.

The lecture on family practice had been one of these, and it had transformed my path as a physician. A series of lectures by professors from the Wharton Business School had opened my eyes to the profound challenges of funding, accounting for, and managing our nation's healthcare. The lecture on death and dying that had helped so much when I'd been confronted by that dying woman during my medicine rotation had also been among these offerings. That lecture was the only time anybody had provided me concrete advice for how to give patients or their families bad news. "When telling a family their loved one has died," the speaker explained, "it is helpful to begin by recounting the medical events that led up to the moment of death. Place these in as positive a light as possible, because your words will be repeated by family members over and over again as they move through the days, weeks and years after what to you will be a brief and awkward conversation." Good advice, I thought at the time. I only hope I will remember it when the time comes. A shudder ran through me at the thought that someday I may be a bearer of such news.

I did eventually return to take microbiology. But while I peered into microscopes and Petri dishes in the morning, instead of studying pathophysiology in the afternoon I'd chosen this elective course on stress and disease. The core content of pathophysiology that had been left out of the first year curriculum, including gastroenterology, gynecology, hematology, and oncology, I would have to teach myself and absorb through my clinical rotations and electives. I trusted in the maxim that anything truly important we would see over and over again. This had proven to be true, but had also left me with significant catch-up challenges in medicine, surgery, and gynecology.

Now sitting in the front row of Dr. Herman's class on stress, his wiry, tense, and restless movements confirmed my suspicion that experts make a life's work of their own unresolved issue. "To understand the human experience of stress and its impact on health and disease," he continued, "we must first understand hunter-gatherer societies." Now he had lost me. "Our physiology, including our hormones, evolved over hundreds of thousands of years of our species' history living in hunter-gatherer societies. Agriculture only developed within the past 10,000 years. This is a very short time by evolutionary standards. The technology we have come to take for granted as an essential part of our lives and that has so dramatically changed the pace at which we live, dates at the longest from the start of the Industrial Revolution, a span of less than 200 years. Our physiology can't possibly have evolved over just these few generations. Instead, we are living in an industrial and now post industrial society with a physiology finely tuned by the evolutionary process to life in a hunter-gatherer society." Now he was beginning to make sense.

"Studies of existing hunter-gatherer societies have demonstrated that, contrary to our prejudices, these societies existed in a state of abundance, not scarcity. The result was that we evolved in a setting of long periods of leisure punctuated by activities of great intensity, such as hunting and battles. Our entire stress response system, including a complex cascade of hormones and other biochemical activities you've studied in other basic science courses, was designed for short term stresses with long recovery periods. During these relaxation periods, our body's biochemistry had time to return to a healthy baseline." Peering around the classroom, again he paused, this time to emphasize the importance of what was to follow. "The problem is that recovery cannot occur in an hour, or a day, or even a weekend. Our hormones require at least two weeks to return to baseline. Today, we move from stress to stress like a ball in a pinball machine, with no time for recovery. The result is a collective lifestyle at odds with our own physiology." With this insight I felt I'd been given another key to understand the experience, the health and illness, of my future patients—and my own.

After several weeks of examining the studies of hunter-gatherer societies and the physiology of stress Dr. Herman had described that opening day, we turned our attention to understanding the impact of touch, being touched by another person in a non-sexual manner, on our hormone levels, rates of disease, and sense of wellness. I was surprised to learn that one of the most potent predictors of mortality in the elderly, meaning who would live and who would die, was whether or not the person had a pet they could touch. Studies had proven that physical contact with a soft, furry pet was one of the most powerful predictors of longevity in the elderly. The research on touch gave physiological support to my intuitive sense, already taught to me so clearly by June Warner, my anxious COPD patient, and others, that the physician's laying on of hands could be a powerful tool for healing.

Finally, we examined the institution of marriage. The data was clear—married men had lower rates of virtually every major illness, from heart attacks to strokes to cancer, than unmarried men. For women, there was no significant difference between rates of most diseases for married women vs. unmarried women. Dr. Herman hypothesized that marriage reduced stress for men but not for women. All I knew was that something about the institution of marriage worked for men in a way it did not work for women. For a budding family physician, one who hoped to treat the whole family, this was an important insight.

*   *   *

"Ahem," the professor in the monogrammed, starched, long, white coat cleared his throat. "Judging from the lot of you," he began in an officious tone, "I am absolutely sure by this point in your training you have all looked in hundreds of ears." He paused, using the moment to turn his piercing gaze on each of us one at a time. "I am equally sure you have never once properly examined an ear in your life. Today, that will change." Imperiously turning to a middle-aged woman in a crisp white uniform he barked, "Nurse, bring in the first patient."

The five of us, all male third-year medical students, stood stiffly in our rumpled short white coats along one wall of the large examining room. The nurse returned, escorting a strikingly attractive woman in her twenties to a special examining chair. Her full, dark eyes scanned the room, opening widely as she took in our motley group standing awkwardly in a row no more than five feet across from her. It seemed she had not anticipated an audience for her exam. As she climbed up the step and settled into the raised chair, I noticed her slender figure, professional blouse and skirt, and well manicured nails.

The nurse announced, "She is here for hearing loss."

"One begins the proper examination of the ear by carefully studying the pinnae,"[15] the professor opened, addressing us and ignoring the woman in the examination chair. "What should the examiner be thinking of upon identifying a structural anomaly in the pinnae?" Silence ensued as we shifted uncomfortably, looking down at the floor. I was aware he had not introduced himself nor any of us to the patient. "It's clear none of you remembers your embryology," he sneered.

"Doesn't it mean a possible problem with the kidneys, sir?" ventured the student next to me in a hesitating tone.

"Well, well. At least one of you was awake during my lecture in gross anatomy. Yes, that is precisely the case. The ears and renal system[16] develop at the same time in the embryo, so a structural anomaly of the ear should cause you to suspect a possible structural anomaly in the renal system." In an exaggerated and ceremonious manner he peered closely at each outer ear, including bending them forward one at a time to examine the backs of her ears. "After carefully examining the pinnae, then and only then is it time to pick up the otoscope." He spun on his heels, and in a grand gesture lifted the otoscope off its cradle. "Remember that the auditory or ear canal runs in an S shape, so to clearly view the tympanic membrane—that's the eardrum for those of you who slept through that portion of my anatomy lecture—you must first apply firm traction to the top portion of the outer ear in a superior

---

[15] Pinna is the technical term for the outer ear that extends off of the skull. Pinnae is the plural of pinna.
[16] Renal refers to the kidney and associated structures.

and posterior direction to straighten out the canal, like this." With that he pinched the upper end of the woman's left ear and pulled it upward and backward. She winced and tried to pull away, but his fingers tightened on her ear causing it to blanch. Reluctantly she gave up trying to escape. "You must also be aware that the outer half of the auditory canal has normal skin sensitivity, but the inner half is exquisitely sensitive. Be very careful with the end of the ear speculum in this inner portion or you will cause great pain for your patient."

He leaned over and peered through the end of the otoscope. "Hmmm. I see the cause of at least part of your problem."

"What's that, doctor?" came the woman's anxious reply.

"Your ear canal is completely blocked with cerumen."

"With what?"

"Cerumen. What you would call earwax. Your canal is filled with earwax. This is actually an excellent opportunity for me to demonstrate to our medical students here just how to remove cerumen." Turning to the five of us he went on, "As ENT specialists, we have multiple methods for removing cerumen. My personal favorite is the ear curette. This is a thin, hand-held instrument with a small steel loop on the end that is easily manipulated within the auditory canal to remove wax. You'll see I have several already here, within my reach on this special ENT examining chair." He pointed to an array of four or five curettes of different sizes hanging within easy reach of his right hand. Selecting one in the middle size range, he proceeded to scoop out copious amounts of soft, dark brown wax from the woman's ear. The nurse stood at his side, holding out a piece of tissue paper the professor repeatedly used to scrape the sticky wax off the curette.

At first the patient seemed relieved to have received a diagnosis explaining her problem. But after the first few scoops of wax came out, she began to wince with each new probe of the curette. The professor seemed oblivious. She tried to twist away, but he only pinched her ear more tightly. Tears welled up in her eyes. He had clearly gone beyond the area of normal skin sensitivity into the delicate inner canal, but was so intent on getting the wax out he was unaware of the pain he was inflicting. Her eyes pleadingly jumped from one student to the next. When they landed on me, I felt she was imploring me for help. I wanted to reach out, to take the two steps it would have required to cross to her side and take her hand, but the distance seemed like a chasm. What taboo would I be breaking if I reached out to her? In the moment of my hesitation, the last student on the end stepped forward. He came to her side and slid his hand into hers. She grasped it tightly as tears overflowed down her cheek. I was relieved and ashamed he had demonstrated the courage I had lacked. I vowed never to hesitate to give comfort again.

*   *   *

"You'll have 25 seconds. If you don't get this tube into my patient in that time, I'm stepping in to take over. You don't want her to die on you, do you?"

"No. No, sir."

"Now remember what I told you when we practiced with the plastic dummy. Position yourself at the head of the operating room table. Once she's asleep, extend her neck and place her head back. Separate her teeth with the thumb and second finger of your right hand. With your left hand, place the blade of the laryngoscope[17] in the mouth, moving it back over the tongue. Once you've got it all the way in, make sure you lift straight up—not sideways and definitely not at an angle that leans on anything else. If the laryngoscope puts any pressure on her teeth, you'll snap one of them right off. Then you'll be in a helluva fix. As for the lips, you could just as easily lacerate the hell out of them. So remember to lift straight up. Then you'll have one chance to get a good look down in there. If you see the opening of her vocal cords clearly, place the endotrachial tube[18] steadily down between them up to this mark we made. You must have a clear view and watch the tip of the tube pass between the vocal cords. If you don't have a clear view, pull out the laryngoscope and let me take over. Got it?"

"Yes, I believe I do."

"Alright. She's asleep. I'll blow some air in her lungs using this bag and mask to get some oxygen in her. She's going to need it since I doubt you'll get this on the first try. Nobody ever does. She's all yours." With that, the anesthesiologist stepped away from the head of the table.

I looked down at the face of this attractive young woman who had come in for a gynecological procedure. She seemed so calm, so relaxed, like the lifeless dummy I'd practiced on. I had to remind myself she had just been given an injection to put her to sleep and a muscle relaxing drug that caused

---

[17] A laryngoscope is a metal instrument used to manipulate the tongue and airway of a patient being intubated for anesthesia or for other purposes. The laryngoscope is composed of two parts. The first is a handle that looks like the long end of a steel flashlight. The second is a straight or curved flat metal piece coming off the handle at a right angle called a blade, even though it is not sharp. At the tip of this blade is a bulb that lights up wherever the blade is pointed. Laryngoscopes come in different shapes and sizes.

[18] Endo means within and tracheal refers to the trachea or upper windpipe that extends from just below the vocal cords to the division of the bronchial tubes in the center of the chest. So an endotracheal tube is a plastic tube placed within the trachea. The tube is inserted through the mouth and down into the throat beyond the vocal cords to maintain an open airway during anesthesia or other forms of unconsciousness.

every muscle in her body to become paralyzed. Suddenly I realized that, unlike with the dummy, I had to move quickly. "Extend her neck and place her head back," I said out loud. Taking the laryngoscope in my left hand, I pried open her mouth with the fingers of my right hand. It was dark.

"Slide the blade of the laryngoscope down over her tongue. It will light up her oral cavity," the anesthesiologist said, reading my thoughts.

Her mouth and throat became clearly lit. The hard metal blade slid smoothly over her tongue. "How do I know how far to go in without scraping the back of her throat?" I asked.

"You're about there," came the quick reply. "Pull straight up now. You don't have much time." My left hand pulled up and away with the laryngoscope. "Careful! Careful! You're pressing on her lower front teeth." He grasped the laryngoscope handle over my fingers and guided it up with more force than I'd expected. "What do you see?"

"There it is! I can see her vocal cords." Looking down the throat and straight through the vocal cords of this woman felt beyond intimate. I was actually inside her. The anesthesiologist handed me the plastic tube. Slowly I guided it down alongside the blade of the laryngoscope. But just as the tip was about to pass through her vocal cords, I lost sight of it. Something blocked my view. "I'm not sure the tip of the tube is in the right place."

"Pull it out. Take out the tube and laryngoscope. QUICKLY!" The anesthesiologist immediately stepped in with a mask he placed over her nose and mouth. Squeezing rapidly on a black bag attached to the mask, I watched as the woman's chest rose and fell with each squeeze of the bag. After about five breaths, he removed the mask, deftly slid in the laryngoscope blade, and in a moment he'd smoothly inserted the tube. "Listen for breath sounds in both lungs," he barked as he attached the bag to the tube and squeezed several times. With my stethoscope I could plainly hear loud breath sounds in each side of her chest. He proceeded to tape the tube in place so it could not be dislodged.

"What went wrong?" I asked once the anesthesiologist seemed settled into his routine of watching the monitors and adjusting her IVs.

"You made a typical beginner's mistake. You inserted the tube directly along the blade. That will definitely block your view. If you're right handed, you need to look directly along the blade, but place the tube in from the right side. This will give you enough of an angle that you can see the tip of the tube as it passes between the vocal cords. Once you've done this, you're home free. You almost got this one. Let's see if you can get the next. By the way, I'll let you bag and mask our next patient so you can start to get some practice with this, too. The whole goal of this rotation in anesthesiology is to learn how to manage an airway, especially in an unconscious patient.

Regardless of your specialty, someday you'll need this to save somebody's life."

For the next patient, as soon as she was asleep, he stepped aside. I extended her neck and placed her head back. With my left hand I held the mask over her nose and mouth. With my right hand I squeezed the black bag. Air hissed as it escaped from the sides of the mask and her chest did not rise.

"Press the mask more firmly onto her face." As I did, the hissing of escaping air stopped, and her chest rose. I released the bag and her chest fell. Compressing the bag again, her chest rose, and again it fell. "You've got it. You're getting plenty of air into her lungs. You could do the whole case like that and she'd be fine." A feeling, almost like a tingling, came over me. I was breathing for this woman. Her life was literally in my hands. With every squeeze of the bag I was sending her precious oxygen, keeping her alive. She needed me for every breath she took, or rather I gave. I'd never felt such a connection with another.

"Now try intubating her."

I pulled the mask away, opened her mouth, and inserted the laryngoscope blade. Down it went along her tongue. I lifted the laryngoscope straight up, using the same amount of force I'd felt the anesthesiologist use. Her vocal cords came into view. Slowly I inserted the plastic tube until I watched as its tip passed between the cords. "I think it's in there," I blurted out. The anesthesiologist attached the bag and instructed me to listen to the lungs as he squeezed the bag. I could plainly hear clear breath sounds on both sides.

"Nice job," he said appreciatively.

"I can't believe how good that felt. It was almost as if, as if…" I couldn't finish my thought. Or rather, I didn't want to share the next thought with him. An ill defined awareness grew within me. I had just placed a tube down another person's throat, a person who was paralyzed, helpless, needed my assistance. The sensation was oddly intoxicating. Something about this act felt almost too good. It was the sensation of power—power over another. Yes, the feeling of competence, the experience of breathing for and with another when they could not on their own, was deeply satisfying. But what of this other awareness? Was it real, or was I imagining it? I watched as anesthesiologists went about their business around me, intubating patients smoothly and efficiently, moving on simply to the next task at hand, all the while chatting with surgeons and flirting with nurses. Did they not feel it? Perhaps they had felt it at one time, but in the mundane routine of their day it had lost its pull. Competence, I was learning, is a sense of power to do something, and these anesthesiologists clearly exuded competence. But power, I was also learning, could mean power over another. Was there lurking unconsciously in their professional demeanor also this sense of power over another? I wondered.

*   *   *

"A man under hypnosis walks across a four inch wide steel beam suspended 20 feet above the ground. How can he do this without hesitation or fear?" The professor gazed around the room. "If I place a four inch wide steel beam on the floor, would any of you have difficulty walking across it? Of course not. The physical act is not difficult. What if it was placed six inches off the floor? Again virtually all of you would do it. What is different about raising the bar 20 feet in the air? The physical act itself hasn't changed, but your perception of the difficulty changed. You now perceive a risk of falling and a significant consequence if you fall. Your interpretation of this perception translates into fear. Under hypnosis, we are not able to change our subject's physical capabilities. We are, however, able to change their perception and interpretation of their circumstances. This change is the essence of the hypnotic state."

Adding this course on hypnosis had been an afterthought as I had designed the schedule request for my fourth year. Initially, given my plan to pursue family medicine, I had wanted to learn more about the underpinnings of my chosen field. I had already come to understand that the family medicine approach, the approach that integrated medical issues with psychological and social concerns, was based upon something called the biopsychosocial model of medicine.[19] I had planned these two months for

---

[19] The biopsychosocial model of medicine had been initially developed in psychiatry in contrast to the biomedical model. The biomedical model understood diseases and medical problems in terms of increasingly sophisticated insights into physiology and biochemistry. In the biomedical model, in order to understand the workings of a human being, one needed to understand the workings of its organ systems. In order to understand the workings of an organ system, one needed to understand the workings of its component organs. This model followed such reductionist thinking, ultimately concluding that complete understanding of how a human being works requires understanding its biochemistry and perhaps even the quantum physics of the subatomic particles which ultimately compose that person. In contrast to this biomedical model, and in part in rebellion against its excesses, a new model arose—the biopsychosocial model. This model, based upon systems theory, understands the workings of a human being as composed of simultaneous activities at different levels of organization. An organ system is an example of a level of organization. But it is not only understood by reducing it to its component parts. The organ system is also understood by knowing the properties of this system. It is further understood by knowing how this system interfaces with other organ systems in the whole that is the human being in which the organ system resides. The same is true on a social level. A human being is not only understood by knowing how his or her psychological activities operate, but by understanding how this

an independent study project on this model. Given my ongoing interest in applying psychiatry within family medicine, my advisor had suggested that, while I did my independent study, I also get exposure to the phenomenon of hypnosis. Though I did not understand why hypnosis would be relevant, I trusted his judgment, which had been on target this far.

As the course in hypnosis went on, I had initially been disappointed. Since it was taught by one of the world's experts in hypnosis, I had expected to learn dark, hidden secrets about inducing hypnotic states. I wanted to hear what enabled ordinary people to do extraordinary things under hypnosis. Instead, he explained that hypnosis enabled ordinary people to do ordinary things, but under extraordinary circumstances. He demonstrated again and again how hypnosis teaches us that what we perceive and how we interpret it determine what we believe we can and cannot do. This echoed Dr. Herman's distinction between strain and stress. In other words, we each provide the inner interpretation that gives meaning to our outer experience.

A mundane example he raised caught my attention. "What do we tell patients as we are about to start an IV?" He paused. "Just a little bee sting. Ha! What are your associations with bee stings? Pain. Fear. What have we just suggested to that patient? We've suggested to her that this IV will hurt like a bee sting, and she in turn will amplify her normal sensations during the placement of the IV into an experience of greater pain and fear than need be. You must recognize," he exhorted us, "that hypnotic phenomena occur all the time, not just under formal induction of a hypnotic state. We are all suggestible, and stress creates an even more highly suggestible state. At times of illness, at times of stress such as hospitalization or even a simple doctor's office visit, patients are highly suggestible. Use this to offer them positive suggestions for how to interpret their experiences."

"So what could you possibly say to a patient that's positive as you're about to stick him with an IV?" one student asked.

"I might say something like, 'I'm washing your skin off with a cool liquid. The coolness may make your skin a little numb. You will then feel a bit of a stick, followed by an odd sensation. Then it will be done.' Remember, an 'odd sensation' allows for a different interpretation than a bee sting. One is very negatively charged and the other is neutral."

---

individual functions as part of a family, and in turn, as part of larger social groupings. In this model, when confronted with a particular constellation of symptoms or complaints, the physician is called upon to determine the level or levels of organization at which dysfunction is occurring and to design and implement the best intervention at each of those levels. This intervention may be prescribing an appropriate antibiotic, performing surgery, counseling the patient or his or her family, ensuring the patient has necessary services in the home, ensuring the patient has adequate money to purchase the prescribed medication, etc.

"Aren't you manipulating the patient by using suggestions without his formal permission to place him into a hypnotic state?" another student asked.

"That is an excellent question. At its heart is the issue of power. You already have the power to suggest to the patient. She has given that power to you by virtue of the doctor-patient relationship and her own state of arousal and fear. What will you do with that power? Will you use it unconsciously and negatively, such as by suggesting a bee sting, or will you use it consciously and for the patient's good? Remember, a suggestion is just that, only a suggestion. Listen carefully to the words I used. 'The coolness *may* make your skin a little numb.' The patient still has a choice of whether or not to accept the suggestion."

"I'm very uncomfortable assuming this power," the same student protested.

"Uncomfortable or not, as a physician you will have extraordinary power. The mistake most physicians make is assuming that any doctor-patient interaction is ever just about the technical aspects of care. Every interaction is laced with this dynamic of power. If you are unconscious of this power, human nature will lead you to use it for self-serving needs of which you will not even be aware, and this will be at your patients' expense. If you are conscious of this power, you will have a choice. You can use this power for the good of your patients. The best physicians do exactly that. Or you can use this power to fulfill your own needs, and your patients will suffer. The choice will be yours every time a patient entrusts his care to you."

I thought of the many times I'd uttered the words, "Just a little bee sting." I thought of the control I'd gained by taking a confused woman's pulse. I thought of the calming effect I'd created in a woman struggling to breathe through the laying on of hands. I thought of the ease I had brought to a dying woman simply through the act of listening. I thought of the uncomfortably satisfying feeling of intubating an unconscious patient. And, with a wince, I thought of the pain I'd caused a dying man with lung cancer because I had not stopped at the edge of my own competence. Yes, he was right. As a doctor, even while in training, I had extraordinary power, both power to do things to patients as well as power over them. I silently asked for the wisdom to use this power wisely.

Chapter X

Subinternship

Something Important Became Buried, but Not Gone

"…Adam Spencer, 213B, is an 11-month-old male in for diarrhea and dehydration. Came in last night pretty dry, down about 10%. He's been tanked up and looks pretty good this morning, but he's still stooling out,[1] so you'd better watch his lytes."[2] I was writing furiously as he spoke. "And the last one is Jennifer Beecham. She's a heart breaker. Nine months old, Down syndrome, congenital heart defect, goes in and out of congestive heart failure at the drop of a hat. Kicker is she's never been home. Not one day. Spent her entire life in this hospital. Parents are sweethearts, but boy are they young—like 18. I left you a long note on her. The problem list tops out at 14. You'll spend a lot of time with her. Gotta go. Good luck."

With that he vanished. I looked down at my scribbled notes. Eight patients. I'd never taken care of eight patients before. And these were sick kids—sick and fragile. I'd just "picked up my new service," as they called it when one intern signs out to another because of changing rotations.[3] The problem was I wasn't an intern, at least not yet. I was a fourth-year medical student, and this was the first day of my pediatric subinternship. That meant I got to act like an intern, with all the responsibility, workload, and sleep deprivation. I didn't get paid because it was part of my education, but I worked just as hard. I also didn't know as much as an intern. I looked up from my list of eight patients. Nurses raced around me. The incessant beep…beep…beep of monitors reverberated everywhere. Babies wailed. A band tightened around the back of my neck. I was scared.

"Come on over here, team," a high-pitched voice rang out over the chaos swirling around me. "Time to start rounds." We huddled together on one side of the busy hallway. "My name is Skip. Skip Peterson. I'm your senior

---

[1] Stooling out is a shorthand way of saying a patient is having frequent, large bowel movements. This is common in children with intestinal infections.

[2] Lytes is an abbreviation for electrolytes. This refers to the concentration of key chemicals in the blood, such as sodium, potassium, chloride and bicarbonate, which must be kept in narrow ranges to prevent severe, potentially life-threatening, complications.

[3] A rotation is a way of referring to a period of time, usually one or two months, that an intern or resident spends in a given specialty or clinical service. This was a general pediatrics rotation, which is different from a rotation or service in the intensive care unit, the nursery, a specialty like oncology, etc. The intern or resident "rotates," meaning changes services, in order to get exposure to a wide range of clinical conditions, knowledge, and skills in their specialty.

resident for the month."[4] He was short, round, and already balding in his mid-twenties. But his face was kind. "Sue here is our junior resident. I've worked with Sue before, and you're all in for a treat." He paused, reaching out his hand to a young woman standing next to me. "You must be Mary," he said warmly. "I've heard good things about you from the residency selection committee. Glad you're an intern on my team." Turning to the woman standing on my other side, he said, "Sally, good to see you again. I remember meeting you at the new resident's barbecue last weekend. Also glad to have you as an intern on my team." His eyes turned to me. For a moment I awaited one of the countless humiliating references to which I'd grown accustomed as a medical student, but it didn't come. "You must be our subintern. What's your name?" He reached out his hand.

"Rick. Rick Sheff." I shook his hand. It was warm, though his grasp was weak.

"Welcome, Rick. Is this the first month of your fourth year?"

"That's right."

"Well, that's good news and bad news. It's bad news because you'll learn a helluva lot more medicine over the next year, which would make being a subintern much easier. It's good news because, since you're doing your subinternship in July, you're starting when all the interns are brand new and plenty green. They're probably just as scared as you are." He looked at Mary and Sally warmly. Their wide eyes blinked back confirmation. "This may not be much comfort to all of you, but this is the first day of my third year, so I'm brand new at being a senior resident." He paused, letting this fact sink in. "We're all in this together. So I promise you every day, on rounds and in conferences, we will cover the basics you need to know. I'll always be here if you need me. There is no such thing as a dumb question. I don't want a single patient hurt on my rotation because any of you was afraid to ask me a question. By the end of the month, you'll all know what you're doing a helluva lot better than you do now, and I'm going to see to that."

The band at the back of my neck softened. This month was going to be alright, I thought. Then I remembered night call. Will I be able to do the nights? I'd never worked straight through the night before. Now I was going to be on call, and up all night, every third night. I'd always been able to grab at least four or five hours sleep in the hospital on other rotations, even obstetrics. Did I have what it took? This month I'd find out.

        *   *   *

---

[4] Pediatric residency training is a three-year program. The first year trainees are called interns, second year trainees are called junior residents and third year trainees are senior residents.

"…Sean Michaels, 318B, four-year-old, pneumococcal meningitis,[5] day three of antibiotics. You shouldn't hear from him," Mary was saying. This was sign out rounds. As a group our team went to the room of each patient on our service one at a time. Outside the room their intern gave a report on the patient's status and handed off care of that patient to the intern on call. For tonight, that was me. "This is Angelina Russo, 223A, age five, asthma. Came in early this morning. She's real tight. You should check her as soon as we finish sign out and keep close track of her all night. She may need to go to the unit."[6]

As we finished sign-out rounds, I received a piece of paper from each of the other interns with their patients' names, diagnoses, key information like critical lab results, and anything I needed to check or do during the night. Sign out rounds had taken over an hour. During that hour, my beeper had gone off five times, but I hadn't had a chance to return any of the calls as each of the interns had just wanted to get out of the hospital and go home. I glanced at my watch. It was now 6:20 in the evening. I was five calls behind and had already been notified that my second admission for the night was on the way up from the emergency room. Grabbing a chair at the nurses' station, I started returning the first of the calls. Just then Skip came by.

"How is it going?" he asked.

"My first night on call as a subintern, and it's going pretty much the way I expected. But I hadn't realized how much sign out rounds really puts you behind."

"Have you gotten dinner yet?"

"No! I completely forgot about dinner. What time does the cafeteria close?"

"In five minutes, but sometimes they close early. Better get down there. A night on call without dinner isn't a pretty picture."

"Thanks. I'm there."

Running down the stairs, my beeper went off again. All I had time to do was write down the number and race into the cafeteria just as they were putting a gate up at the entrance. I grabbed a tray and headed to the hot entrees, suddenly feeling very hungry. The chicken breasts had dried out and shriveled from sitting on the steam table for hours. The broccoli looked pale

---

[5] Meningitis is an infection of the fluid around the brain. It is called meningitis because the meninges are layers of tissue that surround the brain and spinal cord, and itis, as previously noted, means inflammation. Pneumococcal means this infection is caused by the pneumococcus bacteria, which is also called strep pneumonia. It is the most frequent cause of meningitis and also causes pneumonia and infections of the blood.

[6] The unit is a way of referring to the intensive care unit or ICU.

and overcooked. It didn't matter. I took them both and a slice of pie. I was about to take a cup of coffee, but realized I wasn't tired. I'd been told there was always coffee to be found on the wards, so I decided to wait until I really needed it. It didn't matter that the food had sat out for so long because wolfing it down meant there was barely time to taste anything.

Racing back up the stairs, two at a time, I suddenly remembered I hadn't checked on the asthmatic, the one in 223A. I blew into her room to take a quick listen to her lungs as a baseline for the night, so I would know later on if she was getting better or worse.

"Hi…" I checked the sign out sheet for her name. "Hi, Angelina. My name is Rick. How are you doing tonight?"

"OK… I guess," came the whispered reply, requiring two breaths for her to get it out. I looked at her for the first time. This five-year-old girl sat straight upright in bed, the muscles of her neck straining with every breath. Her lips pursed together as she tried to force out the air her lungs held trapped inside.[7] Did I see a bluish hue around her lips? I listened to her lungs with the stethoscope. The skin between her ribs puckered in with each inspiration. At first I checked to make sure my stethoscope diaphragm was rotated the right way because her chest sounded so quiet. She was moving very little air. The ratio of inspiration to expiration was one to four, the worst I'd ever seen.

"I'll be right back." I raced to the phone. "Skip, you need to come to room 223 right away. I've got a really tight asthmatic. I think she may need to go the unit."

"I'll be there in a minute," came the remarkably calm reply.

Reviewing her chart while waiting for Skip, I determined she had been admitted earlier that morning after a week of her asthma getting worse. She had received a bolus of intravenous aminophylline and a first dose of IV steroids in the ED. She was getting isoproterenol nebulizer treatments every four hours. She was also being well hydrated with IV fluids and seemed to be on a reasonably high hourly dose of an aminophylline drip.[8]

---

[7] Asthma is a disease in which the airways in the lungs become narrowed because of three simultaneous changes: the lining of the airways swell; the muscles around the airways tighten; and mucous fills up the airways. These combine to make it difficult to get air into the lungs, and even more difficult to get it out.

[8] At the time of my training, this was state-of-the-art care for asthma. Aminophylline, the primary medication for treating asthma at the time, was administered intravenously with a high initial dose called a loading dose or bolus. Then a steady amount was administered in an IV drip calculated as an hourly dose based upon the patient's weight. IV steroids were also used, but these didn't take effect for six hours and sometimes took one to two days to reach maximum effectiveness. A nebulizer is a method of delivering medication as an aerosol or cool steam that can be inhaled directly into the lungs. Isoproterenol, a drug related to

"When did she get her last nebulizer treatment?" Skip asked as he came around the corner.

"Three hours ago."

"What was her last theophylline level?"[9]

"12, and that was two hours ago."

"How many steroid doses has she gotten?"

"Two. She's due for her next one in a few hours."

"Has she had a blood gas since the ED?"[10]

"She had one just before her intern went home. Her pO_2 was 72, pCO_2 36, and the pH 7.44."[11]

"She's got some room to get worse before she's in serious trouble. Let's go see her." He walked briskly but calmly into her room. "Hi, Angelina," he began warmly, picking up her name by glancing at the stamp on the chart hanging on her bedrail. "I'm Dr. Peterson, and I'm here to check on you. Rick and I will be here all night and will be checking on you lots and lots. How do you feel?"

"Not...so...good."

"Why is that?"

---

adrenaline, relaxed the muscle spasm of her airways. It was also believed that giving large amounts of intravenous fluids kept the mucous in the lungs moist and easier to clear.

[9] Aminophylline is converted to the chemical theophylline in the body, and the level of theophylline can be measured in a blood test. The target theophylline level for which we aimed at the time of my training was between 10 and 20, called a therapeutic window. Levels below 10 were thought not to produce significant therapeutic benefit. Levels above 20 could cause seizures and even dangerous heart rhythms. Checking the level of theophylline in the blood allowed for adjusting the dosage of aminophylline to fine tune the treatment for maximum effectiveness without dangerous side effects.

[10] Blood gas is short for arterial blood gas. This is a measure of the oxygen, carbon dioxide, and pH of the blood taken from an artery.

[11] pO_2 stands for the partial pressure of oxygen in her blood, a way of measuring the amount of oxygen carried in blood. 100 is normal, and below 80 is concerning. Below 60 can be life threatening. pCO_2 stands for the partial pressure of carbon dioxide in the blood, a way of measuring the amount of carbon dioxide dissolved in blood. Normal is 40. Levels below 30 and above 50 can be dangerous. pH is a measure of the acid-base balance in the bloodstream. 7.40 is normal. Levels below 7.30 or above 7.50 begin to be life-threatening. At the time of my training, an instrument called a pulse oximeter had not yet come into common use. Today it is used extensively and allows for the continuous measurement of oxygen concentration in the blood with a simple clip on the ear or finger. No needles or puncturing the skin are necessary. However, it does not measure pH or pCO_2.

"My stomach…is upset," came the two-breath reply. She clutched tightly to a small, worn, stuffed bear.

"I'm not surprised. The medicine in your IV can have that effect.[12] How's your breathing?"

"OK."

"Is it getting better, worse, or about the same since this morning?"

"About the…same."

"Let me take a listen to your lungs." He quickly listened with his stethoscope to both lungs. "Thank you. By the way, where are your parents?"

"Momma was here…but she had to go home…to feed the babies." A tear formed in the corner of her eye.

"Is she coming back?" Skip asked gently.

"Don' know." We knew enough not to ask about her father.

"Well, I think you're doing pretty well. We'll keep a close eye on you." Handing her the call button, "You just push this button any time you need anything, OK?"

"OK." She clutched the bear more tightly.

"Good. We'll see you in just a little while."

Out in the hallway, Skip turned and said, "I understand why you are so concerned. I'm concerned too, but based on my experience, I think she's going to turn the corner in the next six to eight hours, but it's going to be a long night because you'll have to check her at least every two hours, and perhaps more often." He clearly saw the worry on my face, and it wasn't about having to stay up all night. "When I was at your stage, I would have been just as worried."

"She's only five. I've taken care of COPDers who looked almost this tight, but they were old and had smoked all their lives. It wasn't like they were going to die young. But Angelina…" my voice trailed off. I'd read about the surprising number of children who still died of asthma every year.

"We won't let her die tonight, at least not on my watch. We can always take her to the unit and intubate her, but I don't think it will come to that. Trust me. Start by ordering respiratory therapy to move up her next nebulizer treatment. They'll balk at it, but treating her after three hours is alright." Pulling out a laminated card, he showed me an algorithm for aminophylline dosing. "Given her theophylline level was 12, there is some room to bump her dose, aiming to get her level to 18. Based on this algorithm, you can bolus her 2 milligrams per kilogram of her body weight, and increase the hourly drip rate from 1.1 to 1.3 milligrams per kilogram per

---

[12] One of the side effects of aminophylline is nausea and vomiting. This can happen at theophylline levels even between 10 and 20.

hour.[13] Then check her theophylline level three hours after rebolusing her and increasing the drip rate."

I calculated the doses based on her weight and wrote the orders in the chart. Later that night, I planned to write the aminophylline dosing algorithm in my ectopic brain book. But right now it was time to start returning calls and get to my second admission.

Two hours later, after I'd examined my second admission, written her admitting orders, and answered those first five calls (I'd also received four more I hadn't gotten to yet), Skip and I met in Angelina's room. "How are you feeling now?" Skip asked.

"OK...I guess."

"Any change?"

"No."

"Let's take a listen." Her lungs still sounded quiet, way too quiet. Inspiration to expiration was still one to four. Every muscle in her neck stood out on each breath. We stepped into the hallway.

"She doesn't look good to me. I'm afraid she's tiring," I said.

"You may be right, but my gut still tells me she's gonna turn this around. When is her theo[14] level coming back?"

"It's due to be drawn in another hour."

"Right now I've got to get to the ICU since I'm senior resident covering the whole hospital tonight, and there's a kid going bad there. That's not the way I want to free up a bed in the ICU for Angelina if we need it. I'll see you back here in two hours. We'll make the call then." He started to walk away briskly and stopped himself. "How did it go with your first two admissions?"

"Number one was a 20-month-old with diarrhea and dehydration. The ED calculated her initial fluid regimen for the first eight hours, so I just continued their orders. I want to go over these calculations with you when you get a minute to confirm them and plan what to do after that. The second was rule out[15] sepsis[16] in a 14-month-old. Her tap[17] in the ED was negative, so we've got her on amp and gent and will wait for cultures to come back.[18]"

---

[13] Dosing for pediatrics is most often given in milligram per kilogram amounts. This is because as patients grow, their weight changes dramatically, and the amount of medication or other substances they need will vary directly with the weight. To determine the actual dose of medication to be given, you would multiply the child's weight by the milligrams per kilogram amount.

[14] Theo is short for theophylline.

[15] Rule out is a short hand way of referring to a diagnosis that is being considered. Results of further tests or observations are necessary to determine if the patient has that diagnosis (it is "ruled in") or does not have that diagnosis (it is "ruled out").

"How's the rest of the house?"

"So far the calls have been pretty simple—Tylenol orders and lab results coming back. Angelina is my only real problem so far."

"Good. Remember, you can always call me if you need me."

"I'm sure I will."

We both raced off in opposite directions.

Two hours later, we were back in room 223. "How are you feeling now, Angelina?" Skip asked.

"Tired." We had just awakened her.

"How's your breathing? Any change?" She shook her head. "Let's take a listen to your lungs again." She dutifully leaned forward. Nothing much had changed, except I heard a few musical wheezes as she tried to breathe out. Skip looked at the bag of medication hanging from her IV pole, and we went back into the hall.

"What did you hear," Skip asked.

"Inspiration to expiration hasn't changed at one to four. It's still awfully quiet, but this time a heard a few wheezes."

"You're right. And that's a sign she is starting to move more air. Her third dose of steroids is dripping in right now. The problem is she is starting to tire. The steroids will reduce the inflammation in her airways and help them open up, but this effect comes on slowly and won't be felt for six to 12 hours. It's a race now between how quickly the steroids kick in and how soon she tires out. What did her theo level come back?"

"17."

"That's as good as it's going to get without making her toxic on the aminophylline, so there's no room to make an adjustment there. We have to hang tight and hope the steroids do the trick. See you here in another two hours."

"Right."

By then my third admission had arrived, so I headed off to see her. The ED told me another was on the way. I hadn't written up the history and physicals from my first two admissions yet. At least the phone calls had slowed down to one or two an hour.

---

[16] Sepsis is a shorthand way of referring to a bacterial infection that has spread throughout the bloodstream. Another term for this is bacteremia.

[17] Tap is short for spinal tap, also called a lumbar puncture. It's the test to determine if a patient has meningitis.

[18] Before starting the antibiotics, samples were taken from the patient's blood, urine, and spinal fluid. These would be incubated in the laboratory for two days in cultures. If the cultures grew a single type of bacteria, we would know this was the one to treat and would modify the antibiotics to attack that bacterium more specifically. If the cultures didn't grow any bacteria after 48 hours, we would conclude the patient had a viral infection and the antibiotics would be discontinued.

Back in room 223, Skip and I watched Angelina as she slept. Even in her sleep her lips pursed together with every out breath, an unconscious attempt to keep her swollen airways open. She must have sensed our presence in her fitful sleep because she opened her eyes and slowly sat up. There were the neck muscles, still straining with every breath. She looked so tired.

"How do you feel?" Skip asked. She stared blankly. "How do you feel now," he asked again.

"Tired," came the response.

"Of course you are. Can we listen to your back again?" She pulled up her pajama top, and Skip and I listened to her lungs. This time the wheezes were definitely louder. For the first time I heard some breath sounds. Inspiration to expiration was still around one to four, but the skin puckering between her ribs was a little less pronounced.

Outside in the hall Skip asked, "What's your assessment?"

"More wheezes, a little better air movement, and somewhat less prominent intercostal retractions.[19] I think she's doing a little better."

"That's my assessment as well. She could still fatigue, especially with how hard her muscles are all working just to breathe. She needs to be checked every few hours the rest of the night, but I think she's turned the corner."

A wave of relief washed over me. All night Angelina had weighed heavily on me. Now she was going to make it, and I felt relief from a tension I hadn't known I'd been carrying. I looked at the clock. It read 2:15 AM. The realization that I'd been working for more than 18 hours straight occurred to me. For the first time tonight, I felt tired. Skip seemed to read my mind.

"Let's get some coffee and go over the fluid calculations for your first admission. How are you doing with the others?"

"I've done the history and examination on all four of them. They've each got admission orders, but I haven't had a chance to write up the H and P's[20] yet."

---

[19] The skin puckering between the ribs is called intercostal retractions. Inter means between. Costal is a term that refers to ribs. These retractions between the ribs occur when the airways are partially blocked. As the chest expands to draw in a breath, negative pressure is created inside the chest. In healthy people, air rushes in through the mouth and airways, drawn in by this negative pressure. In asthma (and also in emphysema), there is so much blockage and resistance in the airways, that not enough air can get in to balance the pressure, so the negative pressure in the chest pulls the skin in between the ribs while it is also pulling in as much air as it can through the airways.

[20] H and P is short for history and physical. It refers to the written or dictated note in the medical record that officially records the patient's history of present illness, past

"As an intern or subintern, the H and P doesn't have to be like the write ups you did as a second- or third-year medical student. You need to put a brief admission note in the chart summarizing the essentials, and then dictate the H and P. Even what you dictate needs to be concise."

"I've got the instruction sheet on what buttons to push, but I've never used the dictating system before."

"The instructions are pretty straight forward. It'll just take a little practice to get used to it. For now, let's go through the IV fluid calculations." Sitting down with our cups of stale coffee, he began, "So your patient's 20 months old. What's the weight?"

"29 pounds."

"We do all our calculations in kilograms."

"Alright. Dividing by 2.2[21]..." I scribbled on a piece of paper. "That's about 13 kilograms."

"And how dehydrated was she? What percent of her body weight did the ED physician estimate she had lost?"

"The note said 10% dehydrated. During my initial Peds[22] rotation they had taught us how to estimate percent dehydration, but I don't remember exactly how to do it."

"We'll go over that on rounds tomorrow. For now let's just go with the ED estimate. So if she's lost 10% of her body weight, assuming that was all water, how much water did she lose?"

"If she's 13 kilograms now, her baseline weight would have been about 14.4 kilograms, so she lost about 1.4 kilograms of water weight, which is 1.4 liters."

"Exactly. You're doing well for almost 3 AM." His reminder of the time suddenly made me aware of feeling tired. "The best approach is to replace half of this amount in the first eight hours and the remainder over the next 16 hours, so it's all replaced in a 24-hour period. How much will she need over the first eight hours?"

"That's 1.4 liters, or 1400 milliliters, divided by two is 700 milliliters over the first eight hours."

"So the hourly IV rate should be..."

"700 divided by eight is...just a little less than 90 milliliters per hour." I looked at the rate ordered by the ED physician. It was 155 per hour.

"Why would it be higher?" Skip asked. I tried to focus on the question, but my thoughts would not come. Fatigue suddenly made thinking an effort.

---

medical history, family medical history, review of systems, physical examination, lab reports and imaging studies, the physician's assessment and plan of care.

[21] One kilogram equals 2.2 pounds. So to convert weight in pounds to kilograms you divide by 2.2.

[22] Peds is an abbreviation for pediatrics. It is pronounced with a long "e," like peeds.

Skip saw I was struggling, so he said, "How many times has she had diarrhea since admission?"

"The nurses said she was having a large loose stool about every two to three hours." Through the fog, I realized the answer. "She's having ongoing losses."

"Exactly. She is still losing water in the diarrhea, plus she has maintenance requirements just from breathing, sweating, and urinating. We'll go over how to calculate these tomorrow, along with her sodium and potassium deficits and replacement. For now assume she will need 65 milliliters an hour to replace ongoing losses. How much will she need over the next 16 hours?"

"We said her total loss was 1400 milliliters...half will be replaced over the first eight hours which are almost up...so she'll need another 700 milliliters over the next 16 hours which is..." I did the calculation long hand on a piece of scrap paper. "That's about 44 milliliters an hour. Add to that 65 for ongoing losses and maintenance and she'll need 109 milliliters an hour."

"That's right. So write that order, assuming the same sodium and potassium concentration in her IV now. We'll go over this in greater detail in the morning. What do you have left to do?"

"I've got to dictate the H and P's and answer the last few pages. Thank goodness they've finally slowed down. This assumes I won't get any more admissions. And we have one more check on Angelina in about an hour."

Sitting alone in a corner of the nurses' station, I stared at the dictation machine. The instructions swam on the page. I took another gulp of coffee. Forcing my eyes to focus, I picked up the hand set that looked like a typical phone receiver with a few extra buttons on it. I punched in the personal dictation number they'd given me and slowly followed each step of the instructions. Finally a high pitched wine came through the earpiece. I pushed the dictation button, and it stopped. "This is Richard Sheff, subintern, dictating an admission history and physical on..." I paused, releasing the dictation button and checking my notes for the patient's name. The high pitched wine returned. I held the dictation button down again and resumed, "Alisha Johnson. Medical record number 923584." I paused, trying to marshal my thoughts. "This is the first CHOP[23] admission for this 20-month-old black female who presents with a four-day history of diarrhea and dehydration." Slowly, I deciphered my scribbled notes and dictated the rest of the H and P into the machine. My beeper went off, and I answered a page to one of the nursing stations. Another lab result had come back, but it didn't need any further action. The nurse was simply notifying me of the

---

[23] CHOP stood for Children's Hospital of Philadelphia.

results. I wrote the test results in the appropriate place on the page the patient's intern had given me at sign out rounds.

I struggled through dictating the second patient's H and P, eyes drifting closed between sentences. Then it was time to meet Skip for one last check on Angelina. She was sleeping. We listened with our stethoscopes through her pajama top, not wanting to wake her. Long, musical wheezes came through, along with louder breath sounds. She was definitely moving air better. Silently we crept out into the hallway where we exchanged a warm smile, knowing she had turned the corner. Skip had been right. He and Angelina had taught me so much about the natural course of asthma and how it responds to treatment. They had also taught me another important lesson. Experience and knowledge added up to clinical judgment— something Skip had and I was just beginning to learn.

*   *   *

"This is the first CHOP admission for this 20-month-old black female with a benign past medical history who presents with a four day course of diarrhea without significant vomiting and approximately 10% dehydration."

Skip smiled at me as I continued presenting the history and physical for my first admission to our attending, Dr. Brookstein. During work rounds this morning with our team, he had provided constructive suggestions for how to improve my presentations, and I could see he was proud that I had already incorporated them into this presentation just two hours later. I was also surprised at how many details I remembered about the patients I had admitted during the night. Feeling a sense of responsibility for them because I was their doctor now, not just a medical student "learning on them," made a big difference. Having to write the note in the chart, think through their problems enough to write admitting orders, dictate the H and P, and present the case on work rounds, resulted in remarkable retention of information about each of these patients. I began to see how the interns and residents of whom I'd stood in awe in my earlier years were able to perform what had seemed superhuman feats of recall. Their feats were not superhuman, but resulted from the focused attention and urgency that come with significant responsibility, plus a hefty dose of repetition.

As I finished my presentation, Dr. Brookstein looked at me closely. "You mentioned the child was 10% dehydrated. How was this determination made?"

Glancing at Skip, I began, "She had dry mucous membranes. Her skin turgor was decreased with mild tenting,[24] but her eyes were not significantly

---

[24] Skin turgor refers to the firmness of the skin. Well hydrated skin is firm, stretches easily, and returns to its original shape quickly. Dehydrated skin is less firm and

sunken." Again Skip seemed pleased. These were the items we had gone over on work rounds that morning, and he wanted his team to perform well in front of Dr. Brookstein on teaching rounds,[25] also referred to as attending rounds. As the attending physician, he was the most senior member of the team and responsible for care for every patient on our service. Unlike a number of the other attending physicians, Dr. Brookstein had a private practice in an office in the community. Most of the other attendings held full-time academic appointments at the medical school and performed research. Dr. Brookstein's focus on a primary care pediatric practice day in and day out meant that he brought a very practical perspective to clinical problems, which we began to appreciate immediately.

"Then 10% is exactly right. I assume Skip went over how to evaluate percent dehydration with you?"

"Yes, sir, we went over it this morning," Skip reassured him.

"Good. Now here's a piece of advice I've been giving residents for years." He looked directly at me. "The most accurate way to determine percent dehydration is weight loss from a known previous weight, right?"

"Yes, that's right," I responded instinctively

"The problem is that each scale is a little different from others. So to be sure, you should weigh the patient on the same scale. That way you will know exactly how much weight they have lost."

"But what if the scale is in their pediatrician's office?" one of the interns asked.

"That is a problem here in the hospital, but not as much a problem in my office. So once you're in practice, you can use this technique easily. Here in the hospital, it would only work if the patient had been seen in one of the pediatric clinics. Then you could take them down to the clinic and weigh them on that scale. Now let's move on. How did you decide the IV fluid rate?"

"Estimating 10% dehydration in a 13 kilogram child, we determined she had lost 1.4 kilograms of free water. Half of this was replaced in the first eight hours and the remainder over the next 16 hours."

---

takes a longer time to return to its original shape. In moderate to severe dehydration (10 to 15% percent dehydration), when the skin is stretched by the examiner, it holds its shape for up to several seconds. This is called tenting because the stretched skin at that time looks like a tent.

[25] Teaching rounds were when the team of residents and medical students presented and discussed cases with the attending physician, the senior physician responsible for oversight of all patient care provided by the team. The focus was both to use these cases for teaching important information to the team members as well as to ensure the attending physician provided effective oversight of the care provided by the team.

"Excellent." Dr. Brookstein eyed me approvingly. "What about ongoing losses."

Once again, information from work rounds saved me. "The patient was having loose stools every two to three hours. We had the soiled diapers weighed and the output tracked. This led to an estimate of 15 milliliters of water lost an hour."

"And how did you calculate maintenance requirements?"

"Skip taught us the formula: 100 milliliters per kilogram per 24 hours for the first 10 kilograms, then 50 milliliters per kilogram for the next 10 kilograms, and 20 milliliters per kilogram for anything over 20 kilograms. For this child originally of 14 kilograms, this meant 1200 milliliters over 24 hours or 50 milliliters per hour."

"I'm very impressed," came Dr. Brookstein's reply. Skip beamed. I couldn't believe I was recalling all of this, especially in my foggy, post-on-call state. The coffee and adrenaline must have really been kicking in. "Now, how about her sodium and potassium deficits."

I paused, glancing at Skip. "Uh...Uh...We didn't have time to cover those on work rounds. Skip said we'd cover them with you during attending rounds."

"Well, finally I have something to teach this crew." He smiled at Skip. "Alright, Skip. Why don't you take all of us through how to calculate sodium and potassium deficits and replacement needs?"

Skip moved to a white board in this small conference room. As he did, irresistible exhaustion washed over me. I could hear Skip's words, but they no longer made any sense. "This patient was down 10%. That's 1.4 liters. What was her sodium concentration, Rick?"

"What?"

"I asked, what was the patient's sodium concentration?"

Somehow the number 138 came out of the mouth that no longer seemed connected to my suddenly useless brain.

"Right. Since normal is 140, this is close enough to normal for us to assume that her losses were isotonic, meaning she lost sodium at the same rate at which she lost water. This isn't always the case. If a child had a sodium of 155, you'd conclude she had lost more water than sodium. If the concentration was 125, you'd conclude she had lost more sodium than water. Am I making sense?"

Everyone else was nodding. Everyone else seemed to understand. But for me, nothing made sense anymore. I heard the words, but they bounced around in my head without any meaning. My nearly closed eyes met Skip's. He nodded with acknowledgment. That was the last question he posed in my direction. It's also the last thing I remember from attending rounds that day.

As we walked out of attending rounds, I shook my head violently, trying to clear it. Skip approached me. "Hit the wall?" he asked.

"Is that what you call it?"

"Happens to everybody. Did you get a shower this morning? That's usually a key to making it through your post call day."

"Yeah. After we checked Angelina, I went back to dictating the last two H and P's. I kept drifting off in mid sentence so they took forever. I did get to bed, after one last page, by five. Got up at 6:30, showered, and felt remarkably refreshed. But nothing seems to be helping right now."

"Just get busy. If you stay focused on tasks that don't require too much thinking, it keeps you going."

I looked down at the page I'd created that listed each patient on my service, now totaling 11. (One had already gone home this morning.) Next to each name were one or more actions we'd decided upon during work rounds this morning—check lab results, call the consultant on a case, gather more information from the parents, consult social services—these were things I could do in this state. One by one, a check mark appeared next to each action item as I plodded through the day.

Sign out rounds seemed to go on forever. Two more cups of coffee had barely gotten me through the afternoon. "Alisha Johnson is a 20-month-old with diarrhea and dehydration." I heard myself saying these words, but as if through a tunnel. "She's just about completed replacement fluids and lytes over the first 24 hours. Her stooling has slowed down, so we've cut back her IV rate. Starting to take PO.[26] She should be quiet tonight and ready for discharge in the morning." I had no idea how these words came out. I could only hope they were accurate.

Mercifully, sign out rounds finally ended. Emerging from the hospital, the oppressive July Philadelphia heat struck me, and nothing had ever felt so good. Relief washed over me. I mounted my old, black three-speed bicycle, the bicycle I'd ridden every day since starting medical school. The ride to my apartment, pumping my legs, sweating up the long hill into West Philly, felt like a journey home—home to sleep.

*   *   *

"I've made plans for us to get together with Rob and Beverly tonight," Marsha called out as I came into our apartment.

"You did what?"

"We're getting together with Rob and Beverly. Dinner and a movie."

"But I'm exhausted. I just finished my first night on call as a subintern. All I want to do is sleep."

---

[26] PO is an abbreviation of the Latin phrase *per os*, which means by mouth. Taking PO means she was starting to eat and drink.

"You will. After the movie." Marsha turned her dark eyes directly on me. "You've just left me alone for the last two days. This is the least you can do. Besides, I don't care if you sleep through the movie. Just don't fall asleep during dinner like you did the last time we got together with Rob and Beverly."

"I can't help it. Last night I barely got an hour's sleep. What do you expect?"

"I expect you to keep your promise. I told you I wasn't signing on to be a lonely medical spouse, or significant other, or whatever I am to you. And you promised you'd keep a balance between medicine and a life outside of medicine, including me. I'm just holding you to that promise."

A long sigh emerged from deep inside me. It felt as if I'd never wanted anything as much as I wanted to sleep at that moment. But I also knew I'd made a commitment, both to myself and to Marsha. The words of that exhausted medical student came back to me now. "Medicine is a bottomless pit. You can put all of yourself into it every day and never fill it up. You are the only one who can decide when enough is enough and walk away." I'd given all I had to give over the past 36 hours. But now something else was calling—my life outside of medicine, outside the hospital. And Marsha was the link. Yes, she was pushing me beyond my limits of endurance. Was it for me or for her? In my state of exhaustion the answer was not clear. She had agreed to be the link with my life outside of medicine. She was keeping her promise, but also demanding from me what I wasn't sure I could give. Was I supposed to keep my promise, or was there a more important promise to myself—to respect my own limits? Tonight I might be able to push myself, but at what cost? Something inside wondered if I had made a promise to Marsha I would ultimately not be able to keep. Then, with the will that drove me through gross anatomy, the same will that had just enabled me to get through my first night on call as a subintern, I pushed aside the fatigue.

"Alright. Let's go, but you'd better drive."

In that moment, something important became buried, but not gone.

*   *   *

"Jennifer has had a fairly stable past 24 hours," I said to the team on work rounds. A nine-month-old infant girl lay splayed out on her open bassinette, arms and legs restrained so she would not pull out her precious IV. An electric overhead warmer radiated heat to control her body temperature. Wires came off in every direction, monitoring her heart rate, breathing, and body temperature. A round cylinder of clear plastic covered her head, separating it from the rest of her body and allowing warm, moisturized oxygen to bathe her nose and mouth. The opening in this plastic hood was just large enough for her neck, so the rest of her body seemed to dangle off the hood by that slender neck. With every labored breath her abdomen sucked inward, reflecting the work her little body had to generate simply to breathe. "Going through her problems...number one is congenital heart disease secondary to an endocardial cushion defect.[27] As we all know, she has not been a candidate for surgical correction. Number two: CHF.[28] Her diuretic, dig, and preload reducers[29] are all maximized. Her liver is down 3 centimeters, which is where it has been all week.[30] At least for today, her CHF is stable. Number three: lytes. Her sodium is 134, potassium 3.4. The potassium has slipped to just below the normal range, so we are upping the potassium in her feedings and her IV. The sodium has remained a little low, but we have been unable to correct this due to her chronic diuretics.[31] When we've tried to increase her sodium intake, she goes into worsening CHF. Problem number four..."

---

[27] The endocardial cushion is a term for the structure at the center of the heart that divides the heart into four chambers, the right and left atria and the right and left ventricles. Jennifer was born with a distortion in this structure that caused significant leakage among her four chambers. At that time, it was not considered correctable by surgery. An endocardial cushion defect is sometimes associated with Downs syndrome, which was also one of Jennifer's problems.

[28] CHF, as described in chapter IX, refers to congestive heart failure, a condition in which the heart muscle is unable to pump strongly or efficiently. Jennifer's congenital heart defect led to chronic strain on her heart muscle, and she easily slipped into exacerbations of her congestive heart failure.

[29] Diuretic is a term for a medication that causes the kidneys to produce larger volumes of urine, effectively draining water out of the bloodstream and the body. This is one technique for reducing strain on a heart suffering from CHF. Dig is an abbreviation for digitalis, a drug used to increase how effectively the heart muscle contracts. Preload refers to the amount of blood that fills the heart before it contracts. In a patient with CHF, the heart muscles have been stretched too far, which is part of what makes it hard for the heart to contract effectively. By reducing the amount of blood that fills the heart, called the preload, the heart muscles become less stretched and contract with greater force and effectiveness. At this time, these were the only treatments available for CHF. Today additional medications are in use.

[30] As described in Chapter IX, in patients with CHF blood can back up into the liver. When this happened for Jennifer, her liver became enlarged. A skillful examiner could feel the subtle edge of her liver in her abdomen several centimeters below her ribs. During the subinternship Skip had taught me how to feel Jennifer's liver edge and to use this finding as a barometer of how well or poorly her heart was pumping.

[31] By causing the kidneys to make more urine, diuretics not only cause excess water to be excreted from the body, but extra sodium and potassium leave in the urine as

I repeated this ritual each morning on work rounds. As I did this morning, I caught sight of Jennifer's mother. Elizabeth, as I'd come to know her, had backed away from her daughter's bedside upon seeing our team enter the room, but hovered, vigilant in the background. Her hopes hung on my every word, as they did each day. Beneath her long, straight blond hair, blue 18-year-old eyes bulged wide as she strained to hear my recitation of her daughter's seemingly insurmountable problems. Jennifer's hospital room, especially the corner of it into which Elizabeth retreated whenever we came, had been home to her for the past nine months.

"Problem number eight: nutrition. She's receiving 110 calories per kilogram of her body weight through the NG tube.[32] Dietary has been continuing to consult on a daily basis. Her weight has increased by 45 grams today, but we are concerned this may be due to fluid retention from her CHF rather than a true increase in body mass, so we are watching this carefully. Problem number nine..."

Simply learning to keep track of all 14 of Jennifer's problems had taught me how to organize my thinking about complex patients in a way no other experience in medical school ever had. Each day Skip or one of the seven or eight specialty consultants looking after Jennifer taught me something new about one of her problems. Almost daily, she required modifications in her treatment regimen. The nurses who cared for her knew her so well; they were our most sensitive bell weathers for her condition, even more so than any of the numbers that came back on her daily blood tests.

Those blood tests were increasingly becoming a problem themselves. Number 11 in her problem list was venous access, meaning she was running out of places to put an IV. She was also running out of places from which to draw blood. Yet every day the lab team would go over her carefully, looking for the best place to drain a few more drops of blood from her. Sometimes they could find a small vein. When that failed, they resorted to the "heel stick." A heel stick, I was learning, meant stabbing her heel with a sharp blade and squeezing the heel until a few drops of blood oozed out. After nine months in the hospital, her heels had been reduced to a cluster of ragged, bruised, raw wounds that never had time to heal. Each morning her

---

well. Since congestive heart failure usually leads to excessive overload of sodium and water, diuretics are universally used in patients with congestive heart failure to correct this overload. However, because they cause the loss of water, sodium, and potassium to varying degrees, use of diuretics can lead to problems balancing these chemicals in the body.

[32] NG stands for nasogastric. Naso refers to the nose and gastric refers to the stomach. So a nasogastric tube is a tube that is placed through the nose down into the stomach. It is used to provide feeding into the stomach or can be used to suck secretions and air out of the stomach. In this case, it was being used to provide feeding.

piercing cry would ring out as the blade sank into her ravaged flesh. She would howl inconsolably as the lab tech mechanically and repeatedly squeezed her battered heel, trying to milk out just a few drops for the day's tests. As I wrote the lab orders for the following morning, each day I had to blot out the image of what those words on the order form meant for Jennifer. Without the information, she would die. But to get the information, I was ordering her torture.

Elizabeth knew this too. She would look on from her corner, as if beseeching the lab tech to stop, but knowing if she did Jennifer might die. Sometimes Jennifer's father, Jim, would join her. His hand, barely the hand of a man, would stroke Jennifer's arm or tender fingers. But because he came less often, his touch was awkward. I watched as his breaking heart reached out to the daughter he'd never been able to hold in his arms. Each day we talked of her fluid intake, oxygen levels, and potassium results, but all the while their bewildered eyes seemed to be trying desperately to understand why God had visited this tragedy on their innocent daughter. Sometimes, when they would let themselves, they'd turn to each other, sharing the unfathomable sadness and fears that threatened to tear their fledgling family apart.

Miraculously, every day Jennifer's heart kept beating and she kept breathing, even though it took every ounce of energy her tiny, nine-month-old body could generate just to carry on these basic necessities. As if this were not enough, another miracle unfolded before me as Elizabeth and Jim faced each day anew with courage and stamina that came from a source I could not name and did not understand.

*   *   *

"This is the second CHOP admission for this 14-month-old boy with developmental delay of unknown etiology[33] who presents with vomiting, diarrhea, and 8.7% dehydration."

"Wait a minute," Dr. Brookstein interrupted me. "Did you say 8.7% dehydration?"

"Yes."

"How do you know this child's dehydration with that degree of specificity?"

"Last night we followed your advice."

"What advice?"

---

[33] Etiology means the cause or source of a problem. If we don't know what has caused a problem, this is referred to as unknown etiology.

"This child had been evaluated in the developmental delay clinic two weeks ago. At that time, according to his mom, he weighed 21 pounds, which is 9.5 kilograms. When he was admitted at 1 AM last night, we arranged for security to open up the clinic office so we could weigh him on the same scale. He now weighed 8.67 kilograms. That calculates out to a loss of 8.7 percent, which we assumed was his degree of dehydration."

Dr. Brookstein seemed stunned. "You actually weighed him on the same clinic scale?" His mouth hung open.

"Yes. Isn't that what you had recommended?" Now I was worried.

"Well, uh, yes. That's exactly what I'd recommended. It's just that I've been making that recommendation to medical students and residents for 15 years, and nobody's ever actually done it!"

I looked over at Skip who was once again beaming at me. We had come up with the idea together, but he let me take the credit.

*   *   *

"...Michael Williams, 224A, in with asthma. He came in last night really tight, but he's opened up and doing better this morning. I've left you a note with my latest exam findings. And Jennifer Beecham." I paused, not knowing what to say to this earnest looking intern who was picking up the service I'd been carrying for the past month. "There is so much to say about Jennifer. She's a 10-month-old with Downs, congenital heart disease, and chronic CHF. She's never been home a day in her life." He looked at me, eyes widening for a moment. "It's a very long story. I've left you an extensive note. She's got a problem list 15 long." Again I paused, a lump suddenly rising in my throat. "Her parents are so young, but they've been hanging in there amazingly well." He looked at me expectantly. "Be good to her. Be good to her and to her parents. I'm going to miss them."

I looked for Skip to say goodbye. He was already starting to call his new team together for work rounds, but stepped aside when he saw me approaching.

"So this is it," he said, reaching out his hand in that warm but soft handshake.

"Yes it is. I survived."

"You did more than survive, Rick. You did very, very well."

"You know you're the reason. I can't thank you enough for all the teaching you did, and for how much you supported me and the other team members all month long."

"The best way you can thank me is to do the same for others when you get the chance."

"I'll try to do just that."

*   *   *

"Jim," I began hesitantly to my supervising second-year resident, "I know this is my night on call as a subintern in medicine, but I was hoping to ask a really big favor."

"What's that?"

"You know I've been an avid fan of the Phillies. Well, my friend slept overnight in line to get tickets to the playoffs, so I got to go to two of the playoff games, which were incredible. Now that they're in the World Series, she slept out again and got tickets for two of the World Series games. I'd never been to a World Series game before. We went to game two over the weekend. All I can say is it was electrifying. Tonight is game six, and the Phillies could win it all tonight. We've got left field bleacher seats at Veterans Stadium, and…" I paused, not knowing how Jim would react.

"Go. Go on. Get out of here. There's nothing happening tonight I can't handle without you. The Phillies in the World Series! Now that's something you can tell your kids about some day."

"Thank you! Thank you so much! I'll call you after the game to let you know when I'll be back to the hospital."

With that I ran out of the hospital, down into the subway, and made my way to Veterans Stadium, along with 64,000 other crazed Phillies fans. As we found our seats, high in the left field stands, the crowd was already cheering wildly. It had been a deliriously exciting season, watching Mike Schmidt, Steve Carlton, and Pete Rose, some of the greatest ever to play the game, bring the Phillies to the brink of a world championship. Tug McGraw added so much passion and drama whenever he came into pitch late in the game. The night was everything I had hoped it would be. As Tugger threw the final winning pitch, I joined the hysterical celebration, dancing and screaming at the top of my lungs.

The crowd poured out of the stadium. We headed to center city Philadelphia where we joined hundreds of thousands of ecstatic fans, dancing in the streets as music blared all around us. I suddenly remembered Jim and found a payphone. "Jim," I screamed into the phone, my voice now hoarse. "They won! They won!"

"Yeah," Jim replied, "The hospital's gone pretty nuts over it."

"I'm in center city. What's going on with the patients? I can come in any time."

"Don't bother coming back tonight. It's been pretty quiet, and I've got the service covered. Enjoy the celebration. See you in the morning."

"Thanks, Jim. I'll never forget you for this."

The Phillies winning the World Series, and my being a part of it by going to two play off and two world series games—including the sixth and final game when they won it all—was the highlight of my internal medicine subinternship. In fact, this subinternship had been somewhat anticlimactic. During my pediatric subinternship I'd learned how to get through a night on call, manage a service full of sick patients, and assume responsibility as the treating doctor, not just a medical student. Also, the pediatric subinternship had exposed me to lots of new clinical information unique to taking care of children. By the time I reached my internal medicine subinternship almost half way through my fourth year, I'd already been exposed to many areas of adult medicine. My fund of knowledge certainly grew during this subinternship, as did my clinical skills, but these were incremental rather than breakthrough steps.

I did learn at least one critically important lesson from this medicine subinternship. It required me to be on call every fourth night, whereas the pediatrics subinternship had required every third. I found taking call every fourth night dramatically more manageable for the long haul than every third night. So during my fourth year, as I began the process of applying to residency programs, I decided to consider only programs that required call every fourth night. At times I remembered that for many previous generations, medical students, interns and residents had taken night call every other night—36 hours on, 12 hours off. How had they done this year in and year out? One answer was that they completely gave up any hope of a life outside of medicine during those years, which then set a pattern for the rest of their lives. But even this did not do justice to the sacrifices made by these previous generations. One of my friends did a medical student rotation through the intensive care unit which required every other night on call. Half way through the month he had awakened at home in the middle of the night, disoriented. He called out to his girlfriend living with him that if she loved him, she should kill him that instant because he didn't think he could take it any longer. Somehow physicians in the past had coped with every other night on call. Some even recalled those years fondly because of the excitement of learning and taking care of patients and the deep connections they'd developed with fellow residents. I looked upon them with awe, and also some sadness. They probably did not even know how much that experience had left deep wounds of which they were not even aware.

So during my fourth year of medical school, I began the process of applying for a residency position in family medicine to begin the next stage of my training. Together with Marsha, we reviewed descriptions of the more than 400 family practice residency programs scattered across the country. Marsha was part of the selection process because we'd decided she would come with me wherever I entered residency training. So we combed the options with two initial filters: which programs had no more than every fourth night call the first year, and where did we both want to live.

Eventually, I settled on my first choice, a residency program at a large community hospital in Rhode Island associated with the Brown University School of Medicine. This program had risen to the top because it combined exceptional training in psychiatry and the social aspects of medicine with excellent clinical experience and training in the more technical aspects of medical care. Unlike the programs in university hospitals, at this community hospital family medicine was the largest residency, allowing the residents to assume major responsibility for all the clinical services, from the emergency department, to pediatrics, to running the intensive care unit. Match day came, on which a computer somewhere matched each graduating medical student in the country with a single residency program. As I opened the envelope, I was excited to find I'd been matched with my first choice program at Brown. I began to look forward to ending medical school and beginning internship and residency training in my chosen specialty.

But to my surprise, the waning months of medical school were not purely a time of happily wrapping up remaining requirements for graduation and saying goodbye to friends. As I prepared to assume a new level of pressure and responsibility, the dreams returned. Though I'd experienced pockets of stressful dreams during some rotations, such as the pediatric subinternship, not since gross anatomy during the first year had troubled dreams so filled my nights.

*   *   *

I came upon a door and knew I was being called to care for a patient on the other side, but the door would not open. Panic began to build as I sensed the urgency that the patient was sick, and I needed to respond. With full force, I threw my body at the door, which finally gave way. I found myself entering a large amphitheater. The stage was at least 50 feet across, and seats rose up in steep rows on three sides. Every seat was occupied. In the center of the stage lay a man, obese, sprawled out on a stretcher. I approached and realized he had recently undergone surgery, as he had a long sewn-up incision from the groin to half way up his chest. Just then he coughed, and the wound broke open from one end to the other. The edges of skin pulled back revealing coiled intestines, his liver, a beating heart. I looked on in horror as his organs began to spill out of the wound. Gazing up at the seats rising around me, each filled by a physician whose face I could not see, I realized they were all there to judge me—judge me on how well I cared for this patient. I sprang into action, reaching out to grasp slippery coils of small intestine as they slithered toward the floor. With each coil I caught and brought back up, two more fell out. The man's face was turning ashen. His breathing became quick and shallow. I stared at the heart as its beating slowed. I threw my body over his, desperately trying to draw the edges of his gaping wound closed. Just then, his heart stopped. I looked in panic from his dying face to the faceless judges surrounding me. From deep within, a guttural noise began, finally exploding. "Noooooo!" I screamed.

I sat upright in bed, chest heaving. No need to ask what had happened. Another dream had come to remind me I'd be starting as an intern in just a few more weeks. Was I ready for the challenge? Whether I was or not, the time was almost here.

Part 2

Residency

Chapter XI

Internship[1]

"Hey, Doc! I'm not angry."

Just as I was heading to the on call room to try to grab an hour or two of sleep, the beeper on my belt went off. I stopped at the nearest nursing station to call the extension number on the beeper display.

"Dr. Sheff," the voice on the other end of the phone began, "Mr. Walters in 217 has just expired. Could you please come pronounce him?"

"What happened?"

"He was a no code[2]. Came in with end stage cancer."

"I'll be up in a few minutes."

No need to rush in responding to this call, I thought. On the way up to the second floor, the phone call played over again in my head. "Dr. Sheff..." The words sounded so natural. I'd heard them many times before when patients had mistaken me for a doctor, rather than just a medical student. But they had always sounded odd—or more accurately, wrong. Suddenly I really was a doctor. Graduation from medical school, that day of celebration and relief, had produced this remarkable change. Suddenly I truly was "Dr. Sheff." Just as suddenly, nurses, respiratory therapists, and laboratory techs turned to me to decide what to do. At this moment, they turned to me to officially usher Mr. Walters into death. I smiled at the absurd thought that without my fulfilling this ritual, Mr. Walters would not be allowed to be dead. The transformation resulting from the addition to my name of just one word, doctor, was truly extraordinary.

With this new word also came a weight of responsibility I had not expected. Tonight being my first night on call as an intern, I was shocked to discover that I was responsible for covering over 100 hospitalized patients for whatever happened to them all night long. That's because for the first

---

[1] Like pediatrics, family medicine has a three-year residency training program. As noted before, the first year trainees are called interns, second year trainees are junior residents and third year trainees are senior residents. (Other specialties have residencies that range from three to five years in length.)

[2] No code meant somebody for whom either the patient or the family had requested "no heroic measures." Normally if a patient's heart stopped or breathing stopped while the patient was in the hospital, a code would be called, and a team of physicians and nurses would descend upon the patient, pumping on the chest, breathing for the patient with either a bag and mask or intubation, shocking the heart, and performing other measures designed to restart the heart or breathing and try to bring the patient back to life. This is the process referred to as "heroic measures."

month of internship I had been assigned to surgery. In this community hospital, being the intern on surgery meant that any patient in the hospital who had undergone an operation of any kind was my responsibility. Just as during my subinternship, I had a more experienced surgical resident backing me up I could call for help. But the first call from anywhere in the hospital for any surgical patient came to me as the intern, and the calls had poured in.

"Dr. Sheff, Mr. Salerno has a headache…" "Dr. Sheff, Mrs. Ramirez can't sleep…" "Dr. Sheff, the patient in 438 is having increasing pain…" "Dr. Sheff, Mrs. Palmero is vomiting…" "Dr. Sheff, Mr. Alfonso has just developed a temperature of 102…" At the end of each call came the same question: "What do you want us to do?" Every call was a problem the nurse expected me to solve—until this one. At least for this call it was the nurse who told me what to do.

I arrived at the second-floor nursing station and was greeted by the late shift charge nurse. "Thanks for coming up, Dr. Sheff," she said kindly, realizing it was well after 2 AM. "This was a patient of Dr. Martinez. End stage pancreatic cancer," she explained as she led me into room 217. There lay an emaciated, elderly appearing man with skin the color of a banana. His mouth hung open. Everything in the room was still. With a sudden flash of awareness, I realized I had managed to get through all of medical school without ever being present when a physician had pronounced a patient dead. My eyes met the nurse's as she looked at me expectantly, assuming I knew exactly what to do. I looked back down at the patient, wondering what he needed of me at that moment. Make sure his heart has truly stopped beating and that he is no longer breathing, I thought to myself. The image of this man, shut in a casket and being buried underground alive because I had failed to confirm he was still breathing flashed before me. I removed the stethoscope that had been draped over my shoulders and placed the tips in my ears. Holding the other end over his heart, I listened intently. Silence. Leaning across his still body, I listened to the base of his lungs on one side, and then the other. Silence. Standing up, I felt for the carotid pulse on one side of his neck. Nothing moved. Just to be sure, I felt for the carotid pulse on the other side of his neck. Again nothing moved. Finally, remembering an image perhaps from an old movie, I brushed my fingers across the eyelashes of his open eyes. No movement. I looked up at the nurse, who seemed quite satisfied.

Suddenly, a short, balding man rushed into the room. His suit was wrinkled and his shirt collar open.

"Are you Dr. Martinez?" I asked, unsure as this was only my second day of internship.

"No. No." He panted, winded from running up the stairs. "I am Mr. Walters' nephew. Is he dead?"

"Yes."

"Oh," he answered uncertainly. Suddenly he brightened, "His wife is on her way in. She really wanted to be here when he died. Would it be OK if you stepped out, let her come in, and then came back?"

"I…I suppose I could."

"Great." With that, he ran out of the room.

The nurse and I looked at each other a little bewildered. While we waited for Mr. Walters' wife and nephew to return, I sat down at the nurses' station to review his chart. He was a 67-year-old man who'd been diagnosed with cancer of the pancreas six months previously. He looked so much older than 67, I thought. He'd undergone several courses of chemotherapy which hadn't helped. The cancer had spread throughout his body, apparently causing him great pain. It had also grown locally in his pancreas, squeezing off the end of his bile duct, the tube that brought bile and other waste products from the liver into the small intestines. With his bile duct blocked by the cancer, bile backed up into the liver, causing bilirubin[3], one of the main components of bile, to build up in his bloodstream. It was this bilirubin that had made him such a brilliant yellow color. He had been admitted three days earlier with the plan for him to die in the hospital. He'd been treated with morphine, and the notes suggested he had either been sleeping or in a coma for most of the past three days.

Just then, a rustling noise caused me to look up. Running down the darkened hall toward me were Mr. Walters' nephew and a heavy-set, anxious appearing woman, presumably his wife. They ran past where I was sitting at the nurses' station and right into Mr. Walters' room. Because of the late hour and my fatigue, it hadn't been until that moment, watching them race toward the dead man's room, that I fully realized what his nephew had asked me to do. He wanted me to pretend that I hadn't already pronounced his uncle. He wanted me to conspire with him in creating a false impression for his wife that she had been there when he breathed his last. It seemed wrong. I had not only been raised to tell the truth, but felt a professional duty as a physician to tell the truth to patients. I'd also been taught in the clearest of terms that it was a crime to falsify a medical record, one that could cost me my medical license.

---

[3] Bilirubin is the waste product that results from the breakdown of red blood cells and the hemoglobin within those cells. Red blood cells age and die in large numbers every day in the healthy body. Hemoglobin from those cells is chemically broken down in the liver into bilirubin, which is then excreted through the bile into the small intestines. When bilirubin can't get out of the liver as bile, it builds up in the liver, eventually spilling into the bloodstream. This can occur when the bile duct is blocked, as in this case, or when the liver stops working effectively, as in hepatitis. Either of these conditions causes the body to become increasingly yellow, a condition referred to as jaundice.

In the hushed, late-night darkness of this hospital ward, poised at a troubling moment of decision, forgotten words began to come back to me. "Every doctor-patient interaction is laced with the dynamic of power," my hypnosis professor had stated. In this moment I realized that insight extended to family members of patients as well. I suddenly found myself with the power to ease this woman's pain at the passing of her husband. But to do so, I would have to tell a lie—or more exactly, participate in creating a false impression. Which was more important, upholding my personal integrity as a physician as well as the integrity of the medical profession by telling the truth, or offering solace to a grieving widow? I recoiled at the thought that I was at risk of adopting the paternalistic, doctor-knows-best persona, something I'd rejected so thoroughly until this moment. My training as a therapist had taught me people heal best when the truth is not hidden. My experience in medical training had taught me the extraordinary courage ordinary people manifest in the face of the truth, especially at moments of tragedy such as the one unfolding before me. I'd come to believe that without truth, true healing was not possible. The power of the truth could not, and should not, be denied.

Yet another kind of power was at play here. Placed into my hands was the power to ease this woman's pain. "You can use this power for the good of your patients," my professor had asserted. "The best physicians do exactly that." Did I agree? Would I use this power to participate in creating a falsehood in the name of healing? "If you are conscious of this power, you will have a choice." Suddenly the clear simplicity of black and white had no place. Shades of gray danced before me.

Then a scene from the past flashed into my mind. I was back in the meeting of the psychiatry interest group presenting my paper on love. A fellow medical student asked, "What's love got to do with psychiatry?" In the fatigue and darkness of the late hour the words became, "What's love got to do with medicine?" Faced with one of the first decisions I would make with my new title of doctor, I knew the answer. Everything. Confronting ambiguity and knowing action was required, I found myself asking a question I had no way of knowing would provide guidance for the rest of my career. At this moment, what is the most loving thing to do? The shades of gray cleared.

I waited a full minute, and then approached the room. Mr. Walters' nephew looked up and, in a hushed tone, said, "The doctor is here." After a long pause, he took his aunt's arm and slowly ushered her to the door. Looking down at the floor and shaking her head sadly, she stammered to herself, "He's gone. I know it. He's gone."

The door closed, and the nurse and I found ourselves once again standing on either side of Mr. Walters' bed. Awkwardly, we both realized there was nothing to do at this moment. It was almost difficult to keep a straight face at this supposedly somber time. I waited a decent interval and emerged from the room.

His wife looked up into my face. "He's gone. Isn't he?"

"Yes, he's gone." She started to sob softly. I desperately sought something comforting to say.

Again words from somewhere in the past came back to me. "When giving bad news, review the history of events that led up to this moment…Remember, whatever words you say at this awkward moment will be repeated by family members and loved ones over and over again for years."

"As you know," I began somewhat hesitantly, "your husband has had cancer for over six months. His doctor treated him with chemotherapy. He tried everything to save him, but nothing could help. He was admitted a few days ago and has been receiving morphine since he came into the hospital. For the past three days he has been sleeping peacefully and died without any pain. I'm very sorry. Please let me know if there is anything else I can do to help."

She looked up at me with wide, sad eyes. Suddenly she threw her arms around my neck. Burying her face on my shoulder she said in a torrent, "No. Thank you. You've done so much already. Thank you. Thank you." With that, she released my neck, and her nephew slowly escorted her back into the room to say goodbye to her dead husband. Muffled sobs and the words, "I knew he was gone…I knew he was gone…Thank God I made it in time…" were all I could hear.

Slowly I returned to the nurses' station. There, on top of his chart, was a blank death certificate. Only I, as the doctor who had pronounced him, could fill it out. For the first time, I began the ritual of completing this public record of a man's death. Cause of death: pancreatic cancer. Date of death: June 22. Time of death…I looked at my watch. 2:58 AM. Though he had stopped breathing over an hour ago, and I'd completed pronouncing him 20 minutes ago, as his time of death I wrote down 2:55 AM. His wife will forever believe she was there when he died.

Perhaps this is what I'd done that had led his wife to thank me so profusely, saying I had done so much? Or perhaps it was leaving her with an image of her husband sleeping comfortably, dying peacefully and without pain after having suffered so much for the months leading up to this, his final night. Perhaps in that moment I had come to stand for all the doctors and all the nurses who had provided care and comfort to her and her husband during his entire illness, and she had poured out her gratitude to me because I was the doctor standing before her. I paused, another thought forming. Perhaps she was grateful for the relief of knowing that, at this moment of utter powerlessness, when she was forced to surrender the one most precious thing in her life, someone else had stepped in to assume control. In so doing I had also brought order and a semblance of meaning to this searing moment by reminding her that everything possible had been done for her husband, and that his death was the natural order of things.

*   *   *

"Hey, Doc!" the man began, leaning forward, repeatedly clenching his right hand into a fist and opening it again. The muscles on the side of his face bulged visibly as his jaw tightened. "Hey, Doc," he repeated, to be sure he had my attention, then pausing, his eyes boring into mine. "I'm not angry," he finally said, exaggerating the slow articulation of each word.

Doc. He had called me Doc. The one other person who had ever called me Doc had been a psychotic young man at the Institute. At the time I had been struck with the thought that I was not really a doctor. Now that had changed. And it took the newly acquired addition to my name one step further. This man wanted something from me, and felt comfortable enough to address me in a familiar, almost intimate way while acknowledging my professional status. All this he unwittingly summed up in that simple word, Doc. Yet a mix of simultaneous, conflicting messages swirled around me. His pumping fist, clenched jaw, and denial of the obvious in his opening statement, "I'm not angry," created a wary tension in the cramped examination room.

What do I say next, I wondered uncomfortably. My first meeting with this short, balding man in his late sixties was not starting out well. It was about to take a turn for the worse.

"What brings you here?" I asked, deciding to begin with an open ended question, as I'd been taught.

"What brings you here?" he repeated in a mocking tone. "What kind of question is that? My wife brought me here, if you really want to know. But what does that matter?" He paused again, peering closely at me. "How old are you? You seem awfully young to be a doctor. Do you know what you're doing?"

He had gone for the jugular, and so quickly.

"I have completed medical school and am now in my residency training as a family physician," I answered truthfully, though defensively.

"So I got me a baby doc," came the derisive response. There was that term again, Doc, only this time it didn't have any of the positive associations it seemed to have when he'd first used it. I tried unsuccessfully to hide my discomfort. "Shiiit." He stretched the word out as if enjoying its sound and the effect he was having on me. "Why did I have the bad luck to get you?"

"Each patient here at the Family Care Center is assigned one of the family medicine residents in training as their personal physician. So I will be your family doctor for the next three years." I paused, wondering if there was even the slightest chance he would still be my patient three years from now. I certainly hoped not. "But our focus today is on you, not me," I said, desperately trying to change the subject. "So how may I help you?"

"Help me? Help me? What makes you think I need help? You need to know one thing about me. I am a plumber, and I've lived every day of my life with a lead pipe in my hand so I could tell any man to go to hell!" He pounded his fist into his palm. Instinctively, I moved my chair back several inches.

"You're here for an appointment with me as your doctor," I said, somewhat exasperatedly, "so what were you hoping would happen today?"

"My wife made me come here. I've been complaining to her about these pains I've been getting in my leg. I guess she got fed up with my complaining and made this appointment. I told her I didn't want to come to no doctor, but she dragged me here anyway. So Doc, what are you going to do about them?" Now as the "Doc" this was my problem to fix, and he wasn't helping.

"I can't do anything about them until you tell me more about the pains," I said, trying to control the edge in my voice. "Where do you feel them?"

"In the back of my leg, here," he said, rubbing the back of his right calf.

"What do they feel like?"

"What do you mean?"

"Are they sharp, dull, cramping, burning…"

"I don't know 'bout these pains. They just hurt."

"What brings them on?"

"I already told you I don't know much 'bout the pains. They hurt. They just come, like when I mow the lawn."

"You've got a push mower?"

"Yeah."

"So you're walking when you mow the lawn?"

"Of course I am," he said, his irritation rising.

"Do the pains go away when you stop mowing the lawn?"

"I suppose so. Never thought about it. Hey, why are you asking all these questions?"

"I'm trying to diagnose what's causing your leg pains. It sounds like these pains come on when you're walking and go away with rest. Is that right?" I was beginning to think he might have a blockage in his femoral artery, like the patient on which I'd helped operate during my surgery rotation.

"Yeah. Yeah. I never thought of it like that, but I guess so."

"Has this been getting worse or staying about the same?"

"Well, I used to be able to mow most of the lawn, but now it comes on about half way through." This told me the problem was getting worse.

"Tell me about your general medical history. Any medical problems?"

"Yeah. I got the sugar."

"Diabetes, you mean?"

"Yeah. That's right."

"How long have you had diabetes?"

"I don' know. They told me years ago I had it, but I didn't pay it no mind, and I'm not about to start to neither."

"Any other medical problems? Heart disease, stroke, kidney problems, emphysema?"

"No. None of those. Just the sugar."

"You smoke cigarettes?"

"Yeah. But you're not gonna give me a lecture about those, are you? You and the rest of the world. I'm damn sick and tired of everybody thinking they can tell a man what to do."

"Sounds like I would be wasting my breath."

"You're right 'bout that."

"How about alcohol?"

"Yeah, I have a beer now and then."

"How much do you drink a day?"

"I already told you not every day. They didn't teach you to listen too good at that fancy school of yours, did they?"

Despite persistent barbs like this one, I finally finished the medical history. As I tried to examine him, his irritation continued to rise. "Roll over on your left side," I said as I prepared to examine his heart. I placed the stethoscope firmly on the outer left side of his chest as I'd been taught.

"Hey, Doc! Are you listening to my heart?" his voice exploded into the stethoscope causing me to tear it out of my ears.

"What did you say?"

"I asked if you thought you were listening to my heart. I'll bet you're so green you don't even know where my heart is. It sure as hell ain't out in my Goddamn armpit."

"I assure you I know more about your heart than you give me credit for," I said through gritted teeth.

With relief I finished the exam and excused myself, saying I'd be back in a few minutes. Outside the exam room, I headed to the end of the hall, entering a conference room where I found Dr. Parker, the preceptor for today's Family Care Center session. A preceptor was either a faculty member for the residency program or a practicing family physician from the community. The preceptor's job was to supervise the work of residents in the clinic that day, and I knew I needed help on this one.

Seeing patients in the Family Care Center under supervision like this was a requirement for all family practice residents. The founders of this new field of family practice had recognized that most of family medicine is practiced in an office setting, not a hospital. They also recognized that caring for patients in an office setting requires a very different set of knowledge and skills than caring for hospitalized patients. As a first-year resident, patients were scheduled to see me two afternoons a week in the Family Care Center. This would grow to three and then four half days in my second and third years, so I would continue to build greater knowledge and skills in the real world of caring for patients outside the hospital. It was great to start seeing my own outpatients like this so early in my training, patients assigned to me as their family physician. But these sessions in the Family Care Center were in addition to all my responsibilities as an intern on the surgery service, or whatever other service I would be on. I could already see that prying myself away from those inpatient responsibilities was going to be a real challenge.

"This guy is so obnoxious," I exploded at Dr. Parker as I walked into the room. "I wanted to smack him more than I wanted to treat him."

"What is he here for?" came the patient reply.

"Calf pain. He's got calf pain that appears to be progressing. Comes on with walking and goes away with rest, though prying this history out of him was like pulling teeth."

"And you think this pain is caused by..."

"I think he may have calf claudication[4] from a blockage in his femoral artery."

"Any risk factors for atherosclerosis?"[5]

"Yeah. He's had diabetes for some unknown number of years and smokes cigarettes."

"What did he do that was so obnoxious?"

"He was so angry from the moment I walked into the exam room. He criticized everything. Nothing I did was good enough. He wouldn't cooperate with giving me his history, and he kept harping on how young and inexperienced I was."

"Congratulations. You've just had your first encounter with a common experience in medical practice—the angry patient. There are some important

---

[4] Claudication means a cramping-like pain brought on by lack of arterial circulation.
[5] Risk factors are conditions that increase the likelihood of developing a clinical condition, in this case atherosclerosis, which is also called hardening of the arteries. The most potent risk factors for atherosclerosis include cigarettes, high blood pressure, high cholesterol, diabetes, and a strong family history of atherosclerosis.

pearls[6] in how to deal with the angry patient, but the most important is not to take it personally. He's probably angry with his past, his current situation, and with lots of people, perhaps even with himself, so don't feel special. It's got nothing to do with you or anything you did."

"That's hard to remember when he so accurately placed the thrust of his daggers right into the heart of my own insecurities."

"Sometimes patients are amazingly astute at that, so much so it can be unnerving. I don't understand why they are often so accurate. They seem to tap into something, but what that is remains a mystery. Back to his leg pain, what do you want to do?"

"I guess we have to confirm the diagnosis. I'll order an arteriogram[7] and plan to see him a week later to discuss the results."

"Sounds like a plan."

Back in the exam room, the plan didn't go so well.

"I don't give a good Goddamn! I'm not going for no test."

"But we can't confirm whether or not you have a blockage in the artery in your leg without the test."

"That's just because you're not a very good doctor."

Now it was my turn for my jaw to tighten as I bit back the insult I wanted to throw back. "Fine. You came in for my advice, and now you have it. You can do what you want with it. But if your leg pains get worse, or you get a sore on that foot that doesn't heal, my medical advice is that you come right back in. Now, about your diabetes…"

"I don't want to talk about the sugar. I already told you I'll do what I damn well please. You've been no help at all. I hope someday you learn how to help some other patient." With that, he got up and stalked out of the exam room.

I looked down at his chart. The name across the top read James Schmidt. I hoped never to see that name on a chart in my hands again.

        *   *   *

"What a great case!" The surgical intern could barely contain his excitement. "You won't see this kind of case very often. It's a guy in his late twenties with a 25-pound weight loss and a rock hard prostate. He's got prostate cancer for sure. You ever felt prostate cancer?" I shook my head.

---

[6] Pearls, sometimes referred to as clinical pearls, are snippets of information that are particularly helpful in working with patients. The term grew out of the value such information holds for the practicing physician.

[7] An arteriogram is an X-ray study in which dye is injected into an artery and a series of images taken to evaluate how the dye passes downstream through the artery. It can be used to diagnose blockages in arteries anywhere in the body, including the legs, kidneys, heart, neck, and brain.

"Then you've got to do a rectal exam on this guy." He ushered me into an exam room in the surgery outpatient clinic. There sat a terribly thin man in his late twenties, hunched over, rocking anxiously on the examination table. A heavy set, older woman wearing a torn dress sat nearby, his mother I presumed. "Hank, this is Dr. Sheff. He needs to do the same exam I just did on you."

"I can' get da strength," came the reply.

From across the room, I guessed Hank was schizophrenic. He moved awkwardly, even in his rocking, and had that unnatural, flat facial expression I'd come to recognize in schizophrenics. He smelled bad, a combination of stale urine and not having showered in far too long. Or perhaps that smell came from his mother. In either case, I did not look forward to getting my face close enough to him to place a finger up his rectum.

"You'll feel my examination finger," I said, trying to say something comforting, though I had the all too familiar knowledge I was only putting him through an unnecessary and uncomfortable exam just for my education. Sure enough, I felt a rock hard mass just inside his rectum. I couldn't get a good sense of where it began or ended, but didn't want to prolong the exam. "Thank you," I managed to get out before leaving the room.

"That's what prostate cancer feels like," the surgical intern blurted out as we stepped into the hall. "I'm going to order the workup: bone scan, chest X-ray, chemistry profile. What a great case," he repeated.

I found myself experiencing an odd mix of feelings. Part of me did feel excited to have felt prostate cancer for the first time. That sensation of a rock hard, irregular mass would forever remain with me, informing every rectal exam I performed for the rest of my career. Yet there was nothing "great" about poor Hank, a schizophrenic who had the terrible luck to contract prostate cancer in his late twenties. And what of his mother, herself barely able to take care of her own hygiene and certainly failing at helping Hank take care of his? "Thanks for letting me feel that," was all I could respond to the surgical intern.

As I walked away, I felt sad for the grief Hank's mother would soon know. And I could only imagine what bewildering thoughts would fill Hank's mind as he underwent test after test, suffered through treatments he would not understand, and ultimately faced death—a death of which his confused mind would make bizarre and unknown meaning. Sad, indeed.

*   *   *

This first month of internship passed quickly, speeded by a never ending workload and busy nights with little or no sleep. I spent considerable time in the operating room. But unlike medical school, the attending surgeons didn't yell. In fact, they were quite kind both to me and their patients. They took time to teach me sound surgical thinking and technique, including how to sew properly, and also how to assess and manage the common problems handled by general surgeons.

One night I was called to the emergency department to assess a23-year-old man with abdominal pain. I diagnosed acute appendicitis. The attending surgeon came in and supervised me doing an entire appendectomy, from opening the skin, to entering the abdomen, to finding and removing the appendix, and finally closing the abdomen and skin. I came away with two major learnings from that night—a profound appreciation for how many things can go wrong in surgery, even in such a simple procedure, and confirmation of my anatomy professor's initial observation that surgery should not be my first choice, not by a long shot.

During this first month, the family practice interns, all 12 of us, began meeting once weekly in an "intern support group." The thought that interns, struggling with awesome new responsibilities, massive workloads, sleep deprivation, and desperate attempts to maintain strained relationships outside the hospital, might benefit from a structured support group was a surprisingly new idea pioneered by family medicine residency programs. In fact, the residency program paid a psychiatrist to facilitate this group each week. During these first weeks of coming together, the group functioned as a "gripe session," with predictable complaints about the amount of work, unreasonably demanding attending physicians, and too little sleep. Perhaps surprisingly, we had few complaints about our supervising second and third year residents who, like my senior resident during that pediatric subinternship, went out of their way to support and educate us. In the face of all I was struggling to manage, this was the one hour each week I looked forward to and appreciated most. It made me feel I was not alone.

*   *   *

"Welcome to my labor and delivery unit." With those words, Mary, a heavy set, ruddy faced nurse in her late fifties, made instantly clear who was in charge. "This is where my girls labor," she went on, showing me one end of the hall with three double rooms, two of which held moaning women in labor. The rooms were a step up from the four-bed labor rooms I'd trained in at the University Hospital, but not by much. "When one of my girls is almost ready, we bring her into the hallway here. That way we can keep an eye on 'er real close." I'd already had experience with this launching pad and could easily picture a woman, writhing on a bed in this open hallway, with Mary periodically pulling back her sheets for the world to see her bulging vagina and make a group decision of when she was ready to pop.

"Then we bring her into the delivery room, here." She waved her hand in the direction of two sterile-looking rooms. We get her ready while the doctor scrubs. You'll learn a lot from our doctors here. They really know their obstetrics."

It didn't take long for me to meet one of those obstetricians.

"How's she doing?" a diminutive man asked in a slightly Hispanic accent as he strode onto the labor and delivery unit.

"That's Dr. Martinez," Mary whispered. "He'll show you what's what in a delivery room. You'll be thanking him for that before your obstetrics month is over." Then, more loudly, "She was 8 centimeters, 80%, and +1 station just half an hour ago when I called you. Judging from the noise of the last half hour, sounds like she's been in transition.[8] Bet she's ready for the launching pad just about now." She paused, appearing quite pleased with her mastery over her dominion. "Oh, and this is Rick Sheff, our new family practice intern on OB. It's his first day."

"Hi, Rick," Dr. Martinez said, extending his hand. I took it firmly. "Welcome to our OB service. Looks like you'll be getting your feet wet pretty quickly." A loud moan emerged from the first labor room, escalating into a scream. "Come on in with me," he said, turning and walking briskly toward the source of the sound. "Well, Janet. From the sounds of it, you're making great progress."

An exhausted appearing woman in her early twenties looked up gratefully at Dr. Martinez. She labored alone. "Thank God you've come. Is it time? Is my baby coming now?"

"Let's check and see," he said, pulling on a glove and motioning for me to do the same. "This is Dr. Sheff. He'll be assisting me with your delivery this morning." She nodded wearily as she brought her knees up and spread them, revealing a cleanly shaven vulva. Mary's handiwork, I presumed. Dr. Martinez's exam lasted no more than three seconds. Indicating I should examine her as well, he stepped aside. As my gloved fingers spread her labia and entered, I could immediately feel the hard surface of a baby's head. "What's the orientation?" he asked. I felt for the soft place on the baby's head, the anterior fontanelle.

"I'm not sure, but thought I felt the anterior fontanelle just behind her pubic bone at the 12 o'clock position."

---

[8] Transition is the stage in labor in which the cervix dilates from approximately 8 centimeters to 10 centimeters. Women in transition often become extremely agitated and complain of increased pain. Once transition is completed and the cervix is fully dilated to 10 centimeters, the mother is instructed to push with each contraction to help the baby descend through the birth canal.

"That's exactly right." I breathed a sigh of relief. "So the baby's OP. "Do you know what that means?"

I searched desperately in the recesses of my memory from the second year of med school when I last did obstetrics. "Doesn't that mean something about labor being more difficult?"

"Yes. Yes it does. The baby's occiput pounds away against the mom's sacrum with every contraction and push." He looked at Janet with some concern. "That means Janet here has been having a difficult labor all night." Janet nodded a weary acknowledgment to the only part of our conversation she understood. "We will help her out. Mary, let's get Janet into the delivery room. No sense waiting on the launching pad. I'll be needing the Simpsons." He and Mary exchanged a knowing glance. "Rick, let's scrub.

As we entered the delivery room, I recognized the sterile environment, much as I'd experienced in medical school. The difference was that in the intervening years I'd come to understand there were other options in birthing. Sarah had opened my eyes to the barrenness of medicalized childbirth. Between that and my interest in family practice, with its family centered approach to medical care, I'd begun to think childbirth should be a more humane and less medical experience. This made what followed all the more disturbing.

Mary had already strapped Jane's legs into stirrups. The anesthesiologist, a burly man in his fifties, had appeared and was busy drawing up syringes and adjusting IVs at the head of the delivery table. As Dr. Martinez and I donned sterile gowns and gloves, he barked to the anesthesiologist, "We'll be using Simpsons." They exchanged the same knowing glance. He then draped Janet with sterile paper leggings up to her groin and a large sterile cloth across her abdomen. Her vulva, shaved and framed in the harsh operating room light, appeared particularly naked and vulnerable. From a long table of steel instruments, he lifted a syringe and drew up medication I recognized, presumably for anesthetizing her episiotomy. He then attached to the syringe the longest needle I'd ever seen. It must have been seven inches long. He placed it inside a thin steel tube with a trumpet like opening at either end. At that moment, Mary tore open a sterile paper container, causing two long metallic instruments to clang onto the table on top of the other instruments. "Thanks for the Simpsons," he replied. With that, he stepped between Janet's widely spread legs. With one gloved hand he probed deeply into her vagina. She screamed with pain. With the other hand he inserted the syringe and trumpet like tube equally deeply in her vagina. She attempted to pull her legs together to resist the pain, but to no avail as they were firmly tied in place. His roughness seemed to cause Janet unnecessary pain. Being so tied down and exposed added to her powerlessness and humiliation. This felt more like witnessing a rape than a medical procedure. I wanted to push Dr. Martinez out of the way.

"We need some more help here," Dr. Martinez said to the anesthesiologist.

"Help is on the way," came the reply.

Dr. Martinez paused, his hands still deep in Janet's vagina while she struggled to pull away. The anesthesiologist quickly injected something into her IV. Within seconds Janet appeared to visibly relax. This enabled Dr. Martinez to probe even deeper into her vagina, injecting half the syringe on one side and half on the other. By the time this was completed, Janet was mumbling incoherently, drowsy with heavy sedation. With a nod from the anesthesiologist, Dr. Martinez cut a huge episiotomy through the opening of her vagina at the seven o'clock position. Blood gushed out, running down into the waiting bucket he'd kicked into place. I watched in horror as he took one of the long metal instruments, which measured well over two feet, and inserted it deep along the baby's head inside the vagina. "Forceps," was Dr. Martinez's one word explanation for what he was doing. Then he added, "You have to be careful where you place the blades or you can do a lot of damage to the baby."[9] My horror mixed with fascination as he inserted the second "blade" along the other side of the baby's head. With a clang of metal, he brought the handles of the forceps together. The blades visibly tightened around the baby's head within the vaginal canal. I recoiled as I imagined the feeling of those metal "blades" squeezing down on that vulnerable head. "Give us a good push, Janet" Dr. Martinez ordered.

Through her drugged fog, Janet could only mumble, "What? What's happening?"

"Never mind," Dr. Martinez said in resignation. "Fundal pressure now!" he ordered the anesthesiologist. Leaning over Janet's head, the anesthesiologist placed two hands on the top of Janet's very pregnant uterus and pushed with his full weight. Simultaneously Dr. Martinez pulled with surprising force on the handles, pulling first directly out and then diagonally upwards. The baby's head slid out through the bloodied vaginal opening, dragging its body behind. I shuddered at the force pulling on its slender neck and the equally alarming force of the metal "blades" in their vice-like grip along either side of its face. "Head's out," barked Dr. Martinez. The anesthesiologist stopped pushing. Dr. Martinez uncoupled the two forceps handles, and the "blades" came away from the baby's head. Long, curving reddened marks extended along both sides of the face. After briefly suctioning the baby's mouth and nostrils, Dr. Martinez delivered the rest of

---

[9] Historically one of the primary reasons for identifying the location of the baby's fontanelle, and hence the orientation of the baby's head, has been to be able to place forceps correctly to minimize the risk of damage to the baby's face and head.

the baby. "It's a boy," he called out jubilantly. No response came from
Janet. "It's a boy," he repeated, looking up at her for the first time.

Janet didn't respond. After a moment of panic, I realized she was fast
asleep. Made "comfortable" by whatever the anesthesiologist had injected
into her IV, she had completely missed her son's birth. She would never
know the wonder of that moment, or what that moment could have been. "I
gave her enough so she won't be any trouble while you sew up the
episiotomy," the anesthesiologist responded to Dr. Martinez's unspoken
question. That meant her baby would not know his mother's touch and voice
until he was hours old. He would receive that shot, the burning eye drops,
and be whisked away to the nursery where his initial mothering would come
from Mary. I shuddered at the thought.

"Thanks," was his curt reply. He then sat down between Janet's legs to
sew up the unusually large episiotomy he'd just cut. "Now Rick, watch as I
check for any tears. After using the forceps you always have to do a
particularly thorough check deep in the vagina because it is so easy to do
damage with the Simpsons, or any other forceps for that matter. See, there's
a 6 centimeter laceration on the upper left side of her vagina. It's bleeding
pretty heavily." Settling in, he began to repair Janet's torn and mangled
genitalia.

As I watched, saddened that Janet had missed not only the moment of her
son's birth, but the first hours of his life, Sarah's words came back to me.
"It's wrong. It's all wrong." I thought of spending the next three years
learning obstetrics from Dr. Martinez, Mary, and the rest of the obstetricians
on this medical staff. If they all practiced this way, I wanted no part of it.
Yet, what choice did I have? Yes, Janet now had a healthy baby boy. Dr.
Martinez, I was sure, believed he had mercifully taken Janet out of her pain.
But what had been lost? Whatever it was held no meaning for Dr. Martinez.
This was how he'd been taught, and it was the legacy he and Mary wanted
to pass proudly on to me. In that moment I recognized I would learn what I
could from Dr. Martinez and the others, but always hold the truth of what
they could not even see.

*   *   *

"Dr. Sheff, this is Martha in the ED. Dr. Burton has a woman down here
who's bleeding and has a positive pregnancy test. He wants you to come
down and evaluate her. And he asked you to hurry since she's bleeding
pretty heavy."

"I'll be right down." As soon as I hung up the phone, I quickly paged
Sue Werner, the second year resident covering pediatrics tonight. This is
how the family practice residency program had arranged to cover pediatrics
and obstetrics in this community hospital—when an intern was on OB/Gyn,

a second year resident was on pediatrics and vice versa. This always kept a junior resident in the hospital to back up an intern.

"Yeah, Rick. What's going on?" Sue's voice was pleasant, confident, even upbeat, though it was already past midnight.

"The ED called. They've got a woman with vaginal bleeding and a positive pregnancy test. Her bleeding's heavy."

"How many weeks is she?"[10]

"They didn't say...or rather I didn't ask."

"If they didn't say, it usually means she didn't even know she was pregnant so she's probably still in her first trimester,[11] but you can confirm that when you get down there. Is she postural?"[12] I was silent, and she knew what that meant. "So now you know the first two questions to ask in a bleeding pregnant patient. There's one more."

Through my late night fatigue, I tried to retrieve a vague memory from my gynecology rotation. Bleeding and pregnancy...Of course she could be miscarrying, but I knew there was something more in the differential diagnosis. Suddenly I blurted out, "Abdominal pain?"

"Right question. What are you looking for?"

"Ectopic pregnancy,"[13] I said with self assurance.

"Exactly. So get down to the ED and start evaluating her, and I'll be there as soon as I can."

As I arrived, Martha, the night supervisor in the ED, approached me quickly. "You'd better get in there, Dr. Sheff. Since we called you she's started hemorrhaging."

I stepped into the examination room and stopped abruptly. A pale Portuguese woman lay on a blood streaked examination table, clutching her lower abdomen. A towel, soaked dark red, lay wedged between her legs as

---

[10] When describing a pregnant patient, one of the most important pieces of information is how many weeks she's been pregnant. A full term pregnancy is 40 weeks when dated from the first day of the woman's last menstrual period.

[11] Pregnancy is considered composed of three trimesters, each covering approximately one third of the normal term of pregnancy. The first trimester covers the initial 12 weeks, the second trimester includes weeks 13 to 26, and the third trimester includes the remainder, namely weeks 27 to 40.

[12] Postural is a shorthand way of referring to a patient whose blood pressure drops significantly when the patient goes from lying down to sitting up or standing. It's a sign the patient is either dehydrated or has lost a lot of blood. If a patient is postural, it usually means something serious is going on that needs urgent intervention.

[13] As noted previously, ectopic means in the wrong place. So an ectopic pregnancy means the embryo has implanted somewhere other than inside the uterus, usually in one of the fallopian tubes. This can be life threatening because as the embryo grows, it can rupture the tube, and a woman can bleed to death quickly from this condition.

she writhed slowly from side to side, moaning loudly. Bright red blood ran down from the examination table to a rapidly enlarging pool of blood on the floor. My God, she's bleeding to death! I thought almost out loud, and I didn't know what to do to stop it. There was no wound to compress, no tourniquet to apply. I'd read about women hemorrhaging during pregnancy, but I had no idea it could happen this fast or look this bad. For the first time in my medical career, my heart raced wildly with the knowledge that this woman's life was in my hands, and it was slipping away. I hated feeling so helpless. The one thing I could do was call for help.

"Dr. Lucino is the attending on call for Gyn tonight. Page him stat. Call Peds and tell Dr. Werner we've got an emergency and I need her down here stat, too."

"I'm already here," Sue said from behind me, her voice still reassuring as she took in the situation. "What are her vitals?" she barked to a nurse at her bedside.

"BP 90 over 60. Pulse 130."

"Is she postural?"

"We haven't had time to sit her up. Do you want us to do that now?"

"No. She's bleeding heavily enough we'll assume she's already postural. What's her IV rate?"

"200 cc's an hour."

"Open it wide. Has she been typed and crossed?"[14]

"We've called for two units."

"Call for two more. And let's get a transfusion going as quickly as the blood is available." She was deftly laying her hands on the woman's lower abdomen as she talked. "In fact, please start another large bore IV, at least 18 gauge, so we can run two lines in at once." A flood of relief momentarily released the knot in my stomach as Sue's confidence and obvious competence instantly transformed the room from a scene of hopeless tragedy into one of crisp medical care. When would I ever be able to do this, I wondered. Sue was only one year ahead of me, but her knowledge and poise were so far beyond my abilities right now. I was afraid I would never be able to do what she had just done. Turning to me she said more softly, "I know this looks like a lot of blood. It is. I can't feel the top of her uterus, so this isn't likely to be placenta previa.[15] She is most likely miscarrying. If

---

[14] Type and cross refers to the activity of drawing a sample of the patient's blood, identifying its type, and testing it against samples of blood stored in the hospital's blood bank. The goal is to find units of blood or other blood products that match the patient's. This is critically important because giving patients blood that is not appropriately matched to their own can result in a life threatening transfusion reaction.

[15] Placenta previa is a condition of pregnancy in which the placenta lies over the cervix, the opening at the bottom of the uterus that leads to the vagina. As

she suddenly let loose like this, there's a good chance she's passing the fetal sac as we speak. Let's examine her and see. If that's not the case, she'll need to go to the OR stat. We may not even have time for an ultrasound if she's too unstable. But before we whisk her to the OR, let's see if there's something we can do for her right here.

Turning to the patient for the first time, Sue said, "Hello, Ms...." She paused briefly as she scanned for her name on the chart hanging on the side of her exam table. "Hello, Ms. Gonsalvo. I'm Dr. Werner, and this is Dr. Sheff. We'll be taking care of you this evening. It's obvious you're bleeding heavily. When was your last period before this bleeding began?"

In a thick Portuguese accent she replied, "I don' know. I don' know. Maybe two mons ago." She continued to writhe back and forth on the exam table.

"We need to examine you. Please stop rocking so we can examine you. We may be able to help you feel better right now," Sue said in a reassuring but firm voice. With the help of the nurses, we got her feet into stirrups and she slid her buttocks to the end of the table. A steady stream of blood poured from her vaginal opening down the end of the exam table to the floor. "Rick, why don't you do this?"

"Alright," I said hesitantly, moving into position between her spread legs. With gloves on both hands, I spread her vaginal lips and slipped the speculum into place. As it opened, a pool of dark blood splashed out of her vagina, cascading down my white uniform pants.

"Sponge out her vagina and try to see her cervix," Sue said quietly in my ear. One after another she handed me what seemed like the largest Q-tips I'd ever seen, one-inch wide wads of cotton tipping on the end of a foot long plastic handle. Each was quickly soaked and discarded. Within less than a minute, I was looking at her cervix. Blood continued to ooze from it, but it seemed distorted, much wider than I had ever seen. "You see the gestational sac," Sue said, excitement in her voice for the first time, motioning towards a two-inch translucent mass deep in her vagina. "We're going to be able to help her. Use the ring forceps[16] to grab that sac. It's stuck in the cervical opening, and if you can get it out, you will dramatically slow her bleeding."

---

pregnancy progresses and the placenta enlarges, bleeding can occur through the cervix. Sometimes this bleeding can be catastrophically life threatening and requires emergency surgery to save the mother's and baby's lives.

[16] A ring forceps is an instrument that can range from approximately eight to 12 inches long that has two metal rings on one end that are separated and brought together by opening and closing the handle on the other end. It is used during surgery and other procedures to grasp tissue and other anatomical structures from a distance. The handle looks very much like the handle of a scissors which allows the user to open and close the blades of the scissors.

My right hand trembled slightly as I held one end of a ring forceps, much as I would hold the handles of a scissors. Spreading these handles separated the two metal rings on the other end. Placing the rings carefully above and below the sac, I squeezed the forceps handles together. With two clicks, the forceps closed, the edge of the sac firmly caught between those two metal rings. "Pull back with slow, steady traction," Sue coached. I instinctively knew I did not want to tear the thin layer of tissue held precariously between the teeth of the rings. Experimenting to find just the right amount of tension, slowly I felt the sac begin to inch toward me. Suddenly, it came free, and with it another gush of blood poured out of her vagina and down my pants. The ring forceps emerged with a two-inch grayish red sac grasped on the end.

"Ten of Pit IV push now and 10 in the bottle," Sue barked to the nurse. Turning to the patient she said, "I'm not sure if you knew you were pregnant, Ms. Gonsalvo, but you're having a miscarriage. With Dr. Sheff's help, you've just passed some of the tissue, and your bleeding and cramps should get much better almost immediately." The woman started to cry in relief. Then to me she said quietly, "Nice job. You just saved this woman's life."

The words were electrifying. As the blood had continued to pour out, the knot in my stomach had not only returned, but intensified. Now it suddenly released. She was going to be OK. And I had helped. Of course, Sue and the nurses had done most of it. In spite of the "Doctor" in front of my name, tonight I felt very much the student. But in this moment, I glimpsed yet another side of the extraordinary power wielded by physicians: the power to help this woman who just minutes before lay dying, and now, through the knowledge and skill of her treating physicians and nurses, was saved. The sensation felt profoundly satisfying. But more than this, it was intoxicating—the kind of sensation that once the moment was over, I yearned to feel again.

So this was the allure of specialties that practice medicine on the edge: surgery, emergency medicine, obstetrics. I had learned so much from Sue tonight about how to assume control over a medical crisis, instill calm and competence, and transform a moment of impending death into a great healing. Yet already I knew these moments were at best punctuation marks in the work of every physician. The day-to-day practice of medicine requires attention to many levels. Diagnostic and technical prowess constitutes only one of these levels. But if a physician is drawn to a specialty because of the intoxicating quality of heroically saving lives, that physician runs the risk of being inattentive to the subtle interpersonal, social, and emotional levels at play in providing ongoing care to patients. A physician enthralled by the "captain of the ship" experience Sue had just demonstrated, with its aura of autocratic command and profoundly satisfying and immediate results, may not readily function well as a collaborative team member in the more typical treatment most patients need, especially those with chronic illnesses who require "care" over time, and are not amenable to "cure." Personally, I found myself drawn to the joys and challenges of both.

*   *   *

"Dr. Sheff," the urgency in the nurse's voice came through the phone line, quickly arousing me from the brief sleep I'd grabbed in the on call room. "A Mary Disalvo just arrived on labor and delivery. She's a multip,[17] and she's moving along really fast. You better get up here right now. She could drop this baby any minute."

"Alright, Margaret, I'll be right up. Who's the attending?"

"Dr. Martinez. I've already called him, but he won't be here for 20 minutes, and I don't think she'll last that long."

"Then page Dr. Werner. Ask her to meet me on labor and delivery immediately. And get the patient into the delivery room. Start prepping her STAT." By then my eyes had come into enough focus for me to see the clock. It was 3:45 AM. I pulled on my pants, jumped into my shoes, and sprinted to the labor floor. As I pushed through the double doors onto the unit, less than two minutes after hanging up the phone, the final wisps of fog cleared my suddenly awake brain. The clanging sound of instruments being rapidly thrown onto the instrument table in the delivery room told me where to head. First, I stopped at the scrub sink for the fastest surgical scrub I'd ever done.

Moments later, I burst through the delivery room door, quickly taking in the situation. The woman's legs were up in the stirrups, and I could see dark hair bulging from the baby's head pushing on her vaginal opening. She screamed almost continuously. As Margaret pulled on my gown she whispered, "Dr. Werner said she was tied up doing an LP on a sick baby. She won't be able to make it. She said to tell you good luck."

A chill ran through me. I had never done a delivery completely alone before. No attending. No back up resident. Only an intern, barely six weeks

---

[17] Multip is an abbreviation for multipara, which means a woman who has delivered two or more children previously. A nullipara is a woman for whom this is a first pregnancy. A primipara is a woman who has delivered one child previously. These differences are important in several ways, one of the most important being that a multipara woman has already stretched her birth canal to deliver a child at least twice, so it stretches more readily during the next labor. This means her labor, especially the last stage of pushing, is likely to move quickly, sometimes requiring only a few minutes.

beyond medical school graduation. My hands began to shake as I drew up the syringe with local anesthetic for the episiotomy. Just stay calm, I muttered half under my breath. Then out loud I called to this woman I'd never met from between her spread legs, "Mary, I'm Dr. Sheff. This is a helluva way to meet. I'm the resident covering obstetrics tonight. We've called for Dr. Martinez, and he is on his way in. I'll be taking care of you until he gets here. Looks like you're about to have this baby pretty soon. You'll feel my fingers at the opening of your vagina while I check the baby's head position...It's LOA."[18] I reached for the syringe. "I'm injecting numbing medicine here for your episiotomy. You'll feel a stick and some burning and then it will go numb." With that I plunged the needle four inches into the thin skin at the opening of her vagina at the seven o'clock position. She screamed. Another contraction began, and she spontaneously starting bearing down, the instinctual urge to push out her baby taking over.

I began talking out loud to myself. "I'm injecting the lidocaine...It's all in..." Grabbing the scissors off the instrument table, I continued, "I'm cutting the episiotomy." The scissors cut a three-inch gash where I'd just injected the anesthetic. Blood poured out. I barely remembered to kick the bucket into place underneath the end of the table to catch the streaming blood. "The baby's coming out...The head is out...I'm suctioning the baby's mouth and nose on the perineum...I'm feeling for a nuchal cord...There is a nuchal cord...I am reducing the nuchal cord...It's a tight nuchal cord...I am reducing the nuchal cord...I can't reduce the nuchal cord...The cord is too tight to reduce... I can't reduce the nuchal cord..." For a moment, panic rose inside as I realized I couldn't get the cord over the baby's head and the baby might die. Then something burst into my head, coming from a place I did not know. "I am going to clamp and cut the cord on the perineum," I continued out loud. "I am clamping the nuchal cord." I pried the umbilical cord just loose enough to place two clamps on the cord close to each other. "I'm clamping and cutting the nuchal cord on the perineum." The scissors seemed to jump from the instrument table into my hand. "I am cutting the nuchal cord between the two clamps...I am reducing the nuchal cord...It's free now...I am delivering the anterior shoulder...Now I am delivering the posterior shoulder...It's a boy! It's a boy!" A lusty cry emerged from his newly filled lungs. I handed the baby to the nurse.

---

[18] LOA stands for left occiput anterior, meaning the occiput, which is the back of the baby's head, is directed a little to the patient's left and facing partially to her front, or around the two o'clock position. LOA is a normal position that doesn't usually portend problems.

Just then Dr. Martinez stuck his head into the delivery room. "How's everything going?" he asked, quickly surveying the situation and surmising from the loud baby cries that all had gone well.

"Baby's just delivered. He's fine. We're waiting for the placenta to deliver," I responded with an authority I did not know I had.

"Good job, Rick," he quipped, and stepped out to quickly scrub. Then he stuck his head back in, "Go ahead and start sewing up the episiotomy once the placenta's out. I'll be right in."

The placenta delivered quickly with a large gush of dark blood. I pulled the stool into place with my right foot and sat down between her outstretched legs. Just then Margaret handed her the baby to hold. "Mary," I said from the end of the table, I need to examine you to make sure you're OK. You'll feel some stretching, and it may feel uncomfortable. You be sure to hold tight to that beautiful baby of yours."

"I will."

After briefly exploring her vagina and cervix and finding no tears, I began to sew up her episiotomy as I had seen others do. Though it progressed awkwardly, by the time Dr. Martinez scrubbed, gowned and joined me, I was well on my way. Dr. Martinez let me continue, coaching me with tips. The moment it was done he turned to walk out of the delivery room. As he did he called over his shoulder, "I'm glad my patient was in such good hands tonight."

It was well past 5 AM when I crawled into the narrow bed in the on call room. My mind would not shut down, even though I had only an hour to grab a little sleep before beginning the day again. I had performed my first truly solo delivery. In the face of a crisis, I had stepped in, just as I'd watched Susan do for the woman who was miscarrying. Inside I had hovered on the edge of feeling completely overwhelmed. Yet, drawing on everything I had learned since starting medical school, and a strength Susan had shown me how to find, I had known exactly what to do.

Each step of the delivery played over and over in my mind—especially that moment. In the moment when a child's life was placed into my uncertain hands, when his normally life-giving umbilical cord was tightening into a deadly noose that would strangle him within minutes if I didn't intervene, information came to me from an unknown source. In that moment I somehow knew exactly what to do. I didn't remember when or where I had learned what to do for a tight nuchal cord that could not be reduced. Prior to that moment, I'm not sure I would even have been able consciously to recall it. But in the moment, I knew. Faced with the most critical, life threatening challenge of my brief medical career, alone with no back up, information had come to me. The precise information I had needed to save this child's life.

Was this the inner experience that gave physicians that awesome aura of competence they exude with such apparent ease? Was there forming within me a neural network of information and skills that was so complex, so vast, that I could not consciously hold it in my awareness? Yet wasn't this same network becoming complete enough that I could feel myself dipping into it and extracting the very information I needed in the face of this or countless other clinical challenges? Or perhaps it had been the disembodied spirit of some sage healer whispering in my ear exactly what I needed to know. Whether the information had come from within me or from some other source didn't seem to matter. As I drifted off into an ever so brief sleep, I felt myself on the threshold of developing that finely attuned clinical judgment I'd so admired in Chip, the chief resident from my pediatric subinternship.

I remembered that first day of medical school, when I'd been so overwhelmed with the sheer magnitude of the information to be learned. That day, and every day since, my will to succeed had been deeply challenged. Not since that first day had I felt such a profound sense of satisfaction. I was becoming competent.

        *   *   *

"Sure hope this a quiet night," I sighed to Amy, the fellow intern who had just finished signing out her patients to me.

"Why tonight more than any other night?"

"Marsha and I have a wedding to get to in Pennsylvania this weekend. It's a seven-hour drive, starting when I get off call tomorrow evening. She's got night blindness, which means all the night driving will be on me. There's not enough coffee in the world for me to do that if I don't get at least a few good hours of sleep tonight. Besides, we haven't had any time together without me being exhausted since I started my internship. Things have been rough between us, and I hope this weekend will give us a little window of good time."

"Hope you get some sleep, and some of that time."

As always, the night became busy almost immediately. Phone calls, two admissions. Then suddenly all was quiet. I had completed the workup and orders on my admissions, caught up on the phone calls, and handled all the follow up on patients from the other interns. Staring incredulously at my watch, I realized it was only 10:30 PM. I decided to grab a nap in the on call room before my beeper went off again.

Somewhere in that fitful sleep I had a dream. I found myself moving slowly, fluidly towards Sarah. My heart opened. I remembered how much we had been in love for more than two magical years in high school. But a disembodied voice in the dream told me, "No. Not her." Then I saw Alisa, my girlfriend in college. Again my heart opened as I moved towards her. Yet again the voice said, "No. Not her." Finally I saw Marsha. As I moved towards her, my heart opening yet again, the voice said, "Yes. This is the one."

Just then the phone in my on call room rang. I looked at the clock—6:00 AM. Miraculously I'd gotten an entire night's sleep. After taking care of the phone call, I stumbled into the shower. As the warm water washed over me, relief at getting enough sleep for the drive that night gave way to something else. The dream came back. "Yes, Marsha is the one," I repeated to myself. In that still half-awake state, I slowly realized in my dream I'd made a decision. With almost a bubbly feeling in my stomach, I knew I wanted to marry Marsha.

That weekend we found the window of time we'd so desperately needed. On the long drive back, I saw a beautiful stream by the side of the road and pulled over. By that stream, I asked her to marry me. She said yes.

*   *   *

"Bed 18: Mary Wilson is an 82-year-old in with urosepsis,[19] on amp and gent. Presented to the ED in shock.[20] That bought her a ticket to the ICU. She's started to stabilize and is awaiting cultures. Tonight you need to watch her blood pressure. If it drops, she may need pressors."[21]

Even though I'd been on call in the intensive care unit all last night, and had been working almost 36 hours straight with no sleep, my mind seemed to come into just enough focus for me to know exactly what to say on sign out rounds to the intern on call tonight. The whole team, composed of a third-year senior resident, a second-year resident, and two interns, made rounds on all 21 patients in the intensive care unit at the beginning and end

---

[19] Uro refers to the urinary tract. Sepsis, as noted before, refers to a condition in which an infection has spread to the bloodstream. So urosepsis is a shorthand way physicians refer to patients who have an infection in their urinary tract which has spread into their bloodstream. It is life threatening.

[20] As a medical term, shock refers to a condition of dangerously low blood pressure that threatens to produce inadequate circulation to critical organs such as the heart and brain. Shock can be caused by many different conditions, though the most common causes are major infections and severe blood loss.

[21] Pressors refers to medications that increased a patient's blood pressure. Some of these medications raised blood pressure by stimulating the heart to pump harder and faster, some by causing the smooth muscle cells in the arteries to tighten up, and some by doing both.

of every day. Most often we were accompanied by a cardiology fellow.[22]
We brought the charts of each patient with us on a cart so we could write
orders and quick notes while at each patient's bedside. As my 36-hour shift
was coming to an end, I couldn't wait to close my eyes. In fact, they
sometimes started to close for me. Still, I had to stay sharp through the end
of sign out rounds, so just before they started I'd grabbed another cup of
coffee to keep me going.

"Bed 20: Andre Gonzalez is a 73-year-old male with a complex set of
medical problems we've been following along with the surgery team.
Problem number one: ischemic bowel.[23] Mr. Gonzalez infarcted a portion
of his small bowel[24] two weeks ago. Since then he's undergone a bowel
resection[25] and has suffered multiple complications. Hence his long list of
problems. Problem number two: nutrition. His bowel has remained shut
down since the surgery, so he's been on hyperal.[26] He's currently receiving
2200 calories per day, including 70 grams of protein. He still seems to be in
net negative nitrogen balance, so we're upping the protein in his hyperal
solution. Problem number three: hypocalcemia.[27] His calcium this morning
was 7.8. Despite our continuing to increase the calcium in his hyperal, his
calcium's have remained low. He's had one seizure we attributed to
hypocalcemia, and his reflexes remain on the brisk side. He's got another
calcium ordered for the morning, but you should check his reflexes at least
once during the night to make sure they aren't becoming more brisk.[28]
Problem number four: seizures…"

---

[22] A fellow is someone who has completed medical school and residency training
and is now in subspecialty training. A cardiology fellow had completed three years
of internal medicine residency and is in the midst of two or three years of
subspecialty training in cardiology.

[23] Ischemia means not enough blood supply. Ischemic bowel means not enough
blood supply to a portion of the intestines.

[24] Infarction means a condition in which a blockage occurs in an artery, and the
tissue supplied by that artery dies. In this case, the blockage occurred in one of the
arteries that supplies blood to the small intestines. That portion of the small
intestines died. This is often fatal if not diagnosed and operated on in time.

[25] Resection means to surgically remove something. In this case, he had a portion of
his small intestines removed.

[26] Hyperal is an abbreviation for hyperalimentation. Aliment means food or
nutrition. Hyperalimentation refers to providing large amounts of nutrition
intravenously.

[27] Hypo means too little. Emia refers to the bloodstream. So hypocalcemia refers to
a condition of having too low a level of calcium in the bloodstream.

[28] Reflexes that are brisk or highly active are an early warning sign of a calcium that
is too low. This is a gross way of assessing a patient's calcium level at the bedside
without awaiting the results of a blood test.

Stuart, the senior resident who ran the intensive care unit team this month, was also a family practice resident, and I could see him nodding in approval. All that training with sick kids during my subinternship was certainly paying off during my rotation in the intensive care unit. I had been dreading this rotation, in part because it was one of only two months during our internship that required being on call every third night, a grueling, exhausting schedule, especially since you were up most of the nights on call. The other reason I had dreaded it was because of how sick the patients were. But to my surprise, I had actually enjoyed grappling with these patients with long lists of life threatening problems. I absorbed every bit of clinical knowledge I could from the cardiology fellow who rounded with us, the medical director of the unit, the chair of cardiology, and the other attending physicians. These more senior physicians had much to teach me, and I found myself an eager learner. Now, more than three weeks into my rotation in the intensive care unit, I was exhausted but remarkably comfortable. The glow of competence that had begun during my obstetrics rotation continued to strengthen during this ensuing month of challenges in the ICU.

My eyes wandered to the two long rows of glass-encased treatment areas—you couldn't call them rooms—each holding one bed, one patient, and countless IV poles, tubes, and beeping monitors. What privacy that could be had came from flimsy curtains pulled partially closed. Nurses moved briskly from one patient to the next, responding to alarms from monitors, or call buttons from patients, or arriving with some ordered medication. In this bustling environment, I felt remarkably comfortable.

As we left Mr. Gonzalez's bedside, I wrote the few orders Stuart suggested and flagged them. Flagging meant folding down the page on which I had written any new orders. It was a sign to the nurses that new orders were there that needed attention.

"Mr. Joseph Harrison is a pleasant 68-year-old man who'd previously been healthy until he presented last evening with a large anterior MI,"[29] I said to the team as we stood outside his treatment area. I had taken quite a liking to Mr. Harrison and he to me while I had done his history and physical. We had bantered together warmly off and on throughout the day. I already knew the names of his children. His eyes had had a special twinkle when he had told me about his newest granddaughter who'd just been born. He made a point of telling me how much he looked forward to holding her

---

[29] MI is an abbreviation for myocardial infarction. Also called a heart attack, a myocardial infarction occurs when a blood vessel leading to a portion of the heart muscle suddenly becomes blocked off. The heart muscle which lies down stream of the blockage dies. Though any heart attack is serious, large heart attacks in which a significant portion of the heart is damaged are life threatening.

when he got home. "He's already Q'ing out,"[30] I went on. "His pressures dropped some during the day but appear to have stabilized." I noticed the senior resident and cardiology fellow looking closely at his heart monitor.

"Rick," Stuart interrupted, "I'm worried about these PVCs. "You all know PVCs are spontaneous electrical impulses in the heart that arise separately from the heart's internal pacemaker and can be a normal finding," Stuart went on for the benefit of the entire team, "but especially in the setting of an acute MI, PVCs like this may portend a dangerous cardiac rhythm, especially V Tach or V Fib.[31] See here," he said, pointing to one particular squiggle on the monitor printout, "this PVC is falling dangerously close to the preceding T wave. Unfortunately Mr. Harrison is at risk for the 'R on T' phenomena[32] which occurs when a PVC falls on top of the recovery phase of the preceding beat. These could degenerate into a lethal arrhythmia[33] at any time. Rick, make sure he gets started on lidocaine[34] immediately. Begin with a 100 mg bolus followed by a drip at 4 mics per minute to be titrated to control the PVCs."

"Will do," I replied, writing the order in Mr. Harrison's chart as the rest of the team went off to the next patient.

Mercifully, rounds finally ended, and I headed for my car. The drive home, though only 10 minutes, seemed interminable as I struggled to keep my eyes open. In my post-call surreal state, I imagined how easy it would be simply to turn my wheel into the oncoming traffic. Then I wouldn't have to

---

[30] Q'ing out is an abbreviation for changes that occur in a patient's EKG over the first several days following the onset of a heart attack. A Q wave is a portion of an EKG tracing that, when present in a specific pattern, signifies that muscle cells throughout the full thickness of the heart wall are dying and slowly evolving into a scar within the heart wall itself.

[31] V Tach (pronounced Vee Tak) is an abbreviation for ventricular tachycardia. Tachy means rapid. So tachycardia means a rapid heartbeat. V Tach occurs when the heart's internal pacemaker is taken over by beats that begin in the ventricles, the lower chambers of the heart. Ventricular tachycardia can be relatively slow and well tolerated or it can be fast (over 150 beats per minute) and life threatening. But its biggest risk is that it can degenerate to V Fib (pronounced Vee Fib) which stands for ventricular fibrillation. This is a condition in which electrical impulses course throughout the heart chaotically. In ventricular fibrillation the heart muscle cells contract completely independently of each other, so the heart stops beating completely and instead appears to quiver like a bowl of jelly. This is the most frequent cause of death from a heart attack or MI.

[32] R waves and T waves are ways of referring to other parts of the EKG tracing.

[33] As noted earlier, arrhythmia is a term that refers to any abnormal heart rhythm.

[34] Also as noted earlier, lidocaine is a local anesthetic used to numb the skin before suturing or surgical procedures. When given intravenously, it affects the electrical activity in a manner that reduces the likelihood of the heart degenerating into ventricular tachycardia and ventricular fibrillation.

go back to the hospital in the morning. This ordeal would end. I shook myself back awake and jerked the wheel back to the right, correcting the slight turn the wheel had taken. With a surge of anger, I recalled some of the research I'd read about the effects of sleep deprivation on physician performance. All the studies seemed to be done on acute sleep deprivation, asking how errors increased in decision making and performing procedures after 18, 24, or 36 hours of continuous work. Though these studies were important, they missed a critical element that also needed to be assessed. Doing one 36-hour shift was hard, and led to deterioration in my or any other physician's performance. But what happened when you had to do it again just days later? And again after that? What happened after three months, or three years, of chronic sleep deprivation? What did such chronic sleep deprivation do to error rates? What had to harden inside for a physician to keep going? What subtle, complex thinking would become less available? I was unaware of any studies that captured the true toll our training system took on physicians and, in turn, on those who counted on them.

As I pulled into my driveway, I wanted nothing more than to sleep. Yet I knew when I walked through the door, Marsha would want to talk, as she always did. If talking failed to keep me awake, somehow our talking would deteriorate into an argument. Marsha's arguments, fierce and impassioned, always managed to keep me awake. Arguing was not my preferred way to end a shift on call, but it somehow kept us connected as best we could during this dark time in our relationship. Well past midnight our argument ended, and I was finally allowed to close my eyes.

The following morning I was back in the ICU by 7:30 to get ready for 8 AM rounds. I began quickly to check on my patients when an alarm suddenly went off. At the other end of the unit, a nurse called out, "Code 99! Code 99!" This was the universal signal that a patient had stopped breathing or their heart had stopped. Almost immediately, the overhead paging system throughout the hospital repeated the call. "Code 99 ICU! Code 99 ICU!" I, along with everyone else, sprinted toward the site of frantic activity. As I approached, my heart rose in my throat as I realized all the commotion was coming from Mr. Harrison's bedside. A nurse on her knees on the edge of his bed pumped on his chest. Other nurses rushed in with the code cart carrying specialized equipment and medications. Stuart already stood at the head of the bed barking orders into the chaos. He quickly grabbed the laryngoscope from the code cart and intubated Mr. Harrison with surprising speed and skill. Just then a respiratory therapist joined him at the head of the bed and began using the bag attached to the newly placed tube to ventilate him.

"He's in V fib," Stuart called out. "Get the paddles. What was his lidocaine drip running at?"

The nurses paused for a minute, looking at each other. "He wasn't on any lidocaine," one of them responded. A chill ran through me.

"What do you mean he wasn't on any lidocaine? Rick, didn't you order the lidocaine last night?"

"Y-yes…Yes I did," I stammered.

"Then why the hell wasn't he getting any lidocaine?"

Another nurse had pulled open his chart. "The order's here, but it was never flagged, so the nurses on evenings never saw the order." My whole body suddenly went cold.

"Stand back!" Stuart fired the paddles on Mr. Harrison's bare chest. His body convulsed, but the heart rhythm didn't change. "An amp of epi[35] and an amp of bicarb.[36] Keep pumping." Moments later he fired the paddles again. "His rhythm's deteriorating. Repeat the epi and bicarb. Bolus him with lidocaine." He fired the paddles again. "He's gone flat line. We're losing him." Another round of medications and shocks followed, but nothing changed. "Anybody have any suggestions?" Stuart called out to everyone in the room. Silence followed. "Call the code. He's gone."

I stood numbly in the corner as one by one people left. In a daze I found myself stumbling over to where Mr. Harrison's chart lay open at the end of his bed. Out of the corner of my eye I glimpsed his now ashen face, distorted by the plastic tube still protruding from his mouth. His limp arms lay at his side. Never would they hold his granddaughter. His dull eyes stared open at the ceiling. Never again would they twinkle with the love for his family. I stared hollowly at the words on the order sheet. There was my handwriting, but nobody had ever seen it. I had forgotten the simple step of flagging the order. Because of my mistake, Mr. Harrison lay dead.

"Let's get ready for rounds, people," Stuart called out. "Rick, you can take care of the paperwork after rounds."

I wasn't ready to go on. I couldn't go on. Rounds was the last thing I felt like doing. Still cold and numb, I could hear the words repeating over and over again in my head. I killed him. My mistake killed Mr. Harrison. If it wasn't for me, he'd still be alive.

"Come on, Rick, we're waiting for you for rounds."

With the familiar force of will I had drawn upon so many times before, I pushed aside the voice in my head. I ignored the knot forming in my

---

[35] Epi was short for epinephrine, another name for adrenaline. Administering epinephrine to a patient in V Fib was thought to make the rhythm more coarse and therefore more susceptible to being shocked back into a normal rhythm.

[36] Bicarb was short for sodium bicarbonate, a chemical that helped correct the balance between acid and base in the bloodstream. This was thought to make it easier for the heart to restart into a normal rhythm.

stomach. Burying my burgeoning self doubt in familiar activity, I joined the team. "Mr. Gonzalez is a 73-year-old male with a complex set of medical problems. Problem number one..."

*   *   *

I sat hunched over, my eyes unable to meet anybody else's. Eleven other interns bantered easily around me as our support group convened. We began the same way every week with a check-in from each of us. John was enraged at another obstetrician for being so callous to a woman during labor. Monica felt overwhelmed with not knowing enough on her medicine rotation. Paul felt exhausted, not knowing how he would be able to keep up the repetitive sleep deprivation. When my turn came, without looking up I simply said, "I have something difficult to talk about."

Henry, the psychiatrist assigned to facilitate the support group, asked, "What is it that's so difficult to talk about?"

"I killed a patient," I replied flatly, tears beginning to well up.

There was silence. Finally Henry responded, "I'm so sorry." And after a pause, "Tell us what happened."

Slowly, attempting to swallow the lump that kept rising in my throat, I recounted how I'd admitted Mr. Harrison, how much I'd liked him, his PVCs, my writing the lidocaine order, my exhaustion, my coming in the next morning to Mr. Harrison coding, and my horror at learning I'd forgotten to flag the lidocaine order.

After a brief moment, everyone started speaking at once. Henry forced order upon the outpouring.

"That wasn't your fault. The nurses should have picked up the order anyway."

"No. It was my responsibility to flag the order," I retorted. "Nothing can take away the fact that I made that mistake."

"You were exhausted. You're more than entitled to make a mistake like that."

"I know how tired I was. But that still doesn't relieve me of this smothering sense of guilt. If I hadn't forgotten to flag the order, Joseph Harrison would be alive right now."

"We've all been placed in an impossible situation by those who run the residency program and the hospital," John said angrily. "Required to take on extraordinary workloads with sick, complicated patients, operating on the edge of what we don't know, forced into chronic sleep deprivation, we're then told any error is our fault. When they try to be understanding, we know they still think it's our fault."

"Even if they didn't think it was our fault, I'm the one who has to live with it," I responded. "It was my fault. I killed Mr. Harrison. I don't know how I'm going to live with this. I went into medicine to help others. And now I've killed an innocent, good hearted, loving man because I forgot to do something simple and basic. No amount of rationalizing is going to make that go away."

We sat in silence together, each struggling to make sense of Mr. Harrison's senseless death.

Our shared silence somehow helped me return to the ICU that afternoon. I returned again the next day, and the next. Slowly I buried my guilt deeper and deeper. I put distance between me and the intolerable burden of responsibility for Mr. Harrison's death. But this scar joined others that had gone before it. The trauma of sawing a woman's skull in half. The overwhelming pressure to memorize inhuman quantities of information. Repeated public humiliation for what I did not know. Knowledge of the pain I had inflicted out of my own ignorance on an innocent man struggling with cancer. Steeling myself against the repeated exhaustion of nights on call followed by days followed by ever more nights on call. Pushing through my exhaustion in the desperate attempt to maintain the semblance of a relationship with Marsha. Patients entrusting their care to me wholly ignorant of how close to the edge of my competency limits I was always working. And now, Mr. Harrison's death, having nothing to do with my competency, but everything to do with my simple, human fallibility. Yes, I was accumulating a graveyard of scars. Cumulatively, they were taking their toll—a toll of which I was only vaguely aware. Yet even this awareness did not allow me to protect myself from the deep and pervasive inner warping I was undergoing through medical training.

In the face of each scarring event, I had found my two most valued assets to be the will to push through it and my mind's ability to rationalize it. Called upon to rise to these challenges again and again—to absorb more scars—my will and rationality steadily grew in strength, just as exercise strengthens a muscle. Yet any muscle exercised to excess produces hypertrophy, a distortion of the muscle out of proportion to how it best performs. So within me, a gradual distortion grew. This hypertrophy of will and rationality would continue to serve me well in medical training. Yet it would not always serve me well in caring for patients. It would serve me even less well as I tried to maintain relationships outside of medicine with Marsha, friends, family, and especially myself. Despite struggling against it, my consciousness of this distorting process was powerless to arrest it. Instead, an insight gradually dawned within me that all physicians in training undergo the same distorting process. This helped explain how bright, caring young people entered medical training committed to helping others, yet emerged so often appearing rude, insensitive, and oblivious to the true needs of their patients and coworkers. These scars, and the resulting hypertrophy of will and rationality, took an ever deepening toll on the most caring and vulnerable recesses within every physician, including me.

*   *   *

The open-collar shirt hung loosely on his gaunt, hunched frame as he approached down the hall. His full beard and dark, curly hair, both unruly today, bespoke a hurried early morning battle with a comb. Sunken eyes and that slightly faraway look told me his night on call had been a difficult one. Yet his determined gait, punctuated by sharp movements of trunk and hands, made clear the fire was still there. With his white pants and telltale stethoscope draped over his shoulders, I recognized the same intern's uniform I'd worn every day for seven months. That my beard and dark curly hair were unruly most days added to our comradeship that helped get me through each day.

"God, you look like hell," I said, as Howie joined me at the nurses' station, awaiting the start of work rounds.

"You think you look any better?" he quipped. "And you're the one just starting a 36-hour shift, rather than getting off one."

"Thanks for the compliment. Must be the workload on this internal medicine rotation." I omitted telling him that Marsha and I had had another late night fight last night. It had started like so many of our fights.

"I'm listening. I really am listening," I had said, struggling to keep my eyes from drifting shut.

"The hell you are. I was in the middle of telling you how lonely I feel, how cut off from family and friends I have to be because you insisted this was where you wanted to go for residency, and all you can do is run away with the excuse of being tired."

"It's not an excuse. I really can't keep my eyes open."

"I know you better than that!" she screamed. "If I started talking about a patient right now, you'd suddenly be wide awake. If some nurse called saying she needed you, you'd run right up there. But any time I start talking about me, how I feel, or anything that's not about you and medicine, you're gone." The more I protested, the angrier Marsha raged. She knew she was telling the truth, and so did I. I hated to admit what was happening to me. "How come you're always one of the last interns to get out of the hospital every night? You get home an hour later than everyone else, sometimes two. You did it again tonight."

"I already told you why I was late tonight. My last patient in the Family Care Center this afternoon had taken longer than I'd expected. That made me late finishing up work on my medicine inpatients before I could sign out." This was true, but didn't tell the whole story. Antonio Gonzales, a 28-year-old Portuguese man, was that last patient. He had come to the Family Care Center following an industrial accident in which his hand had been mangled, eventually requiring amputation at the wrist. Though the surgery had gone well, his surgeon had found his blood pressure a little high, and had referred him to the Family Care Center to have it evaluated.

When I entered his exam room, my eyes were drawn instantly to the stump at the end of his left arm. It was impossible not to stare. Where his hand had been now resided an uneven scar, its angry red color consistent with a wound in the midst of healing. I expressed my sympathy to him for losing his hand, acknowledging this must have been very difficult. He didn't say much in response. Then, through his broken English, we turned to his blood pressure and established that he didn't have any significant risk factors for heart disease or stroke, such as smoking or diabetes. His blood pressure was a little elevated, but not high enough to require immediate treatment. We decided to check it a few more times before determining if he needed medication. We completed addressing the problem for which he'd been sent, but neither of us stood up. I knew I had a full load of patients to finish seeing in the hospital before I would be able to go home. I knew Marsha was holding dinner for me. And I knew I had a choice. I could end the visit now and get home at a decent hour, or I could ask the question that would make me late again.

I seemed to face this choice almost every day. If I would define my responsibility as only addressing medical diseases, I could get through the day much faster. But I could not ignore what I knew. Sitting with each patient, I sensed when there was something else they wanted to talk about, something that seemed to be the real reason they came to see a physician, even if they did not know it themselves. Sometimes this was subtle, requiring time to tease out and understand what the question behind a patient's question was. With Mr. Gonzales, it was painfully obvious. He had not dealt with the loss of his hand. Typical physician training taught us that such situations warranted a referral for counseling. But I knew what would happen if I simply suggested to Mr. Gonzales that he was depressed about losing his hand and should see a therapist. My patients had already taught me it was easier to make a referral for them to get a tube pushed up their rear than it was to get them to see someone to talk about their feelings. Mr. Gonzales would have politely accepted the slip of paper with a phone number of a therapist but never would have called.

So if I knew all this, what should I do in that moment? I remembered back to Dr. Wilson, and decided it was time to play a variation of the medical card.

"Mr. Gonzales," I began, "your blood pressure is a little elevated. As we discussed, the blood pressure itself doesn't require treatment, but we might be able to help you with the cause of your elevated blood pressure." He looked up. I paused as our eyes met. "You have obviously suffered a devastating loss. I am sure it is harder than you have let anyone know." He averted his eyes toward the floor. Did his limited English allow him to understand what I was saying? I waited to find out.

"I am not a man anymore," he eventually responded softly.

"Not a man..." I repeated, pausing to encourage him to continue.

"Without this hand," he raised the stump, "I...I am not a man. I cannot work."

I realized Mr. Gonzales lived in a Portuguese culture in which, if a man did not work, he was not a man. All the workers compensation and disability payment in the world would not change this. Yet it was not time to help him mourn the loss of his hand, and his identity. That, I hoped, would come later. Now, he simply needed to be understood, to be seen clearly by his doctor.

"Losing a hand must feel like losing everything right now."

"I would rather be dead."

"Have you thought of taking your life?"

"No...I would not do that." I guessed that would be an even bigger disgrace.

I had pushed him enough. He had reached the limit of his tolerance for talking about the loss of his hand. "Mr. Gonzales, I would like to see you next week to check your blood pressure again." He nodded. I also knew this would allow us to spend a few more minutes talking about his loss. It was another way to play the medical card.

Taking the extra time with Mr. Gonzales, including discussing him with the precepting physician, had eventually set me back 45 minutes in getting out of the Family Care Center. Sign out to the intern on call had been delayed because by the time I was ready to leave, she was caught up in an emergency. All told, going beyond Mr. Gonzales' blood pressure to address the loss of his hand made me get home almost an hour and a half later than if I had just addressed his medical diagnosis. Yet I made the same choice every day. I could not bring myself to not ask the question I knew a patient needed asked, to not do what I knew they needed done. I could not let myself be less than the best doctor I could for my patients. But this came at a cost.

"I had a patient who lost his hand, for God's sake! That's what made me late," I screamed at Marsha.

"It's always a patient who lost his hand or has leukemia or is dying!" Marsha screamed back. "I can't compete with your patients. Of course it's more important for you to attend to them. But look what it is doing to me. What it is doing to us. I have dreams of showing up in the emergency room at your hospital with leukemia just so maybe you would see me," she sobbed. "I don't know if I can do this with you. I'm not cut out to be a doctor's wife."

I took her in my arms. "I am so sorry, Marsha. I promised you I would never let medicine dominate my life so much that I could not prioritize you, not value our lives outside of medicine. I need you to be my tie to a life beyond medicine. I will leave this residency before I let medicine make me lose you." I had said these words before. Yet when I said them now, I was no longer sure they were true. When Marsha and I had met, I had been able to walk away from my medicine rotation early every day to place our relationship first. I had tried to keep this my priority. But something inside me was shifting. It was one thing to walk away from a student's responsibilities. It was another to walk away from a doctor's. Was this just the pressure of an abusive system driving a young resident to exhaustion with its demands? No, it was more than that. I could no longer just blame the training of residents for what was happening between me and Marsha. I could not walk away from my patients because this was not just a job. I now knew medicine, being the best physician I could for my patients, was a calling. It was my calling. What this meant for my future with Marsha, we would have to wait to find out.

*   *   *

The harsh buzzer blared overhead in my on call room. "Code 99, Emergency Room! Code 99, Emergency Room!" I was in motion before my eyes were open. Instinctually I pulled on my pants, pushed into my shoes, and was out the door in less than 30 seconds. It was only then that I wrestled my brain into some state of wakefulness.

"Where is this code?" I asked myself, still buckling my belt while jogging down the hall. I played the message over in my head…Code 99, Emergency Room…Yes. I was on my way to the ED. As I burst through the emergency room doors, all was bright and frenetic activity. The usual bustle of individuals responding to a code was already underway. Glancing up at the clock, I noticed it was 1:50 AM. I'd found a lull in admissions and calls and had managed to grab a 20-minute nap. With a modicum of gratitude for even that much sleep, I realized our code team had arrived before the patient. We'd received a call from the ambulance. All we knew was that they were en route with a man found passed out, and they had begun CPR on the way to the hospital. We could hear the siren growing louder as it approached. Suddenly, two burly men burst through the ED doors pushing a stretcher carrying a limp figure. One of them pushed down rhythmically on his chest.

"Ready. One…two…three…Lift!" With that they heaved his body onto the treatment table in the middle of the ED code room, a room specially set up to handle the most complex emergency cases. Suddenly the activity rose to a frenzied but organized medley as two nurses cut off his clothing, one placed EKG leads on each extremity, and I took over the chest compressions. A stool appeared under my feet, and I leaned over, placing both palms, one on top of the other, at the base of his breast bone. With each firm thrust I felt his rib cage compress and then bounce back. Every rib stood out on his emaciated body, and I could feel the detailed contour of his sternum under my hands. On the next thrust I felt the sickening sensation of several ribs break under my hands. I had no choice but to continue the rhythmic chest compressions, a crunching sensation now accompanying each thrust.

"We found this guy on a manhole cover," I overheard the EMT tell Harry, the junior resident running the code. "Probably homeless. He was real cold. After all, it's below 20 out there tonight. We couldn't get a pulse or respirations so we began CPR. We've been going at it for about 15 minutes so far."

"Thanks for your help," Harry replied, and the EMTs left. Harry slipped an endotrachial tube into his airway. "Let's do another round of epi and bicarb," Harry called out as we continued CPR. "Still flat line," he said, glancing at the strip of paper running off the EKG machine. "Anybody have any suggestions?"

Dr. Bronson, the attending physician in the emergency department for the night shift who had been observing the code intermittently while he continued to take care of the other patients, spoke up. "I read an article about a handful of cases like this. They came in cold and flat line, but after being warmed up, they developed a rhythm and eventually recovered. The conclusion of the article was that it may not make sense to declare someone who presents profoundly hypothermic and flat line dead until you warm him up."

"What was the temperature the authors used as the target?"

"They gave a range of 94 to 96 degrees."

"So what's this guy's core temperature?"

"Don't know. Let's check it." I paused the compressions while we rolled him over and a nurse inserted a rectal thermometer. We rolled him back and I resumed pumping on his brittle chest.

"It's so low it won't even register on this thermometer," the nurse said a few minutes later.

"I think we have a hypothermia thermometer for just this purpose," Dr. Bronson volunteered.

After some searching, the special thermometer was found and made its way into his rectum. "Eighty-three degrees," came the answer we almost didn't want to hear.

"Alright, team, let's warm this guy up," Harry said encouragingly. "That means heated, moisturized air into his lungs through the tube, warmed IV fluids, and warm enemas."

"You've got to be kidding!"

"Nope. If we just warm his extremities, we pump blood into them and bring back all that lactic acid and other toxins. It's the quickest way to make sure his heart never starts again. We need to warm his core. That's the drill."

"How long are we gonna go at this?"

"As long as it takes to get him to 94 degrees."

Just then my beeper went off. "Dr. Sheff, please call 5708. Dr. Sheff, 5708," it crackled.

"Could you get that for me?" I asked one of the nurses.

"They need you up on Henderson five. A chemo patient's IV came out, and she needs to be tanked up with fluids tonight for her chemo in the AM."

"Let them know I'll get there as soon as I can. Looks like we're gonna be here a while," I replied, glancing up at the clock which read 2:15.

The respiratory therapist hooked up the air heater/moisturizer. A nurse plugged in the bath for heating IV fluids as they ran through the tubing. Finally, reluctantly, one of the nurses showed up with a large enema bag. "What do we do with this?" she asked.

After a brief silence, Harry said, "Insert the tube as far up his rectum as you can get it. Then fill the bag with warmed saline,[37] raise it, and let it run in. After it's been in a while, lower the bag and let it run out. Empty the bag and repeat the cycle." The nurse rolled her eyes, as we all did.

It took some time, but finally the warming process was fully under way. My back began to ache, and I asked to be relieved of the chest compressions. The nurse holding the enema bag quickly volunteered, handed me the bag, and stepped up on the stool. For a moment, I simply stared at the bag in my hand. "Were you going in or out?" I finally asked.

"In," came the reply.

I raised the bag to shoulder height. A slight gurgling noise could be heard as the warm fluid flowed slowly down through the half-inch wide tube and up into his rectum. I wondered how long to wait. I pictured the inside of his colon. Imagined the clear saline mixing with his partially formed stool, making a slurry mess. Imagined this brown mixture warming up the lining of his colon. Imagined heat emanating out to his other abdominal organs.

---

[37] Saline is the term for a solution of water and salt that matches the concentration of salt in the human body.

When it felt about right, I lowered the bag. A lower pitched gurgling sound accompanied the slight vibration I felt as the bag filled.

"Did you have a clamp for this?" I asked the nurse who seemed quite happy to be doing her new job.

"Nope. Just pinch the tube, open the top of the enema bag, and pour it out into the basin there.

For the first time I noticed a steel basin near my right foot. I unscrewed the plug on the enema bag, and out flowed a foul smelling brown liquid. I raised the bag back up. Hands appeared that poured fresh warm liquid back into the bag, and the cycle repeated. Then it repeated again. And again.

In a dazed state, my mind began to wander. I thought of the rotation I'd just completed last month in pediatrics at the big university medical center. It was part of our family medicine training because we needed to be prepared to care for sick children, especially if we planned to practice in an isolated community like the one in upstate New York I'd gone to that had cemented my decision to choose family practice. Much like my subinternship, the rotation at the university medical center required us to care for staggeringly sick children—children with cancer, children with birth defects, children with meningitis, anorexic children, abused children. The on call "room," if you could call it that, had been a bed right on the sick baby ward. The steady beep, beep, beep of countless monitors droned on all night, punctuated periodically by a high pitched alarm, usually caused by one of the leads coming loose from a crying baby but occasionally from a life-threatening emergency. It made for a fitful sleep at best, when sleep was even possible. During that month I'd been unable to get back for our weekly support group and had been isolated from my fellow family practice residents. Devastated by the suffering children surrounding me, I'd sunk into a deepening depression. It didn't help that it was December. Cold and dark when I came to work, cold and dark when I went home.

I thought it would be better now, back among my friends, and it was, at least a little. Yet every night on call seemed to drag on longer, especially this night. My beeper had gone off four more times while I raised the bag and lowered the bag, raised the bag and lowered the bag. Periodically we'd checked his temperature, which rose slowly. 87...90...92...We were getting closer. I glanced distractedly at the clock. It now read 3:20.

Just then I noticed a new sensation. My right foot felt wet. I glanced down at the bag. It seemed different somehow. In my daze I stared, not comprehending what I was looking at. Then I saw the tube. It had pulled off the bag and was running down the outside of my right leg into my shoe. A dull brown streak along the outside of my pants told me what I didn't want to know. Quickly grabbing the tube and pinching it off, I felt my foot squish inside my right shoe.

"Could somebody take the bag from me? I seem to have a little problem here." I was too embarrassed to tell anyone what had happened. I reattached the tube to the bag and handed it to one of the nurses.

"Let's check his temperature again," Harry said just then. A few moments later he proudly announced, "95 degrees! We did it team." Looking down at the continuously running EKG strip he called out, "He's still flat line. Any objections to calling the code?" He paused, hearing only silence. "Code's called."

I slinked away, feeling my right foot squish in a pool of brown liquid with every step. Ducking into the nearest bathroom, I poured the foul smelling pool out of my shoe. I was able to take off my pants and rinse out the brown stain. The same worked for my sock. But no amount of rinsing seemed to help my shoe. Without another pair of shoes in the hospital, I had no choice but to wipe out the inside of my shoe with a paper towel and put it back on.

Feeling lower than I had all year, I made my way to Henderson five. Along the way I stopped in my on call room for a swig of Maalox. The more coffee I downed, and the more nights on call I survived, the more heartburn seemed to become a way of life. I'd never used antacids before, but now found keeping that Maalox bottle in my on call room as essential as recharging my beeper.

"Mrs. Bartlett in 536B needs an IV restart," the woman at the nursing station said, barely looking up. In the dark, semi-private room I found an elderly woman sleeping soundly. I didn't want to wake her, but she had to get as much IV fluid in as possible tonight or her chemotherapy treatment in the morning might poison her kidneys. Glancing automatically along both arms I realized this was going to be a tough one. (By this point in my internship year, I'd come to see every patient, even friends and family, as either an easy stick or a hard stick. Instead of looking into people's eyes when I met them, I'd taken to looking at the veins in their arms.) Her paper thin skin, multiple large bruises, and tiny, fragile veins told me I was not the first person to face this challenge.

"Mrs. Bartlett," I said, gently nudging her shoulder, "it's Dr. Sheff. I need to restart your IV." Through my fatigue I tried to sound as friendly as possible.

"Leave me alone. Go away!"

"I have to restart your IV or you won't be able to get your chemotherapy in the morning."

"I don't want chemotherapy. Why can't you just leave an old woman alone to die?"

I paused, knowing the wisdom in her words. Yet at almost 4 in the morning, it was not my place to question her attending physician's order. "I'm just doing what your doctor ordered."

She rolled over, sunken eyes regarding me with a mixture of sadness and resignation. "I told him I don't want his chemotherapy, but he keeps insisting. He thinks he's helping me. But to a woman of my age...I'm 84, you know..." Her voice trailed off.

I didn't want to do it. I didn't want to be the next person to torture this woman with another needle stick. And for what? To get an IV in that would last another 10 or 20 hours? Then she will have to go through it all over again. I paused, wanting to walk away, to leave her in peace, but knowing my "orders" were to protect her kidneys by sticking her with another IV. "I understand how you feel. I really do," I said, looking into those hollow eyes, made all the more haunting by the shadows of the night. "But if we don't get this IV in now, the medicines might badly damage your kidneys," I finally said, seeking some justification to myself for what I was about to do.

In silence, she put out her arm. I placed the rubber tourniquet first on one arm, and then the other. No good veins. I excused myself and returned with a blood pressure cuff. Using one of the tricks I'd learned this year, I blew up the cuff half way on one of her arms. I knew she would soon start to feel a painful throb as blood entered her arm but could not leave it. "Oooh! Oooh!" she said, as the throbbing increased. But I'd found my vein. Thin, fragile, it would have to do. "Ow! Stop...Stop...You're hurting me," she moaned as I pushed the needle through her terribly thin skin. Using every technique I'd learned from starting IVs in tiny infants, I managed to stabilize her fragile vein and get the tip of the needle into it. I knew her vein was too fragile to push a flexible catheter into, so I was content to use a tiny steel butterfly needle. Leaving the tip lodged precariously in her thin vein, I propped and taped the needle in place, praying it would stay there long enough to get the necessary fluids into her. I left it to someone else to secure a more stable line for the poisonous chemotherapy drugs she'd be getting in the morning. That meant another needle stick for her, but at least I'd fulfilled my responsibility tonight.

"Good night," I said softly. She rolled over in resigned silence, turning away from me as I backed toward the door.

The next morning I sought out her attending, Dr. Simons, a successful and respected oncologist. I shared with him her request to forgo any further chemotherapy. He replied, "Nonsense. She has multiple myeloma. That means she's got a 40% chance of a response to treatment."

"Response? Does that mean remission or cure?"[38]

---

[38] Oncologists usually describe success for cancer treatment in one of three categories. A response means the tumor shrinks in response to the treatment, usually by at least 50%. Remission means all detectable signs of the cancer have gone, and it cannot be identified on any blood or imaging study. Very often, even after

"Neither. She's got no chance of cure and only a small chance of going into remission. But with some luck we can give her eight good months, maybe even a year."

"Is it worth it, at least in her eyes?"

"Look, Rick, I know you're trying to do the right thing here. But nobody should be allowed to die without a trial of chemotherapy. She just might be the one person who goes against the odds. Do you want to be the one to deprive her of that chance?"

As I walked away, the itchy, brown material caking ever harder in my right shoe, I could not get Mrs. Bartlett's haunting eyes out of my head, nor her plaintive voice pleading with me to stop. Dr. Simons, I realized, was exercising another aspect of physician power: the power of information. He knew the statistics from the latest studies on multiple myeloma. I was pretty sure he hadn't shared these with Mrs. Bartlett. Instead, he would probably have said something like, "Using the recommended treatment protocol, which is supported by the most recent scientific studies, will provide you a significant chance of a good response." Hearing this, Mrs. Bartlett, or at least her family, would likely have thought there was no choice but to undergo this treatment. Dr. Simons would have been just as honest to have said, "We can't offer you a cure. We can't realistically offer you any chance of going into remission. But if we poison your body with our chemotherapeutic drugs, bringing you to the edge of death, perhaps even killing you with complications of the treatment, you'll have a 40% chance of living 10 months longer. Along the way you will experience significant nausea, vomiting, and discomfort, as well as likely require several prolonged hospital stays."[39] If her choice had been presented in this manner, I have no doubt what Mrs. Bartlett, at the age of 84, would have chosen.

Dr. Simons would have been one of the first to claim that today's practice of medicine had moved beyond the well intentioned but now discredited paternalism that characterized previous generations of physicians. But I was not sure he fully appreciated the extraordinary power physicians still wield with information and how they present information to patients. Medicine has embraced informed consent, the understanding that no patient should undergo treatment or have treatment withheld without the freedom to choose a personal course of treatment. Without adequate

---

remission a cancer eventually recurs. Cure is used following remission when no signs of the cancer are detected over a long time period. The usual waiting period for labeling a cancer patient cured is at least five years after remission, though some cancers can recur as late as 10 years after remission.

[39] This would have been accurate at the time of my training. Today, treatment regimens for multiple myeloma provide considerably better results, but the underlying principle of how much influence physicians have in how they present the results of research trials remains the same.

informed consent, a physician performing a procedure on a patient would be considered to be committing a battery, in the legal sense of this term. Yet, to make a meaningful treatment choice, patients must be given the information necessary to adequately inform this choice, including the risks and benefits of the proposed treatment and alternatives to that treatment. As I watched this ritual of informed consent played out again and again, I was struck by how much a patient's choice could be influenced by *how* a physician shares information.[40] This opens up the informed consent process to powerful distortion from the biases, or self interests, of physicians. I had no doubt that Dr. Simons truly believed that by insisting on a trial of chemotherapy, he was acting in full concordance with the Hippocratic Oath. But in his zeal to treat Mrs. Bartlett, I sensed something else. His biases, and perhaps even his self-interest in the fees he would receive from her chemotherapy treatments, also played an important role in how he presented his clinical recommendations to her, though he would vehemently deny such an accusation. Perhaps he was simply fulfilling the maxim, "If your only tool is a hammer, all the world comes to look to you like a nail." Data was beginning to show this was true of many medical and surgical specialties. Oncologists treated cancer with chemotherapy. Surgeons did surgery. In fact, the best predictor of the number of surgeries performed in a community was not the clinical needs of its inhabitants, but rather the number of surgeons in the community. I sensed I was witnessing the fulfillment of my hypnosis professor's observation that physicians can use their power, in this case the power of information, for the good of their patients. Or they can use this power to pursue their own needs, and patients will suffer. I did not believe Dr. Simons was conscious of this power nor how he was using it, and Mrs. Bartlett suffered. I believed these were related.

"No one should die without a trial of chemotherapy," Dr. Simons had said, believing he was instructing me in a sacred truth. Yet this elderly woman with whom I'd spent 15 minutes in the darkest hours one January morning, someone I would never meet again, had taught me a different sacred truth. It's OK to stop.

We are exceptionally good at doing more and more for our patients, yet we are remarkably inept at "not doing." I thought of Jennifer, the nine-

---

[40] Today, the Internet has dramatically changed how patients and their families, at least those with adequate education and resources, can access medical information for decision making. Physicians are beginning to experience the Internet as undermining their authority and role, as undermining this information-based aspect of physician power. Still, interpreting clinical information in the proper context often requires a physician's training, experience, and judgment, so this aspect of physician power will continue to heavily influence patient treatment choices.

month-old with Down syndrome and congenital heart disease from my
pediatric subinternship, struggling with every ounce of energy her tiny body
could generate simply to draw her next breath. Why didn't her doctors stop?
Why didn't I stop? What was it about not doing that was so hard? I, as did
all physicians, harbored the hope that if I did just one more treatment, a
patient who entrusted her care to me might experience a miraculous
recovery. Perhaps one day Jennifer would. But the scientific odds against it
were devastatingly high. What was I afraid would happen if I didn't do that
next treatment? Even in the early hours of this dark morning with Mrs.
Bartlett the answer was clear. I didn't want to be the one to exercise the
awesome God-like power of depriving my fellow human being of life by
withholding treatment. But in pushing treatment after treatment, was I not
exercising an equally God-like power of sustaining life against nature's
natural rhythm and timing? The dark shadows of Mrs. Bartlett's face
revealed the untold, hidden suffering physicians inflict in the name of
striving to beat back nature's plan.

As I responded to the next page and trudged on to the next nurse's
station, I realized we doctors are afraid of something else that prevents us
from stopping all this doing. We fear the opposite of doing. We fear simply
being. Stillness is foreign to us and filled with unknown demons. We have
never been taught how just to be with a patient, especially one that is
suffering. We swoop in, assess, diagnose, prescribe, and leave. Nurses stay
by the bedside and witness suffering and death, not doctors. If we perform a
procedure, we race out as soon as it is over, moving on to the next thing we
must do. To stop requires work. It requires the kind of inner work that is not
taught nor valued in medical training. The opposite of stillness is praised,
rewarded, role modeled. Anything else is just weakness, and must be driven
out of physicians in training so they can simply get through their days…and
nights. The gift Mrs. Bartlett had offered was to wake me up to the inner
work that lay before me, the work I must do so that someday I will be able
to stop doing to a *patient* and simply be with a *person,* even though I am a
doctor.

But that work would have to wait. At the moment, exhausted and
drained, I had to draw once again on the familiar, inner reservoir of will
simply to get ready for rounds and a day that would not end soon enough.

*   *   *

"Here we are again," opened our psychiatrist facilitator. The Thursday
noon-hour support group with my fellow interns remained my favorite part
of each week. On other days we had educational conferences. Mondays
were internal medicine, Tuesdays pediatrics, Wednesdays obstetrics and
gynecology, and Fridays psychiatry. Over the course of the year, the

conferences worked through the core clinical content of each of these specialties, so by year's end all of us as interns would be well grounded in the foundations of each of the specialties. But Thursdays at noon were reserved for the interns just to talk. After I had shared my inner hell from killing Mr. Harrison in the ICU, other interns began opening up, sharing more than just gripes. We supported each other through feelings of incompetence and humiliation, frustration with attendings who didn't seem to care, and exhaustion and depression that come inevitably from so much sleep deprivation. We shared glimpses into strained marriages and dating relationships.

"Sometimes this group feels like 12 desperate people on a life boat in the middle of tossing seas. You are all clinging to each other for survival," our facilitator observed.

"That's exactly what it feels like to me," Harry volunteered. "Like last night. I was up most of the night. I thought of you guys doing the same stuff, and how much we're helping each other get through it all, and that's what made me feel better. Of course it didn't keep me from having to stop in my on call room every two hours to catch another gulp of Maalox."

"You keep Maalox in your on call room?" Marilyn said incredulously. "So do I."

"I thought I was the only one who was chugging Maalox," I added.

One by one, we each sheepishly shared that we kept antacids in our on call rooms, downing them with remarkable frequency to keep heartburn at bay. We laughed together a deep, healing laugh. But that night, I still needed just as much Maalox to get through my night on call.

*   *   *

"Hey, Doc, remember me?"

I looked again at the name on the top of the chart. James Schmidt. The name I never wanted to see on a chart in my hands again. I swallowed hard. Having the Family Care Center clinic the afternoon after being on call was hard enough. It meant I would have to finish seeing patients in the clinic, then return to write up my chart notes, check some final lab results, and sign out. I wouldn't get home until well after 8 o'clock. But facing this venomous man again was more than I thought I could handle, especially in my post-call fatigue.

"Good afternoon, Mr. Schmidt. How can I help you today?"

"Like you think you helped me the last time I came in here?" he said derisively.

The image of Mr. Josephson, the manipulative patient I'd seen during my outpatient psychiatry rotation, suddenly flashed before me. Don't let yourself get caught in this guy's manipulations, I reminded myself. "Well, you are back again," I observed, with a slight edge in my voice.

"You're right about that," he said as his hand clenched into a familiar fist.

"So how come you came back?"

"It's the misses. She made me come back. If it weren't for her, I wouldn't be within 10 miles of this place."

"For a guy who carries a lead pipe and looks any man in the eye to tell him to go to hell, I'm a little surprised your wife could drag you here."

He looked up at me, and for the first time a slight smile began to curl at the edge of his lips. "You know women. I'd rather fight a whole pack o' men than tangle with one woman, especially my wife."

"So, since you're stuck with me because of that piece of hard won wisdom, what can I do for you today?"

"It's my legs. They're no good."

"No good? What's going on with them?"

"Those pains. They ain't goin' away. They're comin' on when I do anything."

"You mean the cramping in your calf we talked about last time you were here?"

"Yeah. Yeah. Of course that's what I mean." The edge was back in his voice. "Shit. Guess that's what comes of getting stuck with a baby doc."

"How far can you walk now before the pains come on?" I said, ignoring the barb.

"Not far. Not far at all."

"How about mowing the lawn? It's early spring now, so how much of the lawn have you been able to mow?"

He hung his head down. After a pause he exploded, "I can't mow nothin'! I start and have to stop right there. It's my wife mowin' the lawn now. How do you think that makes me feel?" His hand clenched in and out of a fist furiously.

"Given what little I know of you, I'll bet you can't stand it."

"You're damn right! You gotta do something. You gotta get me my legs back, Doc."

There was that word again, Doc, only this time it came out with a plea. It must have taken so much for this angry man, this Napoleonic plumber, to admit in his own awkward way that he needed something from me. Making himself vulnerable to another man, something I guessed he had never done in his life, could not have been easy. Using the word Doc seemed to make it alright, but just barely. It was time to play the medical card to ease past his discomfort.

"Hop up here on the exam table and let me take a look at those legs of yours."

"Used to be these here legs would carry me wherever I wanted to go," he volunteered as my fingers searched for the pulses in his feet. "I could carry an 80-pound sack of lead pipes up a ladder like it was nothin'. Now look at 'em."

His legs were concerning. The thin skin and pale color told me his circulation had deteriorated since my last exam. His foot pulses were barely palpable on the left now and couldn't be felt at all on the right. Luckily he didn't have any ulcers or non-healing sores on either foot. After a quick listen to his heart and lungs, I stepped back. Sitting in a flimsy johnny, his spindly legs dangling off the exam table, he looked ridiculous. I decided to give him back as much dignity as possible. "Why don't you get dressed? I'll be back in a few minutes and we'll talk about what we can do for those legs of yours."

I had already concluded he needed a bypass operation for his legs, especially the right one.[41] Just for completeness I reviewed his case with the precepting physician in the Family Care Center who concurred. This gave Mr. Schmidt a chance to get dressed so we could have a conversation like two adults. Walking back to his exam room, I decided to make this my standard practice: interview patients while they are dressed, leave so they can change into a johnny, and leave again after examining them so they can change back into their clothes. Then we can discuss my recommendations with them dressed and in a setting of respect. It would take a little longer, but with several exam rooms going at once, I could still make this work.

"Mr. Schmidt," I began.

"Call me Jim," he interrupted. "When someone calls me Mr. Schmidt I feel like I'm back in grade school."

"Jim, last time you were here I had recommended a test, an arteriogram. You were pretty adamant you didn't want to do any tests back then. How about now?"

"Doc," he leaned forward in a conspiratorial tone, "I know it's gettin' worse. You tell me I gotta go through that test, and I'll do it. I'm not afraid of nothin'…except the misses, of course." Our eyes met for the first time. A silent understanding passed between us.

"I am concerned about your legs. You got a problem with the circulation in both of them, right worse than left. If you don't take care of this, you could lose one of your legs, perhaps even two of them."

"Aw, Christ! I can't have that. I'd rather die than try gettin' around without one of my legs. What do I gotta do?"

---

[41] Today he might be a candidate for angioplasty, a procedure in which artery blockages are treated without surgery. But at the time of my training, this technology had not yet been developed.

"You need to have an arteriogram. It's a study in which they put a catheter in the artery in your leg. They enter the artery through a small opening they make in your groin under local anesthesia. Then they inject dye in the artery and see if there is a blockage." I could see him struggling to visualize all this. "Here, let me draw you a picture," I offered. I rolled my stool next to his chair and began to sketch a leg, an artery, and a blockage in the artery on a pad of paper. Then I drew pictures of how an operation to bypass the blockage would work.

"Hey, Doc," he suddenly said brightly, "sounds just like plumbing. You got a good plumber here at the hospital; I mean one who works on legs like these?" He made a disgusted gesture toward his legs.

"Yeah, we got a real good one." I made a mental note to use this approach of drawing pictures as often as possible since it seemed to put Mr. Schmidt at ease and allow him to quickly grasp what I was describing.

"Then let's do it."

"It's possible the arteriogram will show there isn't a chance to bypass the blockages," I said.

"Then what's gonna happen?"

"We'll have to handle the problem just with medications."

"I've always been partial to plumbing, so let's hope your plumber doctor friend has something he can do for me."

"I really hope so, Jim." We shook hands as he stood up. I could feel his steel grip. "You've still got the strength of a plumber in those hands of yours. I'll bet you were a good one."

"Damn good," he responded, his grip tightening even more.

*   *   *

"Dr. Sheff, you've got to come right away. We've got a real problem here with your patient."

"But I'm on a pediatrics rotation now, and you're calling me from a medical-surgical floor."

"It's Mr. Schmidt. He's signing out AMA,[42] but he's only three days post op."

"Why? What's going on?"

"He won't say. I was just in there, and he yelled at me to get out and said he was leaving. Can you come now?"

"I'll be right there." After hanging up the phone I found my junior resident on this pediatrics rotation and explained the situation.

"That's part of the challenge of having your own practice in the Family Care Center. When one of your patients is admitted, you have to act like the

---

[42] AMA stands for against medical advice.

attending for that patient, regardless of whatever rotation you're on. It's especially a pain in the butt when one of your patients goes into labor. Go ahead. Sounds like you've got a hot one. I'll cover your service until you get back."

Walking into his room, I found Jim half dressed, standing next to his bed and pulling on his pants. "It's about time you showed up. They told me I couldn't leave until you got here. What took you so long?"

"Jim, what's going on?"

"It's those damn nurses. They won't leave me alone. They're as bad as my wife."

"What did they do?"

"Everything. They did everything, and did it all wrong!" he screamed. Suddenly his expression changed. His ruddy face became ashen.

"Jim, what is it?"

"I...I...my chest." His hand reached up to the center of his chest.

"Jim, are you feeling pain or pressure in your chest?"

"Yeah. Godamnit. Yeah."

Stepping out into the hall, I called to the nearest nurse, "We need an EKG stat in 219, and oxygen, too." Stepping back in the room I moved quickly to the bedside. "Jim, sit down." He mechanically sat on the edge of the bed. "Now lay back, Jim. I've got to figure out what's going on here."

"Alright...alright," he muttered.

All was activity around us. The EKG machine arrived, and I helped the nurses attach the leads. An oxygen mask appeared. "Get some nitroglycerin,"[43] I barked to the nurses swarming around. I want it in him as soon as we get this EKG tracing." Looking down I could already see his EKG showed the early signs of a heart attack. "Jim, I need you to take this nitroglycerin under your tongue." Turning to a nurse next to me I quietly said, "Call the admitting resident in the unit. Jim's got to get there ASAP. He's infarcting right in front of us."

Later that evening I visited Jim in the ICU after finishing with my patients on pediatrics.

"I sure made a mess of things, didn't I," he said sheepishly through his oxygen mask.

"Yes, Jim, you did," I agreed, settling into a chair near his bed. We sat in silence for a while.

---

[43] Nitroglycerine is used to relieve chest pain that is due to not enough blood flow to the heart, which is the problem in a heart attack. It's primary mode of action is to relax the veins that bring blood back to the heart. This decreases the amount the heart stretches and also lowers blood pressure, both of which result in the heart needing less energy to contract.

"I didn't mean to be so much trouble. It's the nurses. They…"

"Jim, it's best if you just let it go. You need to rest now."

"You're right. I do get kinda wound up, don't I?"

The silence continued. Eventually I said, "You're in good hands here tonight. I'll check in on you in the morning."

"Thanks, Doc." I placed my hand on his arm, and he let me keep it there for a moment before pulling away.

Five days later, a frantic call came again. "Dr. Sheff, it's Mr. Schmidt. Ever since he transferred out of the unit he's been demanding to go home. Now he's threatening to sign out AMA again."

"I'll be right there."

"You can't stop me this time," Jim said menacingly as I walked into his room. "I'm fed up with this place. I want to be in my own bed. My wife can do everything for me these incompetent nurses are doing."

"You realize you are only five days away from a heart attack. You're still not healed from that, not to mention your leg surgery. And your diabetes is far from any kind of good control."

"I don't care. I'll be fine. And if I'm not, that's my problem, nobody else's. Just sign the damn forms so I can leave."

"You know you're leaving against medical advice. It's my professional opinion that if you leave now, you are at significant risk for complications from your heart attack, including death," I said sternly. "I can't be responsible for any complications you may suffer."

"What do you know anyway? You're only a baby doc. I'll be fine. You'll see." His eyes narrowed, glaring at me as if he had that lead pipe in his hands.

"I'll take care of the paperwork," I bit off the words, doing my best to contain the anger surging up inside. "It will say you take full responsibility for leaving the hospital against my medical advice, and that I've explained to you the serious risks leaving now creates for you, including that you may die."

"If I die, I die. It's better than being in this hell hole of a hospital!"

"You'll need to see me in the office in one week, assuming you live that long."

"Whatever you say, Doc, but I ain't guaranteeing I'll be there…either way," he added darkly.

"The nurses will give you your prescriptions before you leave. Be sure to follow your medication instructions carefully." There's little chance of that, I thought. "Call me immediately if you develop any renewed chest pain, shortness of breath, or swelling of the ankles."

I turned quickly and left. Racing back to my responsibilities piling up on pediatrics, I looked down and noticed my right hand clenched in a fist. Were Jim and I that very different, I wondered. Day after day, night after night, I, too, was building up a reservoir of anger and resentment. It grew larger every time I reached down into that seemingly bottomless pool of will, drawing on it again and again in the face of yet another challenge. Jim had simply provided me with one more. Deliberately I relaxed my right hand, releasing the fist, and cleared my mind to focus on the sick five-month-old baby who needed me on pediatrics. It never occurred to me that my own growing reservoir of pent up anger would someday explode.

*   *   *

"Hurry up! She's seizing again. Get the oxygen," I called out to the nurse in the doorway. Looking down at the young woman, her face contorted, limbs shaking, I hoped this one would be brief like the others. Indeed, as the nurse strapped on the oxygen mask, her flailing limbs fell back to the bed. Christiana lay motionless in that familiar unresponsive state seizure victims fall into immediately following a grand mal seizure.

What is going on here, I wondered. The consulting neurologist and my teaching attending had both concurred she was on the best anti-seizure medications. Her blood levels for these drugs had been right in the target range. So why was she still seizing? At attending rounds we had discussed the most recent review article on refractory seizures and concluded we were doing everything recommended. This being the last month of my internship, I didn't like being stumped by a young woman in her late teens because we couldn't stop her seizures. If we were doing everything right medically, maybe this wasn't a medical problem.

"When a patient doesn't fit into a recognized diagnosis or when normally effective treatment doesn't work," our junior resident, Mark, explained, "this should trigger you to look beyond purely medical diagnoses and consider a psychological or social diagnosis. In this case, we are providing the highest doses of the most effective anti-seizure medications, and she is still seizing. So we have to consider the diagnosis of pseudo-seizures."

In a seizure, brain cells begin firing spontaneously in an intense, local pattern. If this disordered electrical activity spreads throughout the brain, it is called a grand mal seizure. A pseudo-seizure occurs when a patient has all the outward manifestations of a typical seizure, but is not having a true seizure. The patient falls down, shakes, and appears to be unconscious. But if you ran an EEG during a pseudo-seizure, you would see normal electrical activity, not the bizarre and erratic activity of a brain gone haywire.

"I pulled this article on pseudo-seizures," Mark continued. Let's all read it tonight and discuss it on rounds tomorrow. Rick, why don't you order a psychiatry consult as well."

As I read the article that night, despite my fatigue I became increasingly excited. Pseudo-seizures didn't occur simply because a person consciously "faked" a seizure. Instead unconscious psychological processes interacted in complex and unpredictable ways with physical predispositions to produce the pseudo-seizures. Nobody truly understood why these occurred. According to the article, the greatest treatment success came from establishing a strong doctor-patient relationship over time, not from any particular medication or intervention. This complex interplay of the psychosocial and the physical was precisely the kind of challenge that had attracted me to family medicine. I returned to the hospital the next day hopeful that we would finally be able to help Christiana.

"No. No. No. I don't need to talk to some shrink!" Christiana protested, when I told her of our plan for a psychiatric consultation.

"Look, Christiana," I tried to explain, "your seizures aren't getting any better. You had two more last night. We think the seizures may be triggered by something emotional, not just physical."

"So now you think I'm crazy," she replied, hurt in her eyes.

"I didn't say that," though I was beginning to think it. "Why won't you talk to a psychiatrist?"

Something dark came over Christiana's face. After a long pause she replied, "I can't tell you."

We had struck the mother lode. "You can't tell me what?"

After an even longer pause, during which she clearly struggled, she finally said, "I didn't think I could tell anybody. I want to tell you, but I'm afraid." Then suddenly looking desperately into my eyes she added, "You're the only one I can trust."

Like a powerful drug, her words produced in me an exquisite and irresistible high. She was about to open her secret to me and only me. Trust. Vulnerability. A strong doctor-patient relationship. These were the best medicine for her condition, at least if she truly had pseudo-seizures. She was letting me in where she had never let anyone else before. This was exactly what I had hoped to find by choosing family medicine as my specialty, melding the medical and psychological. Now I was living it, and helping Christiana as nobody else could because I stood at this unique interface.

"If I tell you, I won't have to talk to a shrink?" she asked hesitantly.

"I still think it will help for you to talk to a psychiatrist."

"I won't. I won't talk to anybody but you."

"Alright," I gave in, knowing it was best if she at least told someone what was really going on. "If you tell me you won't have to talk to a psychiatrist."

Her voice dropped to a whisper. "If I tell you, they'll kill me."

"Who will kill you?"

"My family."

"Who in your family?"

"My father. My brothers."

"But why? What is it they don't want you to tell anyone?"

She looked around the room furtively, as if to make sure nobody else could hear, then leaned forward and whispered, "They hurt me."

"What do they do to you?" She looked away in silence. "Do they have sex with you?"

"Oh no! Not that."

"Then what do they do?"

"They hit me," she said in a barely audible voice.

"They hit you," I repeated. Then, with firm authority, "You know they have no right to do that. Nobody has a right to hurt you."

"It's easy for you to say that, but it won't stop them. They do it to everybody."

"Who's everybody?"

"My mother. My sisters. Everybody."

"Look, Christiana, this is something we can make stop. I can set it up for you to talk to a social worker..."

"No! No! I already told you I can't talk to anybody. And you can't tell anybody else. If they found out I told you this much, they'd not only kill me, they'd kill you, too."

For a moment I considered she may be telling the truth. My hands felt completely tied. Christiana was a service case, meaning she had no attending physician of her own, so the teaching attending was now her doctor. He was a bright internist, but had no interest in anything that wasn't strictly medical. Once we started talking about the possibility of pseudo-seizures, his only suggestion was to get the psychiatric consultation to deal with the "psychosocial stuff," as he called it. Now I had been entrusted with critically important information, but told I couldn't tell anybody else. The attending wasn't going to be any help. Do I violate her trust and tell Mark or the psychiatrist consultant? I thought of the training I'd had as a therapist before medical school, and now the training I'd had all year in our psychiatry conferences. Perhaps I could truly help her without violating her trust. I decided to try. I would invite her family in so I could talk directly with them, both to get more information and to start to build their trust in me as well.

The only time a family meeting could be arranged was when the men got out of work, which turned out to be an evening after a night I'd been on call. I'd been up for most of the previous 36 hours, so I grabbed one more cup of coffee, a swig of Maalox, and headed to the meeting room. As I walked in, it didn't take long to figure out who the patriarch was. The family members were arranged with all of them facing in a semicircle toward one chair. In it sat a diminutive man. He couldn't have been more than five feet four inches tall. His face was leathery, as if he worked outdoors, and his hands oversized and strong as they shook mine, perhaps a little too firmly. He looked directly into my eyes in a way that made me uneasy. The same happened with his two sons. They introduced their wives, but the women would not meet my eyes. We all sat down. Was I seeing extra makeup around the eye of one of the women hiding a bruise? From this distance I couldn't be sure.

"Thank you all for coming this evening," I began brightly. "As you know, Christiana was admitted several days ago with seizures. I was hoping you could shed some light on what you know about her seizures." I was playing the medical card as my entrée.

"I tell you what we know," her father began in a thick Hispanic accent. "We don' trust you." He said it firmly, leaving no doubt he had already reached his conclusion.

"Why is that?" I asked, taken aback.

"Christiana, she call us. She tell us what is happening here at the hospital. What you are doing to her."

"What is it she is telling you," I asked, becoming alarmed.

"She tell us you keep her against her will. She say she want to come home to her family where she get good care. In here, she say, she afraid and not want to take the medicines you give her. She say they make her sick. She say the medicines make her have those…those…seizures as you call them."

I was shocked. I also suddenly felt out of my depth. How could I have been so blindsided? Why would Christiana feed this misinformation to her family? Why would she tell me just the opposite? I completed the family session, gathering some additional information, but knowing the most important information I needed would have to come from Christiana.

"What else could I tell them," she pleaded with me. "You don't know them. They are suspicious of everybody. I had to show them I was siding with them."

""Frankly, Christiana, I don't know who to believe. You had three more seizures since we spoke yesterday, and I don't know if they are real or just in your head."

"They are real. I'm not doing anything to make myself have seizures, I promise you. You have to believe me. And you have to get me out of there. They…They…They tie me up."

"They do what?"

"They tie me to the bed…with a chain…they chain me to the bed so I cannot go out without one of them."

"They chain you to a bed?" I said incredulously.

"Yes. Yes. But you can't tell them I told you. It would be very bad for you."

"Look, Christiana, I don't know what you expect me to do if you tell me these things but then tie my hands by saying I can't tell anybody else. If you are in danger, if you are being abused, you need to get out of there. You're 19, so you're no longer a minor. I don't have any authority to make them let you leave. Only the police and the courts can do that."

"No. Even they cannot get me out. My family will find me wherever I go. I will never get away. I will have to go back home." Dejection settled across her face.

Later that day I received an emergency page. "Dr. Sheff, Christiana was just found in the bathroom. There's blood everywhere. She's cut her wrists."

"I'll be right there."

"I...I...don't remember anything," she was saying as I walked in.

"What are her vital signs?" I barked.

"Blood pressure 110/60. Pulse 100."

"Draw a STAT CBC to see how much blood she's lost," Then turning to Christiana sympathetically I said, "You really didn't see any way out, did you?"

"I don't know what you are talking about, Dr. Sheff. I don't remember anything."

"You cut your wrists. Do you remember doing that?"

"No," she said looking down at the fresh bandages on her wrists. "I don't remember cutting my wrists. I wouldn't do that. I am Catholic and would never do such a thing..." Her voice trailed off.

It fell to me to notify her family. Now that she had attempted suicide, she needed to be transferred to a psychiatric hospital. Though I was profoundly relieved someone with greater skill, knowledge, and seniority would be taking over her care, I did not relish telling her family. Her brother took my call.

"Dr. Sheff, we are very angry with you. We send my sister to your hospital to get better, and she is worse. Now you say she cut herself. She would never do that. I know my sister, and you must have done something very bad to her. She will never go to a psychiatric hospital. We will take her home."

"Manuel, this is out of my hands. Now that she has shown she is a danger to herself, she has to be admitted to a psychiatric hospital. If she resists, the state says we will have to commit her."

"Dr. Sheff, listen to me carefully," the icy, threatening tone in his voice unmistakable. "Our family will never allow my sister to go to a psychiatric hospital."

"I understand why you feel that way, but I have no choice. The law says we must commit her. If you try to stop us, the police will escort her to the psychiatric hospital."

"I am afraid you do not understand, Dr. Sheff. The police, they cannot stop us. We have a very big family." He paused for emphasis after each of these last three words. Then the phone went dead.

When the ambulance arrived to transfer her to the psychiatric hospital, I called hospital security. "I'm afraid her family may make trouble. They have the potential to get violent. Is there some way we can have a show of force from hospital security?"

"Yeah, I'll be there," came the less-than-reassuring answer.

Her family had gathered in her room. The ambulance crew rolled the stretcher down the hall, followed by one lone hospital security officer, an elderly, thin, white-haired man who carried no weapon. Some show of force. Luckily, her family backed out as the stretcher was rolled into her room. Just then one of her brothers pulled me aside, his face inches from mine. My heart started to race. "Dr. Sheff," he said menacingly, "you made my sister sick. You and me, we have a problem." Then, forcing his face directly into mine he spat out, "I will see you later!"

That night I called the phone company to unlist my phone number. I wasn't sure what good it would do, since Christiana's family knew where I worked. For the next several weeks, I looked warily over my shoulder every time I walked to my car in the hospital parking lot. At home I virtually sprinted from car to front door. Gradually my vigilance waned.

What did not wane was my new found respect for this "psychosocial stuff." I had fallen into a trap. Christiana had convinced me I was the only person she could trust and, hence, the only person who could help her. The experience was like a drug, a seductive yet toxic mixture that played on my passion to heal and an ego hypertrophied by repeated experiences of rising to meet every challenge of medical training. As I neared the end of my internship, I felt I'd not only survived, but that I'd mastered being a doctor. I finally felt competent. I knew I could handle anything medicine threw at me. It was a source of strength, yet it also put me at risk and my patients at risk, but I couldn't see this—until now. Now I was forced to sit with uncertainty. Had Christiana been telling the truth, that she was a victim of domestic abuse? Or had she been delusional, disturbed, and perhaps manipulative. I would never know. Stung with the realization that I'd allowed myself to go beyond my competence, I had to accept that perhaps I had made Christiana worse. I could also see how easy it would be for me to be seduced into going beyond my competence again and again once I was out in practice.

Even though well aware of the power of transference[44] and countertransference[45] from my previous counseling training, I had missed the signs that countertransference had begun to color my interactions with Christiana. I had wanted to rescue her, and this had clouded my judgment. Again, the words of my hypnosis professor came back. "If you are unconscious of the power in the doctor-patient relationship, human nature will lead you to use it for self-serving needs of which you will not even be aware, and this will be at your patients' expense."

I had fallen into a soon-to-be familiar trap of a patient telling me, "You are the only one I can trust. You are the only one who can help." As I talked about this experience in the intern support group and with other mentors and teachers from the residency program faculty, they were able to help me see that, beyond their seductiveness, these words should have served as a red flag warning. They both drew me in and tied my hands from doing what was best for Christiana. I should have required that those with greater expertise be called in. I should also have set limits on what I would hold in confidence, letting Christiana know that I would violate her confidence if I thought her safety or the safety of others was at risk. In the rich, fertile interface between the medical and psychosocial in which Christiana and I were treading, I had allowed my desire to help, and her seductive trust and vulnerability, to manipulate me into doing the wrong thing. Every physician loves the feeling of being the one doctor to help when all others could not. In every physician's career, these moments do happen, and they are deeply rewarding. But such experiences inevitably set us up for the trap of "You're the only one...."

*   *   *

Several weeks later, I emerged from the hospital, my last day of internship just completed. With a cautious glance over my shoulder for Christiana's brothers, I walked slowly to my car, savoring the moment.

---

[44] Transference is a term that refers to a common phenomenon in therapeutic relationships, such as a doctor-patient relationship. Transference occurs when the patient "transfers" to the therapist feelings they have toward another important person in their life, such as a father or mother. These feelings toward the therapist are often far more powerful than would seem appropriate to the present therapist-patient relationship because they are fueled by the history of the previous primary relationship.

[45] Countertransference is term that refers to the reverse phenomenon in which the therapist or doctor feels feelings toward a patient that are "transferred" from a previous primary relationship, making these feelings out of proportion and distorted by the history of the therapist or physician's primary relationship.

Exhausted, spent, I basked in a new found lightness. Internship—that overwhelming, endless challenge of sleep deprivation, work load, performance pressure, and terribly high stakes for failure—was now behind me. I had done well. I had earned the respect of my fellow residents and teachers, expanded my medical knowledge and skills exponentially, and provided compassionate and competent care to many, many patients. Just then Mr. Harrison's face drifted before me with that twinkle in his eye, followed by the searing image of his still, pale figure, lying lifeless after the futile code caused by my error. Somehow in the course of this year I had found a place for even this experience of shattering failure and shame. Another rite of passage, I thought, just as gross anatomy had been, my first night on call as a subintern, and countless other moments. Every day of this past year I had been tested. I'd been pushed repeatedly to find my depth of will, to renew my personal commitment to caring service to others in the face of my own self deprivation. With great satisfaction, I realized both my will and commitment had been found to be enduring. As I turned to look back at the hospital that had served so much as my personal crucible these past 12 months, I knew that these would continue to endure. The strength I'd discovered within myself this year, the capacity to give to others when I thought I had nothing left to give, to continue to offer my competence and caring, would endure no matter what professional or personal challenges lay before me.

In that same moment I was not as aware of the other truth—that the scars would endure as well. The hypertrophy of will, the submerged anger and resentment, the walls I'd built around vulnerability and feelings in order to go on, and a sense of entitlement because of the sacrifices I'd made, these too would endure.

Just as I had during gross anatomy, I again entered unwittingly into a deep, unspoken bond with all other physicians who had gone before me and would come after me. Until we fundamentally change how we train physicians, all of us will continue to carry these same strengths and scars as the legacy of our internship year. Those who depend upon us for their care will be the recipients of both.

Chapter XII

Junior Residency

One by One, the Dying Were Teaching Me
What They So Desperately Needed

A powerful odor assaulted my nostrils as I entered the cramped exam room in the Family Care Center. This particular combination of stale urine and the odor of not having showered for too long felt strangely familiar. On the exam table sat Hank, appearing even more emaciated than when I'd seen him almost exactly a year ago in the surgery clinic. Was his mother wearing that same torn dress?

"Hi, Hank," I said, surprised he had survived his ordeal with prostate cancer. "How are you doing?"

"I can' get da strength," he answered, rocking as he had before.

Turning to his mother, I said, "I'm Dr. Sheff. I don't think we were properly introduced the last time I saw Hank. What is your name?"

"Shirley."

"I presume you're Hank's mom."

"Yeah."

"What can I do for both of you today?"

"They sent us here from emergency," came the resigned reply.

I looked down at Hank's chart, noticing it contained a copy of a recent emergency department visit form. "Please excuse me while I look over his report from the emergency room." It read:

> *29-year-old schizophrenic male with 30 lb weight loss over past year. Workup in surgery clinic, GI clinic, and oncology clinics all non-diagnostic. Now complaining of vomiting and further weight loss.*

At the bottom of the page it concluded:

> *Exam and labs non-specific.*
> *Assessment: Vomiting and weight loss of unknown etiology. No primary care physician.*
> *Plan: Refer to Family Care Center.*

So I was the plan. Specialists in surgery, GI, and oncology couldn't figure out what was causing Hank to lose weight, so they sent him to me. "What happened to Hank's diagnosis of prostate cancer," I asked Shirley.

"Don' know. They done sent us for every test they got. We never did get no answer." I could tell she didn't think I would be of any more help than the dozen or more physicians they'd seen over the past year. "Them records should have it all," she added, pointing to a four-inch thick file on the counter next to the sink.

The nurse who ran our team in the Family Care Center had had the foresight to request that his medical records be sent over for this visit. "Hank, Shirley, I hope you don't mind if I take a few minutes to review what you've already gone through so we don't have to put you through anything unnecessary or repetitive."

"That would be a change," Shirley replied glumly.

"Can' get da strength," Hank added, as he continued to rock on the exam table.

Hank's medical record, like every medical record, was padded thick with every form, lab report, nursing note, order sheet, and physician note recorded at any point in his care. Interspersed among these were the nuggets that allowed me to trace the thread of Hank's year-long ordeal. There was the initial emergency room visit for vague abdominal complaints and weight loss which had led to his referral to the surgery clinic. I read with a sense of irony the surgical intern's note which documented his certainty that Hank was suffering from metastatic prostate cancer, a moment in which I had actually participated. When the workup was negative for prostate cancer, the intern, upon recommendation of his attending surgeon, had referred Hank to the GI clinic. There he underwent the usual battery of GI tests, including a repetition of his myriad blood tests plus upper GI and barium enema X-ray studies. These were followed by upper GI endoscopy and colonoscopy. I cringed at the thought of Hank and Shirley, already marginal in handling basic daily hygiene, struggling through separate bowel preps for these procedures, including the repetitive, explosive, uncontrollable diarrhea the heavy dose laxatives of these preps induced. What bizarre meaning did his already disoriented and hallucinating mind make of these experiences? Finally, when these studies all proved negative, he was referred to the oncology clinic, because such profound weight loss most often occurred as a result of metastatic cancer of some kind. There he underwent a bone marrow biopsy and further tests, many of which repeated for a third time studies ordered by physicians in the previous two clinics. The sum total of findings was persistent mild electrolyte abnormalities consistent with vomiting, but no clear diagnosis. This would have been bad enough, but I found several notations that every time Hank needed blood drawn he went into a confused panic, requiring four people to hold him down simply to draw his blood.

And now Hank and Shirley sat before me, expecting to be treated the same way with the same results. What did I have to offer that experts in surgery, intestinal disorders, and cancer did not? I began with the one thing I felt they needed most.

"I am so sorry you have been through all of this," I said, setting the huge chart aside. "Just going through that many medical visits and tests must have been miserable. But to come out with Hank still sick, you as his mom worried, and still no diagnosis is horrendous."

Shirley looked at me oddly. "You're the first doctor we done seen that said what I been thinking. Like you actually care."

"I'm not sure what else I can offer to you and Hank after his illness has stumped so many excellent physicians, but at the very least I can let you know that I really do care. I'm prepared to serve as his family physician, helping him and you navigate the healthcare system through whatever it takes to try to get you both the help you need."

Shirley started to cry. "I been so scared. I don' know what I'd do if I lost Hank." As much as Hank had been a lifelong drain on Shirley, had kept her from having what others would consider a normal life, she had built her life around the task of caring for him. She wanted nothing more than to continue to do so.

I had offered my empathy. I had committed to stewarding them through the medical maze as his primary care physician. What else could I offer that might help? Just then words from Dr. Davis, the instructor in my medicine rotation, popped into my head. "If you learn nothing else from me this month, I want you to come away knowing that a good history will tell you far more about your patient's diagnosis than any exam finding or laboratory tests."

"Alright, let's begin by going over exactly what Hank is experiencing," I began. "What actually happens when he feels sick?"

"I try to fix him foods I know he likes. I try real hard. But no matter what I make for him, he throws up. And he keeps losing weight. He ain' nothing but skin hanging on bones now. You see him."

"How long after eating does Hank throw up?"

"I don' know. Seems like not too long after."

"Hank," I said turning toward his perch on the exam table, "Sounds like you throw up after eating. Tell me what happens when you eat."

"I can' get da strength."

"Does the food taste good to you?"

"Yeah. Yeah. It taste good." Finally a different answer than something about getting strength.

"So you like eating it?"

"Yeah. Mom, she cook good."

"Then after you eat, what happens?"

"Tell the doctor how you have to throw up," prodded Shirley.

"Where do you throw up," I asked, trying to get as specific and concrete as possible.

"In the bathroom," he replied.

"What happens in the bathroom?"

"I can' get da strength."

"What do you mean you can't get the strength?"

"I can' get da strength. Then I throw up."

"You can't get the strength, then you throw up," I repeated, trying to understand what was going on medically at that moment, what physiologic mechanism triggered Hank's repeated episodes of vomiting. "Tell me exactly what happens once you go into the bathroom."

"I can' get da strength. Then I throw up," he repeated.

"Do you do anything to make yourself throw up?"

"I do this," he said, motioning that he inserted two fingers into the back of his throat.

"You mean you put your fingers down your throat to make yourself throw up?"

"Yeah."

"Why?"

"To get da strength."

Suddenly it made sense. Hank had a psychotic delusional system in which he had the repeated experience of not having enough strength and believing that inducing vomiting would somehow give him back the strength. This was the diagnosis that had escaped all those other specialists.

Before sharing this hypothesis with them, I examined Hank thoroughly to make sure I wasn't missing something else. Terribly thin, with an abdomen that sank inward, he showed the signs of chronic malnutrition, but no other medical conditions. On his rectal exam I once again felt that rock hard mass. This time I forced my examination finger to trace its full contour. I could feel its sides, so I knew it was five to six centimeters across, and it extended upward without ending. A prostate would not do that, I now knew. Instead, I recognized the feel of impacted stool in his rectum, the result of chronic poor nutrition. The pieces all fit.

With an accurate diagnosis, I could formulate a treatment plan. It began with alerting Hank's psychiatrist to the nutritional effects of his delusional thinking. The psychiatrist then adjusted his medications, and we collaborated in attempts to modify his self-induced vomiting behavior. A nutritionist helped Shirley with a series of dietary and cooking changes to maximize the nutritional content of what foods Hank kept down. I prescribed a regimen to reduce his constipation and stool impaction. Hank continued to live in his psychotic delusion in which he struggled to find strength. His vomiting persisted, but at a reduced frequency. He put on about 10 pounds and then leveled off. And Shirley, grateful for an end to her worst fears (at least for now), seemed content to carry on as she had for so many years, making the day-to-day care of Hank the focus of her existence. I saw Hank periodically to coordinate and follow up all aspects of his care and reinforce the behavioral changes that had reduced his vomiting. Neither Hank nor Shirley ever improved their personal hygiene.

Why had so many intelligent, well trained physicians failed to help Hank, I wondered. Why had tens of thousands of dollars been wasted in misdirected diagnostic efforts? And why had caring physicians repeatedly ordered the same blood tests, when every single effort to draw blood literally tortured Hank? Initially I thought these failures were a result of the overspecialization of medicine, resulting in a focus on individual diseases, not the whole patient. Though true, this was only part of the answer. The clinics had been staffed by residents in training, like myself. Because Hank was on Medicaid, with few social and economic resources, he received care from physicians who were learning on him while they provided his care. They were supposed to be supervised by attending physicians, as I was in the Family Care Center, but this supervision varied in quality and intensity. This allowed a surgical intern to misinterpret a physical finding, leading to his chase for the diagnosis of prostate cancer. Contributing to Hank's mismanagement was the over dependence of all his treating physicians, residents and attendings alike, on our increasingly sophisticated and expensive laboratory tests and imaging studies. They had been seduced away from the perennial medical wisdom that listening to the patient was the best diagnostic tool we will ever have.

The final ingredient in this tragedy of errors was each physician's well intentioned focus on finding something they could treat using the medical arsenal at their disposal. Just as with Mrs. Bartlett's oncologist, these physicians appeared to be playing out the maxim that if your only tool is a hammer, everything comes to look to you like a nail. Each of Hank's physicians looked at him through the lens of their specialty, in search of something they personally could do to treat Hank, especially something that would cure him. Ultimately, Hank's condition had no cure. Instead it could only be managed over time. Modest changes in his medications and behavior stabilized his condition, improved the quality of life for Hank and Shirley, and stopped the useless drain of healthcare resources that were being spent in pursuit of an illusory cure for his condition. Hank and Shirley needed caring, not curing.

As a family physician, I was being trained to look at the whole person, to work within the context of the doctor-patient relationship, and to be flexible in moving across the spectrum from cure to caring, selecting the best approach to meet the needs of each patient and family. I had already come to learn that the best medicine will be practiced when every specialty adopts this approach. Hank and Shirley had served as my teachers in this effort, but they should have also served as teachers to the other physicians who had tried to help them. Unfortunately, because of how fragmented our healthcare system is, those physicians never got the chance to learn what Hank and Shirley had to teach them.

*   *   *

"Full moon tonight. Hope you're ready for a long night on call," Mary chided.

"Isn't that full moon thing a myth? I thought studies had debunked it," I replied.

"I don't care what some study showed. I've been working as a nurse in this ICU for 17 years, and I can tell you from my personal experience, more often than not when the full moon's out, this place is jumping. Stuff happens on a full moon you don't see any other time. Let's talk tomorrow morning. You'll see."

Just then my beeper went off with the extension of the emergency department. When the ED paged the junior resident in the ICU, it meant one thing: they had another potential patient for the unit. "We've got Sam Wilson here, a 73-year-old guy with four hours of substernal chest pain and 3 millimeter ST segment elevation[1] in the anterior leads.[2] He's been treated with nitro and morphine and is now pain free. Blood pressure and rhythm are stable," the ED nurse reported.

"Sounds like he's bought a ticket to the unit with an anterior MI.[3] I'll be down to evaluate him as soon as morning rounds are over,"[4] I responded.

---

[1] ST segment refers to a portion of an EKG tracing. As explained in the previous chapter, an EKG is described by referencing specific parts of the tracing, labeled Q wave, R wave, S wave, and T wave. Portions between these parts are referenced by the waves on either end. Hence the ST segment is the portion of the tracing between the S wave and the T wave.

[2] A standard EKG has 12 leads. Each lead provides a view of the electrical activity of the heart from a different perspective. The leads are labeled I, II, III, AVR, AVL, AVF, and V_1 through V_6. Leads V_1 through V_4 "view" the front part of the heart and are referred to as anterior leads.

[3] When the ST segment is elevated, it usually means acute injury to the heart muscle, most often from a heart attack, which is also called a myocardial infarction, or MI. The ST segment elevation in the anterior leads means that this patient is experiencing blockage of a major artery that supplies the anterior portion of the heart muscle. The anterior portion includes a large component of the left ventricle, the part of the heart responsible for pumping blood out to the entire body. An anterior MI or heart attack is usually more serious than a heart attack in other portions of the heart muscle.

As the junior resident in the ICU, my responsibilities not only involved caring for my own patients in the unit, but also evaluating each patient another physician wanted transferred to the ICU from anywhere else in the hospital. It didn't matter if the physician was a fellow resident or a senior attending. When called, I had to determine if the patient was sick enough, whether or not they were likely to benefit from the intensive services available in the unit, and if we had an available bed for them. Sometimes, especially when the ICU was full, it meant making weighty judgments about which patients would get ICU-level care. In other words, I was charged with deciding who would be given the resources for the best chance of surviving and who would not. I could call my senior resident, the cardiology fellow, or the ICU medical director for input, but most of the time it came down to my clinical judgment, the very quality I had admired so much in Skip, the senior resident during my pediatric subinternship. This month as admitting resident for the ICU was intense, on-the-job training which was rapidly honing that clinical judgment.

"Hi, Mr. Wilson. I'm Dr. Sheff, admitting resident for the ICU." I greeted the affable man with a ruddy, round face, thick neck, and full head of white hair. "Sounds like you've had a helluva day already, and it's not even 10 AM."

"Young man, you are absolutely correct. But that morphine your nurses gave me seems to have given a nice glow to the day anyway," he said through a smile that hinted he was more nervous than he wanted to let on. "You can call me Sam," he added warmly. I liked him already. "This heart attack thing isn't too serious, is it, Doc?" There was that word again, Doc. It meant he wanted something from me, in this case reassurance.

"Sam, a lot of people have heart attacks, and the vast majority of them recover just fine." I gave the reassurance he wanted without being inaccurate. It didn't seem to be the time to go into potential complications and dangers of his type of heart attack, unless he specifically asked about them. "We're going to show you our very best hospitality in the ICU to keep that glow of yours going. But first I have to bother you for some medical details. When did your chest pain begin this morning?"

---

[4] My training occurred shortly prior to the development of thrombolytic therapy and routine angioplasty for blocked heart arteries. When I trained, treatment of a heart attack involved oxygen, nitroglycerin, and morphine. These did nothing to open up the clogged artery, but did stabilize the patient. If the patient was stable after this treatment, there was little urgency to transferring them to the ICU. Today, this patient would be treated as rapidly as possible with medication aimed at dissolving the clot that is blocking his artery, called thrombolytic therapy.

The beeper had already gone off three times by the time I finished performing Sam's history and physical. I wrote some quick admitting orders to the unit and hurried back upstairs.

"Ms. Martinez in bed 17 just bottomed out her pressure. What do you want us to do?" Mary greeted me as I came through the unit doors.

"Hang dopamine at 8 mics per minute and titrate the dose to keep her blood pressure over 90 systolic."

"I thought you'd say that. So I've already got the dopamine mixed and ready to go." Mary was the kind of nurse who made any resident better, if only they had the humility to let her.

Just then my beeper went off. "Code 99 room 347, Code 99 room 347."

"Hold the fort down 'til I get back, Mary," I called out over my shoulder as I sprinted through the ICU doors. Another responsibility of the second year resident in the unit was to head up the code team, regardless of where in the hospital a code occurred. Moments later I dashed into room 347. "What do we have here?"

"Eighty-four-year-old, status post right hip fracture, pinned three days ago, now suddenly unresponsive. No pulse. No blood pressure."

A thin elderly woman laid completely naked on the bed, a nurse by her side pumping on her chest and a respiratory therapist pushing air in and out of her lungs with a bag and mask. "She's a full code?" I asked before we did much more.

"Yup. She was healthy, living alone at home before she fell and broke her hip."

"She's in V Tach,"[5] I called out in the authoritative tone of the team leader, looking down at the EKG strip running continuously off the machine in front of me. "Hold the pumping. Any pulse with this rhythm?" I pressed firmly in her groin, feeling for a femoral pulse. "No pulse," I announced. "Resume pumping and prepare to shock."[6] By then an intern had arrived and was charging the paddles.

"Everybody off!" called the intern.

Her frail body convulsed with the flood of electric current. "She's deteriorated to V Fib," I informed the room. "Load her 75 of lidocaine, start her drip at two, give her an amp[7] of epi and an amp of bicarb and get ready to shock her again. Where's anesthesia?"

---

[5] See footnote 21 on page 244 for an explanation of discussion of V Tach and V Fib.
[6] Many people are familiar with TV and movie depictions of a cardiac arrest in which the patient is shocked. This technique applies a predetermined amount of electric current through the heart. The goal is to cause all the cells of the heart to react electrically or depolarize at the same time. If this happens, it gives the best chance for the body's natural pacemaker to take over and restart a healthy rhythm.
[7] Amp is an abbreviation for ampule. An ampule contains a standardized, preloaded amount of a medication.

"They called from labor and delivery. They've got to finish an epidural and won't be here for about 10 minutes."

"We need to tube her," I said moving to the head of the bed. The respiratory therapist pulled over the code cart, opening the intubation drawer. I grabbed an adult size curved laryngoscope and a number seven endotrachial tube. "Give her four good breaths for me," I asked the respiratory therapist. He did so and then pulled the mask aside. I repositioned her head and neck as I'd been taught, opened her mouth, slid in the blade of the laryngoscope. As I lifted straight up, being careful to avoid pressure on her few remaining teeth, her vocal cords popped into view. I watched as the tip of the tube passed cleanly between them. "I think it's in. Give her some puffs." With relief I heard breath sounds in each side of her chest. "Shock her again."

"Everybody off!" the intern called. Again her body convulsed.

"I've got a slow sinus rhythm!" I called out, excited for the first time. "Give her 0.5 of atropine.[8] Hey guys, she might actually make it." This was a rare one. Less than one in five patients that coded in the hospital survived the code. Less than one in 10 made it to discharge, and many who did were neurologically damaged. "Call the unit and tell them she's coming up. Let's just hope she didn't knock off too many brain cells during her period of poor blood perfusion to her brain." She was still unconscious, so we'd have to wait to find out.

I'd been back in the unit for only a few minutes when the beeper went off again. "Dr. Sheff, you remember Dorothy Johnson, the real nice lady with the MI you sent us two days ago. You know, Dr. Patel's patient. She's developed renewed chest pain. The intern's with her now, but it looks like she's infarcting again."[9]

"Shit," I said under my breath. I had grown to like Dorothy a lot during her stay in the unit. "We've only got two more beds in the unit, but it looks like she's just bought one of them. I'll be right down."

Minutes later I entered her room. Amidst the commotion I went straight to her bedside and took her hand. "Hi, Dorothy," I said as reassuringly as I could.

"Oh, Dr. Sheff, I'm sorry to be such a bother," she said through her oxygen mask.

---

[8] Atropine is a medication that increases heart rate.

[9] Infarcting again is a way of saying she was having another heart attack, another myocardial infarction. It means she had developed a blockage in an additional coronary artery and was damaging more of her heart muscle.

"No bother. If you missed me so much, there were other ways you could have visited me in the ICU without developing chest pain." She chuckled. It helped the fear I could see in her eyes.

Back on the unit, things quieted down for a while, so I could catch up. I had to do the history and physical for Sam and the transfer acceptance note for Dorothy. My five other patients needed attention as well. It was mid afternoon when the code buzzer went off again. "Code 99 ICU! Code 99 ICU!" I was already standing in the middle of the ICU. Where the hell was the code? Suddenly I saw nurses running toward one of the beds at the end of the unit. I ran too, with a sickening realization that they were all headed to Sam's room.

"What the hell is going on?" I demanded.

"He was talking to me, then suddenly went into V Fib and passed out," the nurse pumping on his chest yelled over her shoulder. Respiratory therapy was already setting up the bag and mask at the head of the bed. Another nurse had pulled in the defibrillator and was charging up the paddles.

"Shock him as soon as the paddles are charged," I ordered.

"Stand back!" the nurse called out.

He convulsed. I looked anxiously at the monitor. "Sinus rhythm!" I announced with relief. "Have we got a pulse with this?"

"Good pulse here," called out another nurse.

"What...what's happening," Sam muttered to everyone's relief.

"Sam, you gave us a bit of a surprise here," I said, leaning over his face which was now being fitted with an oxygen mask. Then to the nurses, "Had he thrown any PVCs prior to this?"

"Not to speak of."

"Bolus him with 100 of lidocaine and start his drip at 3. Let me know if he shows any more arrhythmias or PVCs." Then turning back to Sam, I asked, "Did you feel any chest pain before this little episode?"

"Not that I can recall," he said confused. As his mind cleared he added, "I was thoroughly enjoying that great hospitality you had promised. Mary was being real nice and all. Next thing I remember is looking up at you. Don't take it personally, Doc, but you're no match for Mary when it comes to looks," he said, trying to smile to cover the dawning realization that his heart attack was more serious than he'd been trying to make it out to be. Then in a more hushed tone he asked, "What's the low down on what I got, Doc?" Now he was ready for the straight story.

"You've had a sizable heart attack," I acknowledged. With a pad of paper I sketched out a diagram of a heart and the three main blood vessels supplying it. I made it clear he had blocked off the largest one. I drew a close up of a blocked coronary artery, describing the damage to his heart muscle downstream from such a blockage. Lastly, I explained the healing process and the potential complications he might face during that time, especially the rhythm problem he'd just had.

"Doc, I got lots to live for. I may joke around, but I got a great wife. My kids are great too. And the grandkids, they're something special." He was suddenly serious. "I need you to take good care of me. I'm not ready to check out, you know."

"I'll do my damndest, me and the whole team. You're a good man, and I like you." I looked directly into his eyes. They were clear, with a hint of tears at the edge. "I'll do everything I can to help you get through this."

"I'm counting on you, Doc," he said in a hushed whisper.

At sign out rounds, the cardiology fellow went over Sam's EKG strips very carefully with the team. "See this here. He's throwing an increasing number of PVCs, and their timing is concerning, close to the R on T phenomenon that can trigger V Fib again. You better switch him to procainamide. I don't think the lidocaine is holding him, not with the size of his anterior infarct."

"Will do," I said, writing the new orders and carefully remembering to flag them for the nurses.

After sign out rounds, I was kept plenty busy fine tuning patient after patient throughout the unit. It seemed everybody needed something that night. I barely had time to get to the cafeteria before it closed, bringing my tray back to the unit and taking a bite or two whenever I could. Around 11 PM, I got another call from the ED for a possible admission.

"What do we have here?" I asked, approaching the ED physician.

"This is a judgment call. Got an 81year-old lady here who's damn sick. Septic, probably from a UTI.[10] Problem is she looks like she's been badly neglected. Emaciated, dehydrated. She's been living with her daughter, but I can't believe the condition she's been allowed to deteriorate into. Her protein stores are probably about zero, so she's gonna have a helluva time fighting off any infection, let alone sepsis like this. It's your call whether she goes to the unit for everything we've got or to a regular medical floor for just antibiotics and fluids."

"She a code?"[11] I asked.

---

[10] UTI stands for urinary tract infection. This can be a simple bladder infection or a serious infection involving the kidneys as well. It is a common problem in the elderly that can lead to infection spreading in the bloodstream.

[11] Being a code meant that a patient warranted receiving all the heroic measures we were capable of providing if their heart stopped or their breathing stopped. It was an abbreviation of Code 99, the announcement used in this hospital to signify that somebody needed the services of the code team for a life-threatening complication, usually a cardiac or respiratory arrest.

"Haven't discussed it with the family. Left that up to you."

This was going to be a tough call, especially for the last bed in the ICU tonight. In the moments it took to walk from the desk in the ED to her treatment room, I quickly organized my thoughts to help her family work through the decisions before them, critical decisions that had to be made under great pressure in the next few minutes. She was clearly sick enough to deserve care in the unit, but after so much neglect, and with severe malnutrition, she was at risk for a long and stormy ICU course. This, I knew, was a euphemism for the horrific ordeal I'd seen other patients suffer. Tied down to a bed, tubes thrust into every orifice, lying helpless in the face of repeated needle sticks and suction catheters probing deep into her throat, triggering convulsive gagging, her skin breaking down into gaping, oozing sores—these were the ministerings of well intentioned doctors and nurses in our ICU. Mercifully we'd sedate these poor patients to lessen the pain and disorientation. But all too often the net result was a mocking distortion of nature's plan, prolonging the dying process while multiplying the patient's suffering many fold. And who paid for this futile care? All of us. Daily I was saddened to watch tens and even hundreds of thousands of dollars spent on these worn and broken bodies trying desperately to give up the ghost, but we wouldn't let them. With every order, with every day spent in the ICU, they consumed resources that could be spent so much more fruitfully on prenatal care, immunizations for children, and countless other needed services that provided far more proven benefit. America maintains the most expensive healthcare system in the world, 40% more per capita than our nearest competitor. Yet we are no healthier than other Western countries. Thirty-one countries have a longer life expectancy. Thirty-three have a lower infant mortality rate.[12] And our citizens are far less satisfied with their healthcare than residents in many of those other countries.

My thoughts were interrupted by the sight of a gaunt, bone dry, elderly woman lying on a stretcher, her sunken eyes staring blankly at the ceiling. Her breathing was shallow and rapid. Her skin pulled tightly over bones that pressed unhappily into the stretcher supporting her rigidly distorted frame. At her side stood a well dressed woman in her sixties, rubbing her hand. "Mom, don' die. Mom, please don' die," she repeated over and over again.

"Excuse me. I'm Dr. Sheff, the admitting resident for the intensive care unit. I'm here to evaluate your mother. How long has she been sick?"

"She was fine. She was fine until tonight." I found this hard to believe. "She just didn't seem herself tonight, so I brought her here."

I could tell I wasn't going to get much helpful history from her. "Could you please excuse us so I can examine her?" The daughter stepped out.

---

[12] Based upon US Census Bureau's International Database 2000.

After a quick exam, I completely agreed with the emergency physician's assessment. I went to find her daughter.

"Your mother is gravely ill from a very serious infection," I began. "She looks like she's been ill for some time," I added, looking her daughter in the eyes. Something in those eyes told me I was right. She knew she hadn't taken adequate care of her mother for a long time. I paused, wondering how to convey the full complexity of the decision she was about to make. "I'm afraid your mother is very close to dying." Her daughter let out a sound, a mixture between a gasp and a sigh. "We have lots of technology we can use to try to treat her. She will need intravenous fluids, antibiotics, medications to support her blood pressure, and she'll probably need to be placed on a breathing machine. We'll have to get some nutrition into her, because without nutrition she won't be able to fight off this infection. Even if we did all this, she has only a small chance of surviving, though it is still a chance. We would make her terribly uncomfortable in the process, and it's not clear to me she is mentally with it enough to understand why we were doing all of this to her. She may have suffered some brain damage already and is at risk of more brain damage, even if she does recover from her infection." I paused, Mrs. Bartlett's sad eyes floated before me, pleading to be allowed to die from her cancer. "The other choice we have is to treat her, by that I mean treat her aggressively with antibiotics and intravenous fluids, but not to use all the invasive technology at our disposal. I'm concerned that using all the technology we have will only prolong her suffering with no guarantee it will help her in the long run." I paused to see if she was taking any of this in. The daughter's eyes just stared blankly ahead. "Did she ever express a preference for what she would want done in a situation like this?"

"No. We never discussed it," she said, wringing her hands. An avoidable tragedy, I sighed. A family discussion of such questions in advance would have provided so much insight and relief to this impossible moment.

"Are there any other family members who need to be brought into this discussion?"

"Dad's been long gone. I'm afraid I'm her only daughter. I had a sister who died just over a year ago, one she had lived with for years. Since then she's lived with me." About a year of neglect is what I thought I was looking at.

"Well, without her previously expressed wishes to guide us, we have to decide what to do, and we have to decide it pretty quickly. I'm afraid we don't have much time."

In that moment I realized the two of us stood at the nexus of a primary cause of the runaway costs, disappointing results, and poor satisfaction that defined our American healthcare system. With all our technology and wealth, there was so much we *could* do for her mother, but it wasn't clear we *should* do all of it. Should we commit her to suffer in our ICU, to consume vast amounts of precious healthcare resources, to hover suspended on the edge of death for some undetermined time? And for what? For the terribly slim chance she might somehow survive, with or without her brain intact. Yes, that was it. It was a matter of chance, of an uncertain future. There truly existed a small chance she might survive all we would do to her, that she might return to some semblance of life as she knew it before this illness. Who was I, just because I was a physician, to deprive her of this chance, no matter how small? Was I to sit in judgment of who deserved a shot at this slim chance, even if it meant draining valuable resources from others in need? No, this was a decision, the ethicists now told us, that should be left to the individual patient. But her voice was silent, unable to tell us her choice in this all important moment. Neither had she communicated her wishes in advance to her family. So we were forced to place this intolerable burden on her next of kin, on her daughter, the very same daughter who had neglected her for so long, yet clearly loved her nonetheless. All the ambivalences of their mother-daughter relationship danced in the air.

Unfortunately, they did not dance long enough, for I could tell this woman had no tolerance for the nuances, the subtleties of the decision weighing on her. She could not see that, if her mother truly was so close to death, the most loving decision might be to admit her to a regular medical floor where we could limit our interventions to antibiotics and fluids. If she turned around with this treatment, we all would have done a good thing. If not, if she died from this serious illness at the age of 81, she would have passed on in relative peace after living a full life. Given what I'd already seen in the ICU, this seemed the more humane, the more loving of the options before us.

But this was not to be. "You have to do everything. You just have to do everything," her daughter blurted out.

"Even though we can do everything, I'm not sure…"

"I love my mother. You said there was a chance she could recover. If there's any chance, any chance at all, you must do everything you can." She said this with true earnestness and concern for her mother, though it seemed disingenuous after months of neglect. The subtleties that should go into this decision—the statistical chances of recovery, this woman's guilt over neglecting her mother, her ambivalence at the continued burden her mother would cause if she recovered, the other possible uses of the financial and medical resources her care would consume, and the torturous nature of the treatments we were about to commit her to undergo—these washed over this distraught daughter as she clung to my words that there was a chance her mother could recover, no matter how slim that chance was. With a sigh I reconciled myself to where the conversation was headed. This daughter was her next of kin, and we had to do whatever she decided or pursue a court order to do otherwise.

"What if her heart were to stop or her breathing stop? What if she were to die? Would you want us to shock her, to put her on a breathing machine, and to pump on her chest?" I could almost feel those brittle ribs cracking under my hands from the chest compressions. "Or would you want us to keep her comfortable and allow her to die in piece and dignity, with as much comfort as possible?" I was trying to slant the decision the best I could.

"I said do everything," she responded indignantly, "and that means everything. That's what she deserves."

"I'm not sure you understand the ordeal you are committing her to..."

"You asked me what should be done, and I've told you." Her eyes were less resolute than her words.

"So you want us to do everything, using every intensive care resource we can?"

"Yes." My hands were tied. I had done my best for her, but now I had to do my best to her.

"By the way, what is your mother's name?"

"Gloria. Gloria Sackett."

With a sickening feeling in my stomach I returned to her treatment room. "We're doing it all to poor Gloria," I announced to the nurses. "Let's tube her. Call surgery to place a central line. Put in the Foley catheter. Go with amp and gent. Let's start her IV at 125 an hour and watch her closely to see how she tolerates this. Her body hasn't seen that much liquid in a long time."

Back in the unit I stopped in to see Sam. He was resting comfortably. "Hi, Dr. Sheff. How's your night going?"

"Busy so far. Thanks for asking." I looked warily at his heart rhythm on the monitor. He was still throwing a few PVCs, but nothing alarming. "Any more chest pains?" I asked, laying my hand warmly on his shoulder.

"None whatsoever. You folks are doing a great job. I'm in good hands." He smiled, seeming to enjoy the touch of my hand on his shoulder.

"Good. I'll check in on you later."

"You do that," he said, pleased by the extra attention.

Moments after I left Sam's room, the code buzzer went off again. "Code 99 ICU! Code 99 ICU!" I froze, looking warily around for the source of the trouble. Nurses again began racing toward Sam's room, as did I.

"He's in V Fib again," called out one of the nurses.

"Damn it! Let's shock him," I ordered. He convulsed but remained in V Fib. "A round of epi and bicarb. Then shock him again. How 'bout his lytes?"

"Here's the latest set from two hours ago." They seemed OK.

"I've got to draw a blood gas."[13] I felt for the arterial pulse in his groin with each chest compression. Inserting a needle directly into the pulsating femoral artery, I was relieved to see the blood come out fairly bright red, as it should from an artery. "Send that off for STAT analysis. Prepare to shock him again." Another convulsion followed. "The V Fib is coarser now. Another round of epi and bicarb. We're not stopping until we've done everything." A few minutes later the blood gas came back. "His pH is 7.22 and his bicarb 19,"[14] I called out. "His oxygen's 90[15] and his CO_2 is high. Give him an extra bolus of bicarb. If he's too acidotic he won't be able to get into a healthy rhythm with the shocks. Once the extra bicarb has circulated for a minute, shock him again. The bag and mask aren't ventilating him adequately. Looks like I'd better tube him as well to blow off that extra CO_2 and keep his oxygen level optimized."

His thick neck was difficult to extend. It took two tries to get the tube in, but eventually we confirmed it was in his trachea. "Shock him again." His thick limbs jumped off the bed. I anxiously waited as the monitor recovered from the shock to see what his heart was doing. "Back in normal sinus rhythm!" I shouted.

"Got a pulse," one of the nurses added. "A good one."

"Sam, Sam," I approached the head of the bed with a mixture of concern and relief, "Sam can you hear me?" Bewildered, Sam's eyes wandered, unable to find the source of the words. His jaw chewed on the tube that made it impossible for him to speak. "Sam, you had another close one," I said. "But you came out OK." He grunted. His eyes finally came to rest on my face. This time they were pleading. "Sam," I said looking directly into his eyes. "You had another episode, a serious one. We were able to help your heart again." His mouth attempted to form the words "Thank you." I could feel his gratitude, but also his fear. I had never felt anyone so entrust his very life to me. "I know, Sam. You're in good hands, and I'm gonna do everything I can to keep it that way." I had the nurses page the cardiology fellow who made some recommendations to adjust his arrhythmia medications, which I ordered.

---

[13] Please see footnotes 10 and 11 on page 195 for a discussion of what an arterial blood gas is and what the pO_2, pCO_2, and pH mean.

[14] Normal pH is 7.4. It is usually controlled in a remarkably tight range from about 7.36 to 7.44. Normal bicarbonate is around 24. In this situation, the patient had accumulated too much acid from poor blood circulation during the code, making his pH drop. He also had accumulated too much CO_2 because we couldn't blow enough air in and out of his lungs with the bag and mask. The CO_2 converted into acid and lowered the pH further.

[15] The normal oxygen level on an arterial blood gas, also called a partial pressure of oxygen, is 100. Levels above 80 are acceptable, and no benefit is gained by raising the level over 100.

The rest of the night continued the same pace. Of the 21 patients in the ICU, almost every one needed attention at some time. First one nurse would call about a patient's change in vital signs, then an arterial line would clog and need to be replaced on another, then an alarming blood test result would come back on a third. As I stood at one bedside, I noticed the setting full moon out one of the windows. Just then I was called to yet another bedside. I felt very much like a ball in a pinball machine, bouncing from one bed to the next all night long. In the face of this, I adopted an unflappable tone and gallows humor. Bring on the next challenge. I was at the peak of my competence. Nothing could faze me, or so I thought.

A hint of light appeared on the horizon. This night of the full moon was almost over. Just then the code buzzer went off again. I didn't even wait for the overhead announcement. I sprinted to Sam's room. The nurse was already pumping on his chest. The familiar chaotic pattern of V Fib ran across his monitor. "Paddles!" I yelled. We shocked him. Still V Fib. "Epi and bicarb." Another shock. "Give me a blood gas syringe." Out came the blood, not quite as red and more concerning. "Another round of epi and bicarb!" My composure was fraying. We had been coding Sam for 30 minutes, but his rhythm remained in V Fib, deteriorating minute by minute. "Get me a second blood gas to see where we are." Ten more minutes went by as we continued pumping on him, injecting him, shocking him. His rhythm had become flat line. A nurse handed me the latest blood gas result. pH 6.93, oxygen 180, CO_2 32, bicarb 8. He was fatally acidotic. No wonder we couldn't get his heart to beat again. In desperation I ordered three more amps of bicarb and ventilated him faster to blow off more CO_2 to try to raise his pH. He remained flat line. With more sadness than I had felt in a very long time, I asked, "Does anybody have any suggestions?" Silence. "Any objections to calling the code?" Silence again. "The code is called."

Everyone backed away from Sam. I approached the head of the bed, laying my hand on his forehead. Silently I apologized for not being able to do more for him.

Trudging slowly to the nurse's station to begin the paperwork and the call to his family I dreaded, I suddenly felt exhausted. It was an exhaustion beyond the long night's work. I had the distinct impression that I had spent most of the previous 24 hours wrestling, wrestling for Sam's life, wrestling with Death. I had never before felt Death's touch so directly. With the distortion and simultaneous clarity a sleepless night induces, I sensed that the Angel of Death had hovered over Sam since his heart attack began. He had been beside both of us in the emergency room, but our orders and skillful nursing care held him at bay. He had tried to take Sam with his first episode of V Fib, but again we beat him back. I felt I'd personally wrenched him out of Death's grip with his second episode of V Fib. But now I had the powerful sensation of having wrestled hand to hand with the Angel of Death during Sam's long, last code, and that I was no match. In spite of my best efforts, in spite of everything medicine had to offer, Death claimed Sam. It was the natural order of things.

As the rich glow of sunrise filtered into the ICU, Sam's death forced me to accept what Gloria's daughter could not. There was a natural order of things, and it included death. Her daughter clung to the hope that death could somehow be beaten back, and with it the false hope that if it could be beaten back for just a little while, perhaps it could be beaten back forever. This delusion, shared by so many in America, fueled countless, futile efforts at the end of life. While I had already learned there certainly were times to do all we could to intervene in illness and heal when healing was possible, I was slowly learning there was a time to shift our focus from fighting back death to helping patients die well. I had been trained to *do* and *do* and *do* ever more for those who came under my care, even those whose natural life had come to an end. When would I be trained in the wisdom, compassion, and acceptance to stop all this doing and simply *be* with those who are dying? With a tired smile I realized that, one by one, the dying were teaching me what they so desperately needed.

*   *   *

"Gloria Sackett is an 81-year-old female who presented last night with urosepsis, shock, dehydration, malnutrition, respiratory failure, and coma." I recited on rounds that morning. "Problem number one: sepsis." As I glanced down at Gloria, I noticed she already looked different. Her blood pressure had hovered dangerously low all night. I had increased her IV rate from 125 to 300 cc per hour, and she had required pressors to support her blood pressure. The bone dry quality of her skin had started to give way to a doughy softness as fluid poured into it, a problem called third spacing. One space is considered the area inside her blood vessels, and another space the area inside the individual cells of her body. The third space, as it is called, is the area between and around the cells. Proteins in the bloodstream and cells are critical for holding enough fluid within them. Since she was so malnourished, her protein levels were too low to hold much fluid, so most of the fluids we poured into her veins leaked out into this third space and didn't get into her cells. Her serious infection made this worse.

By the sixth day, she was not recognizable. Her face and body had blown up like a balloon, and her paper thin skin stretched to the point of tearing. We had tried to feed her through a nasogastric tube into her stomach, but her digestive tract was not prepared to receive nutrition. Even the liquid nutrition we gave her ran right through, producing copious diarrhea. The diarrhea poured out so fast the nurses could not keep up with cleaning her, so her torn, raw flesh lay in her own excrement despite our best efforts. As the nurses rolled her to pull away the soiled sheets, I saw the green, putrid liquid clinging to broad patches of denuded skin. I winced with the knowledge that through our attempts to treat her, we were pouring her own stool on open skin sores. We replaced the tube feedings with intravenous feedings, an expensive undertaking that required frequent blood tests and changes to her regimen. Yet obtaining blood specimens from her became increasingly difficult and painful, as her skin swelled to the point of making her veins impossible to find.

Unfortunately our antibiotics had started to work, so she was no longer in danger of dying from her blood and urinary tract infections, at least not immediately. We had managed to stabilize her enough so we would have to continue to treat her. She awoke from her coma just enough to moan in pain with every needle stick, every dressing change, every attempt to move her grossly distorted body. None of this surprised me. It is the tragedy I foresaw in my brief but critical discussion with her daughter. I estimated Gloria would linger like this for weeks before she would finally die a tortured death. A hundred thousand dollars of expenses and three weeks later, Gloria mercifully gave up.

Meanwhile Dorothy Johnson initially improved slowly. Since she had suffered a second heart attack, more of her heart muscle was damaged, and she slipped into mild congestive heart failure. Upon further evaluation we found she had diffuse hardening of the arteries from diabetes and high blood pressure that had not been well controlled for many years. There was not much we could do to reverse the many blockages in her arteries. Yet she was cheerful, as were her four children, a son from New Hampshire and three daughters who lived nearby. I grew close to all of them.

One morning Mary came to find me. "Dorothy says she isn't feeling well."

"What's going on, Dorothy?" I asked lightly as I walked into her room. At a glance I could tell she was in trouble. Pale and sweaty, she could barely catch her breath. "Dorothy, are you having any chest pain?"

"Dr. Sheff, I didn't want it to happen again…"

"Get a STAT EKG. Bump her oxygen to 100 percent rebreather mask. Give her nitro sublingually right now, and get Dr. Patel on the phone immediately."

"I can't breathe, Dr. Sheff. I can't breathe."

"Page anesthesia for a STAT intubation and get the code cart. I don't know if we have time to wait for anesthesia to get here, so let's set up for a crash intubation."

Anesthesia did arrive in time, for which I was thankful because I didn't look forward to intubating Dorothy while she was awake. Leave that to a trained anesthesiologist, I thought. Watching her panicked eyes dart from me to the anesthesiologist as he coaxed the endotrachial tube past her vocal cords was hard enough. She looked at me imploringly, but could say nothing because of the tube.

"Blood pressure's down to 70," Mary called out.

"Hang dopamine starting at 8 mics and titrate it to a BP of 90 systolic."

Several hours later Dr. Patel, her cardiologist, came in. "She's infracted one more time. We're barely able to get her BP over 80 with the dopamine," I informed him.

"It looks like her heart's shot. Given the damage she's sustained, it can't pump any harder or faster. Switch her to norepinephrine, the strongest alpha agonist[16] we have. It will clamp down her blood vessels and give the best shot at supporting her blood pressure so she can perfuse her brain and kidneys. That's what she needs to stay alive long enough to see if her heart will recover from this latest insult," he said matter-of-factly. Then going into Dorothy, he patted her on the arm and said, "We're changing your medicines. I'm sure you'll do fine." She grunted around the tube.

Dorothy's eyes implored me, while her mouth worked around the tube suddenly filling her throat, forcing air in and out of her weary lungs. She strained to say something I could not understand. Frustrated, she finally gave up. When her daughters arrived, she again tried in vain to speak. In desperation we gave her a pad, but she couldn't focus. The very act of raising her hand to write exhausted her. Finally we made out the scrawled words, "I love you all," the last words she was able to communicate.

For four days after her latest heart attack, Dorothy hovered in this state. The norepinephrine sustained her blood pressure just enough to keep her alive, but at a cost. The blood vessels to her extremities clamped down so hard that her hands and feet turned black. The tips of her fingers and toes began to slough off, leaving desiccated stumps, no longer able to feel loving touch. Yet mentally, she remained completely alert and increasingly frustrated at not being able to communicate with her family.

On the fifth day, her blood pressure started to slip. Unlike most physicians, Dr. Patel knew when to stop. "No reason to keep pushing," he concluded. "Wean the pressors and let her go." He turned to leave the unit.

---

[16] All blood vessels have two classes of receptors called alpha and beta. Receptors are sites that bind to chemical messengers in the bloodstream and react to these chemicals. When stimulated, both alpha and beta receptors cause blood vessels to constrict or tighten. Beta receptors on the heart also cause the heart to pump harder and faster. Agonist refers to something that stimulates a receptor. So an alpha agonist stimulates the alpha receptors on the blood vessels, causing them to constrict. Dopamine is a mixed alpha and beta agonist, while norepinephrine is a pure alpha agonist. So dopamine stimulates the heart as well as tightening the blood vessels. Norepinephrine only tightens the blood vessels.

Though this seemed the most humane thing to do, something didn't feel right. It did mean we would stop pushing to keep her alive, stop goading her frail heart to keep going when it so clearly wanted to finally rest. We would sedate her, and she would die peacefully. Wasn't this the kind of compassionate death in which I was coming to believe? What was nagging at me? In response, the question I'd formulated more than a year ago early in my internship came back. At this moment, what is the most loving thing to do? I knew what was wrong.

"Dr. Patel," I chased after him, "she's been trying for days to communicate with her family. Since she is going to die anyway, would it be possible to extubate[17] her so she can talk to her family before she dies?" This was a bold suggestion coming from a resident still in training to an attending cardiologist, but I knew it was what she would want, what she needed to die well.

"Do what you like," he said dismissively.

With her children's quick agreement, this became the plan. One by one her daughters joined together at her bedside. Dorothy's blood pressure continued to slip, and it wasn't clear she would survive until her son's arrival. "Bump the norepinephrine and add dobutamine," I instructed Mary, who in all her years in the ICU had never seen anything like this, never seen a patient extubated knowing she would only have minutes to say goodbye before dying.

Finally her son, Bill, arrived. As the children stood by, I leaned over her. "Dorothy," I began slowly, "you know you've had another heart attack. Your heart is now too weak to keep going." Her eyes widened, but she understood. Again she tried desperately to say something around the tube. "Dorothy, I know how hard you've been trying to communicate with your children." She nodded. "They want to hear what you have to say. They also want a chance to say goodbye. If it is alright with you, I will remove this tube so you can say whatever you want to each of your children." Again she nodded, a look of deep gratitude coming over her face. We both knew she was dying. With astounding speed of insight, she accepted this shattering truth, and in an act of exceptional courage wanted nothing more than to share a few loving words with her children in whatever time she had. This was in my power to grant her, a power thrust into my hands as a physician, just as my hypnosis professor had predicted. Thankful to be able to use this power to help her as I sensed she so desperately wanted, I removed the tube and stepped back.

---

[17] Extubation is the act of removing the endotrachial tube from a patient who has previously been intubated.

First Marie approached and knelt next to her mother's head. With soft eyes and quiet words, Dorothy said something only she could hear. Marie let out a moan and buried her head on her mother's shoulder, pouring out what needed to be said for mother and daughter to part in peace. Each sister took her turn. Then Bill knelt next to her. With her last lucid moments, mother and son exchanged final words. Dorothy's eyes closed as she seemed to drift off to sleep. All eyes turned to the monitor over her bed which captured each slowing heartbeat. Her chest rose and fell. Then, in great peace, her heart simply stopped.

I hugged each member of the family and left them to their grief. I knew with certainty that I had just done something important. I did not know at the time that allowing Dorothy and her family these precious minutes to say what needed to be said and then to say goodbye would remain one of the finest things I would ever do in my career as a physician.

*   *   *

Amidst the late night fights and stressful nights on call, Marsha and I married. Stealing 10 days, we had a beautiful wedding and flew off to Bermuda for the honeymoon, a much needed oasis to reconnect and commit to our future together. Upon arriving, we found the beach on which we thought we'd reserved a cozy bungalow had been washed away in a hurricane 10 years earlier. The brochure just forgot to mention this fact, or change its picture. The October weather in Bermuda was cold, damp, and windy. After three days, we cancelled the rest of our stay and returned home. There, in our cozy apartment, we did the best we could to nurture our love and prepare to return to the lives that strained it so.

*   *   *

"Angela, how do you feel about your mom having another baby?" The six-year-old looked down at the floor in uncomfortable silence, every now and then stealing a glance at the video camera. "How about you, Jolina?" I asked her four-year-old sister brightly. Another uncomfortable silence followed. Though I saw both of these girls for their regular medical care, during which we joked together easily, today something else was going on, and they didn't want to talk about it. "Well, Maria, Tom," I said, turning to their mother who had come to me for prenatal care and her boyfriend, "I can see why you thought we all needed to talk about this baby before it arrives." I wasn't sure if it had to do with the coming baby, or something about Tom. I sensed they feared him. For the first time I considered he may even be molesting them. Or was it just that they worried about losing their mother's attention and affection to Jim and a new baby. In either case, this family counseling session was clearly not the setting in which to get at that information. It would best come out in my ongoing pediatric care for the girls and prenatal care for Maria. In fact, I had to admit the family session was not going well at all. I winced as I thought of Catherine, our faculty psychologist, kindly but incisively criticizing my counseling technique. But that's what I wanted. That's why we videotaped these sessions in the Family Care Center. I really wanted to improve my counseling skills so I could better weave them into the day-to-day practice of family medicine.

Just then somebody knocked on the door and simultaneously my beeper went off. "Dr. Sheff, they need you urgently in the emergency room."

"Please excuse me. I apologize, but an emergency's come up. Maria, we'll continue talking about this next week when I see you for your regularly scheduled prenatal visit." I also knew I wasn't getting any more helpful information in this session with Tom there. The emergency call was a welcome interruption, until I found out who it was about.

"We've got your patient, Jim Schmidt, here," the emergency room physician started. Not him. I thought I'd wiped my hands of Jim the last time he signed out against my medical advice six months ago. I hadn't heard from him since. "When he came in, he gave you as his doctor. That's before he coded."

"Coded! What happened?"

"He developed some shortness of breath at home. When he showed up here, he seemed to be in mild congestive heart failure with a few rales in both lung bases. Within minutes his lungs filled up completely up with fluid in fulminate pulmonary edema,[18] and he went into respiratory arrest. I've never seen that happen so fast. We've tubed him. Don't see any clear cardiac damage on his EKG, but the unit's already accepted him. They want your input on what to do."

"I'll get over to see him when I finish my Family Care session. For now, the residents in the unit should treat him for congestive heart failure and do the rule out MI protocol."

Later that night I went up to the ICU, not happy to have to tell Marsha I would be home late again. There was Jim, already extubated. Seems he'd responded dramatically to the initial treatment, and his heart failure had

---

[18] Pulmonary edema is the result of congestive heart failure in which the heart doesn't pump blood to the body effectively, so the blood backs up into the lungs. Pulmonary refers to the lungs, and edema refers to the inappropriate accumulation of fluid in the tissues. In essence, the lungs become "wet," like a sponge soaked with water, so they are stiff, don't expand and contract well, and don't transmit oxygen to the blood effectively.

improved quickly, though the tests showed he'd had a small heart attack again.

"Dr. Sheff," he said softly through his oxygen mask, "sorry to bother you like this. Looks like I got myself in a heap of trouble again."

"Looks like you did, Jim. What happened?"

"It's the misses. We got into a row over that damn table."

"What damn table?"

"Her table in the living room. I hate the thing, but she's so damned attached to it. Drives me crazy. Knocked my shin on it again tonight. Well I was tellin' it to her good when I suddenly didn't feel so well, like I had trouble breathin' and all. That's when she took me to emergency. I didn't want to go, but here I am."

I just shook my head. "Jim, what are we going to do with you?"

"Don't know. But I'll be good this time."

Four days later the call came from the head nurse. "Dr. Sheff, Mr. Schmidt says he's signing out AMA. Says he's fed up with this place. He's giving my nurses a helluva time. You'd better get over here right away."

"I'm supervising an intern working up a woman with severe vaginal bleeding. I'll get there as soon as I can."

"What the hell took you so long?" Jim greeted me as I entered his room. "They've kept me in this jail forever waiting for you to get over here! Guess ol' Jim just isn't that important to Mr. Baby Doc."

I ignored the barb. "Jim, you had another heart attack four days ago. You came in in congestive heart failure, unable to breathe. It's not safe for you to go home so soon."

"Safe or not, I'm going. Just do your damn paperwork and get me outta here."

"If you leave now you know you may die."

"I know I'm *gonna* die. It's just a matter of when. But it ain't gonna be in here. Sign whatever you have to so I can leave."

"Gladly," I bit off as I walked out of the room.

Three weeks later, he was back, again in pulmonary edema and coded. Again he was intubated and then quickly extubated. Again he had a small heart attack. Again he was contrite that first night. As an added complication, he appeared to have suffered a small stroke during the code. This time the call came on the sixth hospital day. He was signing out AMA. "Let him go. Have him sign the AMA papers and just let him go," I replied wearily.

And so began a repeating pattern. Over four months Jim presented a total of six times in congestive heart failure that quickly developed into fulminate pulmonary edema requiring intubation and treatment for a small heart attack. In fact, his nickname in the ED became "Flash," because of how quickly his pulmonary edema progressed. By the sixth time he knew the routine so well that as he gasped for air on the stretcher in the ED he called out, "Code 99! Code 99!" before passing out.

The cardiology consultant referred to his problem as "stiff heart syndrome" in which Jim developed global ischemia, meaning not enough blood supply to his heart, which in turn prevented his heart from relaxing so it couldn't fill with blood adequately before the next contraction. We tried every combination of medications we could, with the same result. Each time his symptoms were triggered by a fight with his wife, usually over the same table in the living room. He became enraged, which caused his adrenalin to rise, which triggered his ischemia, which triggered his flash pulmonary edema. And during each of the six hospitalizations, somewhere between the fourth and seventh day, a nurse would do or say something that ticked him off so much, he would fly into another rage and demand to sign out AMA. It didn't matter what service I was on—pediatrics, orthopedics, OB, medicine—the call would come in. "Flash is here again." Everybody eventually knew what that meant.

When he came in for the sixth time, he'd developed a diabetic ulcer on his right foot just like the ones I'd seen so often during my surgery rotation in medical school. It looked infected, possibly down into the bone, which meant he might have an infection that could lead to amputation if he didn't get the right care. This upped the ante when the predictable call came that he was signing out AMA. I was completely exasperated.

"Jim," I said bursting into his room, "you are obnoxious and obstreperous! If you sign out of here AMA you're going to lose that foot."

"I ain't lettin' nobody take off no foot!" he shouted.

"Then you will go home to die!" I shouted back.

"Fine by me," was his only reply.

"Alright, Jim. I'm putting this in the chart. You know you're going home to die. There's no sense bringing you back into the hospital again. I'll put you on antibiotics by mouth, and I'll see you at home until you die."

"That's just the way I want it," he snapped back. I stormed out of Jim's room, relieved to put an end to this infuriating cycle.

A week went by, and I realized begrudgingly I had to do the house call on Jim. At the end of my day on a Wednesday, I rummaged through the Family Care Center and found a not so little black bag that contained everything I might need for a house call. My preceptor pointed out that the residency program had just bought a portable scale for checking weights on homebound patients, so I loaded the sizeable black bag and scale into my car and made my way to Jim's house. About 10 minutes from the hospital, I found myself in a neighborhood of small, single-level houses, each with a tiny, well kept lawn. A blue collar neighborhood, it appeared, in which the owners took pride in their very modest homes.

Pulling up in front of Jim's house, I immediately noticed the lawn he'd first told me about when describing his leg symptoms. Shockingly small, it gave me a whole new appreciation for the severity of his leg cramp symptoms. The cement outer wall of his house showed signs where Jim had patched it over the years, but now had fresh cracks he'd been unable to repair recently. I could sense his frustration and humiliation at his impotence to maintain his own home in this prideful community. When the doorbell rang, I heard hushed angry voices inside and the sounds of scurrying feet.

His wife, a plump woman in her late sixties with a puffed up hairdo, opened the door. Was it my imagination, or had she had her hair done for this visit? "Come in. Come in, Dr. Sheff," she said effusively. "We're so glad you came. Would you like some tea? I made you some tea and some of my cookies," she added, pointing to a cup and plate on a low lying wooden table in front of a faded, green, living room couch. The infamous table, I thought to myself.

"Hey, Doc," Jim's voice called from a bedroom off the living room, "sorry I can't get up to welcome ya proper. When the misses gets done fussing over you, come on in here." I stepped to the door of his bedroom. "Doc, what you called me that day in the hospital when you were so mad...You called me obnoxious and ob...ob..."

"Obstreperous."

"Yeah. Yeah. That's it. I had to go home and look it up. And, Doc, you're right. That's exactly what I am. Obnoxious and ob...ob...obstreperous. You nailed me right."

"Glad you and I finally see something the same way," I said wryly. A big, toothless smile lit up his face.

"No. You were right to be so mad. I deserved it. Now go on in and enjoy some of Ellie's cookies. But watch out you don't break a tooth on 'em." He was still smiling.

It seemed the dutiful thing to honor Ellie's hospitality, so I went over to the couch. I bumped my shin on the table trying to squeeze onto the couch.

"I'm so sorry, Doctor," Ellie fussed, "Jim's always after me to move that table."

"He has given me an earful about this table. It seems awfully small to warrant such a fuss."

"That damn table's what I stubbed my foot on what gave me this ulcer," Jim called out from his bedroom.

"I can see why. It is a bit of a tight fit." Then turning to Ellie I added, "Is there a reason you keep the table so close to the couch? It does pose a bit of a hazard for Jim, with his poor circulation."

"Do you think so, Doctor Sheff? I'd be happy to move it if you think it would help."

"Yes, I believe it would."

"Damn, woman. You never moved it after my tellin' ya to every day for the last 10 years. The Doc comes in here one day, and suddenly the damn table's movin'," Jim bellowed.

I used that as an opportunity to shift my focus to Jim, specifically his foot and his heart. Moving into the bedroom, I pulled out my stethoscope and a blood pressure cuff from the black bag. His lungs were clear and his heart unchanged from the previous week. The foot ulcer looked a little less red and perhaps a little drier. I redressed the bandage, showing Ellie how to do it. With the scale we established his baseline weight and compared it to their home scale. I wrote another prescription for his oral antibiotic and said I'd see him again in a week.

One week later, as I came through their door, Jim called out, "I can't believe you're here to see this obnoxious and ob...ob..."

"Obstreperous?"

"Yeah, this obnoxious and obstreperous plumber." He seemed truly touched that I had come back.

And so was born our weekly ritual. Late every Wednesday afternoon, at the end of my day, I would gather the large black bag and scale and head off to Jim and Ellie's house. I'd have a few cookies and then examine Jim. Sometimes his foot ulcer needed debridement, cutting out any dead tissue, so I started bringing scissors and forceps on the house calls. At some point I realized his ulcer was very slowly healing and that he was not likely to die from this infection. So I shifted to ensuring his heart and diabetes medications were well adjusted. Since he required aggressive diuretic[19] medication, I had to follow his kidney and chemistry numbers closely. This led to me drawing blood from him each week, and then, based upon his vital signs, weight, exam and lab results, making minor adjustments to his medication regimen.

About the sixth week I visited Jim, Ellie and I were having our usual brief conversation over the cookies when Jim called out, "Ellie, you're bendin' his ear with your silly chatter. He's got serious work to do. Let him be."

"Now don't you mind him," Dr. Sheff. "He gets a little out of sorts, but he doesn't mean most of what he says, 'specially when he gets in that state."

"In what state?" I asked.

"You know, when he's had his beers."

"He told me he drinks occasionally, and only a couple of beers at a time when he does." She suddenly got quiet, with a wary glance at the bedroom

---

[19] Diuretics are medications that cause an increase in urine output. They are part of the mainstay of treatment for congestive heart failure.

door. The conversation switched to other topics. As I finished the cookies, I insisted on taking my plate to the kitchen at the back of their little home. Once in there I quietly asked Ellie, "You said something about Jim and his beers. Exactly how much does he drink?"

"Oh, I couldn't say," she said evasively.

"Ellie, I need to know the truth if I'm going to help Jim."

"It gets the worst when he starts to fighting. Like about the table. When he gets worked up, that's when he fights back at me with the salt shaker and his six pack."

"You mean he intentionally takes in extra salt and beer?"

"Yes. That's what triggered each of those visits to emergency. He'd start yelling about something, and if I said anything back, he'd grab a box of pretzels and his beer. Next thing you know, he'd downed a six pack and we're off to the hospital."

Once again I had the dawning realization of putting together the pieces of a puzzle. Jim's "flash" pulmonary edema was not about a stiff heart syndrome. It was about sudden salt and fluid overload Jim intentionally caused to get back at Ellie as part of their never-ending fight. As Jim's true alcohol intake came out, I realized his recurrent pulmonary edema was an atypical presentation of occult alcoholism. I and all the other consulted specialists had missed the diagnosis until this moment. Once I started seeing him weekly at home, I could make the correct diagnosis, and these bouts stopped.

I realized how much I had learned about Jim, Ellie, their relationship, and the interface of their lives together with his illness through making a handful of house calls. As a result, I'd helped them make substantive changes that improved Jim's health, some as small as getting them both to agree on moving the living room table, some as large as breaking a destructive cycle with alcohol. I'd also come to understand Jim's fierce pride and his profound frustration at how much of his identity he'd lost through his illness. This allowed me to reach in and connect with him on a completely different level. As a result, his incessant and infuriating attacks on me vanished, though occasional barbs persisted, usually with that smile. Perhaps most importantly for Jim, by simply going out of my way to come to his home, I'd showed that I cared about him, something no one in a position of authority had ever done. In the process, I discovered that house calls were an unmatched, rich source of critical information and opportunities for helping my patients. Why didn't physicians make house calls any more, I wondered.

This led to my inquiring at the Family Care Center about reimbursement for the house calls I was making. Medicare paid very little, certainly nowhere near enough to warrant the time they took. I had the luxury of continuing the house calls because I was employed as a resident, not worried about how much revenue I had to generate to cover my practice overhead. But physicians in private practice would lose money on any house call they did, which is the primary reason house calls have become mostly a thing of the past. Some family physicians still make them, I found out, but only a small minority. Hardly any other specialties still made house calls. A very great loss, I thought.

On one of my drives out to Jim's house I entertained myself by calculating how much money I was saving the healthcare system with these house calls. Each of Jim's six admissions had cost approximately $6,000,[20] for a total of $36,000. If my house calls saved one admission every six weeks, a conservative estimate given Jim's track record, I was saving the healthcare system on the order of $54,000 in a year. For that, Medicare reimbursed $22 per house call, each of which took about an hour round trip. If I saw Jim weekly for a year, that would total $1,144 over a year. If Medicare tripled what they paid for a house call, up to a whopping $66 for an hour of a physician's time, think of how many unnecessary hospital admissions would be prevented, at a savings of millions and perhaps even billions of dollars. But house calls aren't valued by Medicare and other third party payers,[21] certainly not as much as highly reimbursed diagnostic tests, like CAT scans, or high tech admissions like Jim's stays in the ICU.

With each weekly visit to Jim and Ellie, I realized that we, both physicians and patients, were letting insurance companies and other third party payers make critically important decisions for us. Increasingly I heard from my patients that if something wasn't covered by their insurance, it *couldn't* be done. The idea of paying for healthcare services they personally valued out of their own pockets was becoming a relic of the past. For physicians, facing the challenge of how best to spend their scarce time in a busy practice, why would they spend that time on visits, treatments, and procedures that were poorly reimbursed compared to the time and resources they consumed? Even if physicians wanted to do the right thing, the pressures of rising practice overhead and reduced reimbursements from Medicare and managed care forced them to look to their bottom lines. So over time, physicians tended to perform these poorly reimbursed services less often. An alternative strategy was to perform more of these poorly

---

[20] Please note that all of the numbers in this calculation are estimated costs from 1983. To adjust these to reflect accurate numbers for 2008, all the numbers should be approximately tripled.

[21] Third party payer refers to any insurance company, government agency, or employer that pays for healthcare for an individual or a family. The patient and provider (physician, hospital or other entity providing care) are the first and second parties.

reimbursed procedures, such as office visits, in higher numbers. This led to shorter office visits and the use of allied health practitioners, such as physician assistants and nurse practitioners, to provide the less well reimbursed services. The net result was patients spent less time with their physicians. More highly reimbursed procedures, which were invariably the most technical, over time came to be performed more often by physicians. Because of these distortions in how healthcare is reimbursed, our entire healthcare system, including patients, physicians, and third party payers, was systematically devaluing physician time, thinking, and caring, while raising the value of technology. In other words, we were letting those who collect our money in the form of taxes and insurance premiums, and then redistribute it to pay for healthcare, determine what will be valued and what will not. Whatever is valued will be performed more often. What isn't valued will not. Physician house calls were a precious casualty of this inexorable trend. They were being replaced by an important service, home care provided by nurses and other non-physician practitioners, but still there was a loss in physicians no longer spending time in their patients' homes.[22]

My mind wandered to lectures I'd attended during medical school on the financing of healthcare which were provided on an optional basis by business school professors. I envisioned the graph they had shown of the percent of gross domestic product that went into healthcare. Prior to 1965 it had hovered at the 4-5% level. But since 1965, the beginning of Medicare and Medicaid, it had grown dramatically to over 11%.[23] This meant we were increasing spending on healthcare at two to three times the rate we were increasing expenditures on most everything else, and nobody knew when this trend would level off. Most of this expenditure seemed to be going to the high tech services: expensive hospital stays, more procedures performed by physicians, and ever more expensive instruments, drugs, and tests.

I had an image of President Eisenhower as he left office warning the country to beware of the military-industrial complex. He had been right, as

---

[22] These trends were beginning by this time in my training but accelerated powerfully over the last 20 years.

[23] By the early 1980's, the percent of gross domestic product (GDP) spent on healthcare had grown to 11%. Over the next decade it would grow to 14%. During the 1990's, as our economy grew so rapidly and managed care generated some controls on rising healthcare costs, the percent of GDP spent on healthcare leveled off temporarily, though the actual dollars spent on healthcare grew dramatically. Now, in the first decade of this century, as growth in the economy has slowed, healthcare expenditures continue to grow, which means we are seeing another increase in the percent of GDP spent on healthcare. We are currently at 16% with greater increases predicted. As the baby boomers age and require healthcare in increasing numbers, this trend is expected to accelerate, with no end in sight.

the 1960's and 1970's had seen unprecedented growth in the percent of GDP going into military expenses. But nobody had warned us of the growth of the medical-industrial complex. Technology companies, such as GE, Siemens, and Hewlett-Packard, and the large pharmaceutical companies were turning out ever better, though far more expensive, equipment and other innovations impacting almost every aspect of medicine, from imaging, to medications, to tools for invading the human body. As specialization in medicine proliferated, each specialty developed its unique and expensive tools. The fact that such tools led to billable procedures was not lost on physicians, hospitals, and an increasing number of free standing surgical and imaging centers. Hospitals were expected to keep up with the latest technology, costing billions of dollars every year.

Consumers demanded access to these latest innovations in the oftentimes inaccurate belief they were getting "better" healthcare if it was more technological. Sometimes it was better, sometimes it was just more expensive and invasive, and sometimes it was actually worse. Plaintiff attorneys got into this game on both sides. If a patient didn't have a test or procedure that *may* have helped them, a physician was held liable. I watched this lead to defensive medicine, in which physicians ordered all tests and procedures that *might* be indicated, practicing "CYA" medicine, meaning they were "covering their ass" by ordering marginally indicated tests and procedures. But an important and predictable result followed. The more we did to patients, the greater the chance our interventions would have untoward side effects or adverse outcomes. And plaintiff attorneys cashed in on this as well. So physicians were caught either way. If they didn't order the latest technological testing and treatment for their patients, they were perceived as withholding care. If they did, they were applying often unproven technology that produced additional risks to patients. Even proven technology was not uniformly effective and still carried sizable risks to patients from side effects and errors.

The risks of growth in technological healthcare were well described in a book I'd read in medical school by Ivan Illich that coined the term medical nemesis.[24] Nemesis was the punishment visited upon the ancient Greeks for the sin of hubris, the sin of trying to act godlike. His thesis, with which I now wholeheartedly agreed, was that as our technological capabilities grew, allowing us to more and more powerfully intervene in the natural course of illnesses in godlike ways, negative consequences would inevitably follow. The suffering I personally put Jennifer through at Children's Hospital and Gloria through in our ICU came vividly to mind.

---

[24] Illich, Ivan. *Limits to Medicine: Medical nemesis, the expropriation of health.* Marion Boyars Publishers, London, 1999.

The technological drive of the medical-industrial complex, combined with rising consumer demand and a predatory malpractice environment, was creating a cascade of more sophisticated technology, greater costs, lower satisfaction with healthcare, and only marginal improvements in health for most of our citizens. As I sat munching cookies with Ellie and laying on hands with Jim in his own bed, I was saddened by my visceral sense of what we were all losing.

*   *   *

Our second year residency group had been augmented by the addition of an unusual couple: Tony and Hillary.[25] This husband and wife, both physicians, had been residents in the program several years earlier, but had dropped out because the husband, Tony, had developed lymphoma, a form of cancer in the blood. Actually this was a recurrence of a lymphoma he'd had 10 years earlier. That cancer had initially been treated and gone into remission.[26] But then it recurred during their second year of residency, and they had left the program to find the best treatment for his lymphoma.

Unfortunately, the state of medicine at that time held little promise of cure for his particular form of lymphoma. Rather than accepting this likely death sentence, they had scoured the world for potential treatments. Finally, they had settled on an unusual set of what would be called alternative treatments: diet, herbs, and homeopathy. He had kept strictly to this regimen, and, to the surprise of his physicians, the disease had gone into remission. With this success Tony and Hillary had gone on to have an adorable son, and were now returning to complete their training as family physicians.

I looked at this couple skeptically from the moment I met them. They were extremely bright, exceptionally well educated, and seemed quite sane. I couldn't understand how they could consider such alternative treatments when Tony's very life was at stake. One day, I decided to ask the question that had been on the tip of my tongue since meeting them.

---

[25] Please note that a number of details of this couple's story have been changed to protect their privacy.

[26] As explained in the previous chapter, remission is a term that refers to a state in which a patient with a disease shows no evidence of that disease when examined and tested. Being in remission is different from being cured. It is never precisely clear when someone whose cancer has gone into remission is considered cured. Whereas a period of five years in remission is often a rule of thumb for considering a patient cured, as Tony's case proved, patients can be vulnerable to a recurrence years later.

"You seem extremely bright and well educated—Stanford undergrads, Johns Hopkins medical school—you don't actually believe any of this alternative medicine crap, do you?"

"As a matter of fact we do," came the bemused reply.

"But homeopathy, for goodness sake. Homeopaths take a substance and dilute it one part in 10 to the 10th, one part in 10 to the 20th, or even one part in 10 to the 30th." Then, borrowing a phrase I'd heard from my pharmacology professor that first year of med school, I gave what I believed would be the crushing conclusion to my argument. "Why based on Avogadro's number we all know there's not a snowball's chance in hell there is even one molecule of that substance in the diluted solution. So this has to be just quackery." I folded my arms in defiant satisfaction. I had studied intellectual history and the history of science as an undergraduate. I had studied epistemology and philosophy at Oxford. I was on very comfortable ground and sure that I would prove them wrong. They continued to look bemused.

"Your argument appears to make sense," Hillary replied, "but unfortunately you've fallen victim to the sophistry of scientism, not real science."

"Alright. I'll go with you on this one," I responded, warming up for a good argument. "What is scientism, and how does it differ from real science?"

"Real science is the development of hypotheses which are then tested against observable phenomena. The hypotheses are either disproved by the data or supported by the data. If supported by the data, a hypothesis can, at best, be said to be well supported. It can never be completely proven."

"I'm with you so far. That's consistent with my study of epistemology and the history of science. But you haven't said what scientism is."

"Scientism is the use of the trappings of science in defense of dogma."

I paused. "Say more," was all I could respond.

"You referenced Avogadro's number, which is part of our understanding of how matter is made up of molecules. We understand the logic of the argument you made, but it had a basic flaw."

"Which is?"

"You assumed matter, the substances all around us, is *only* made up of molecules."

"But that's because they are."

"Could it be possible that matter was also made up of essences, or perhaps of energy, in addition to or as part of molecules?"

"The theory of essences has been disproved by modern science over a century ago."

"Has it? How do you know?"

I thought of all the chemistry and physics courses I'd taken, all the textbooks I'd read, all the experiments conducted by the giants of science over the past 400 years. I wanted to count on these as rock solid foundations for everything I believed. After all, I still considered myself a scientist and a skeptic. Science had to hold the answers to our most fundamental questions about matter. It had to. So I responded, defending the absolute truth of what I knew about the nature of matter. It was composed of molecules, which in turn were composed of atoms. Our theories predicted how these molecules would interact, and these predictions had been proven true. We were even using electron microscopes to see smaller and smaller particles. Every time we did, they confirmed the atomic and molecular nature of matter.

"All of what you say is true," they responded. "These are the trappings of science—hypotheses, experiments, theories. You are showing us how everything you believe about how the world is made up, how it works, reinforces everything else you believe. There is much faith in all you profess to believe." I bristled at the insinuation that what I knew of science was based on faith. "But you haven't disproved anything about homeopathy. You've only reinforced the dogma of conventional medicine." My bristling turned to anger at their accusation that the medicine in which I was training was in any way associated with dogma. "Real science would consider the empirical evidence, not just how one knowledge claim about Avogadro's number could be used to discredit another knowledge claim about the effectiveness of homeopathic treatments. It would seek instead to understand homeopathy in its own context and to ask how hypotheses about homeopathy could be well formulated and tested."[27]
"But what if homeopathy is based on a claim that is not testable? It then could not be scientifically proven or disproven," I retorted.
And so the argument raged well into the night. In fact, it raged for six months. Together we sparred about science, knowledge, what we think we know, what we believe and have faith in, and what, if anything, we truly know. At the end of six months, three things had happened. First, I had two new best friends. Second, I lost the argument. By this I mean that I grew into a deeper understanding of the value and limits of science as it is carried

---

[27] I recently came upon research demonstrating that the impact of antibodies can be enhanced by progressive dilutions in water, supporting the underlying dilution principle of homeopathy. See E. Davenas *et.al.* "Human basophil degranulation triggered by very dilute antiserum against IgE," *Nature*, 1988; 333(6176): 816-8. Also see McTaggart, Lynne. *The Field: The quest for the secret force of the universe.* HarperCollins Publishers, Inc., New York, 2002 (59-73) for an interesting discussion of the controversy surrounding this research and additional research demonstrating the relationship between molecules and electromagnetic energy that also provides experimental support for homeopathy.

out. I came to realize that the medicine I'd been taught was only partially based on science. It was still very much an art. I also came to realize that so much of what I thought was true about health and illness, about what makes people sick and how we try to heal them, is founded on limited and deeply flawed science. This did not take away from the miraculous advances gained from our understanding of germs and their role in disease, the great strides that resulted from effective vaccinations and medications, and the extraordinary advances in surgery and diagnostic technology. I valued all of this and more about our scientific medicine. But I came to see the medicine in which I was training, allopathic medicine as it was called to contrast it with other approaches which were not based in pharmacology and surgery, as possessing only a part of the truth about health and illness and what constituted effective treatment. Other healing disciplines, like massage therapy, homeopathy, acupuncture, naturopathy, herbal medicine, nutrition, and chiropractics, also held some truth. Perhaps so did energy healing, though this remained quite a stretch for me. Each of these disciplines developed its own coherent, mutually reinforcing theories and knowledge claims, just as did the medicine I'd been learning. Some of these theories seemed extremely farfetched, given what I thought I knew about the human body and disease. I realized I did not have to accept their theoretical explanations to consider the possibility that their diagnostic and healing modalities might offer some degree of effectiveness.

The third thing that happened was that my world expanded. I saw all healing disciplines engaged in an exercise of the blind men and the elephant. Each discipline held a piece of the truth about health and illness, about how to keep people healthy and heal them when they are sick. Each discipline made the same mistake of concluding this was the entire truth. Conventional medicine, allopathic medicine, held a very large piece of this truth—but not all of it. I owed it to my patients, to my future career in which I sought to become the most effective physician I could, to explore these other healing disciplines. As I embarked on this exploration, I was shocked to discover how polarized all these different disciplines were. Each fought with the others for legitimacy, patients, and ultimately for a share of the healthcare dollar. Allopathic medicine fought vehemently to discredit all other healing modalities. Any that couldn't be discredited would be absorbed under the hegemony of allopathic medicine. It was a sorry and demeaning battle raging for power and money, not for truth and healing, even if the participants believed otherwise. True science should instead embrace the entire elephant.

I committed to treat all healing disciplines with the same simultaneous appreciation and skepticism that I now brought to allopathic medicine. So I and my fellow residents organized an independent study group to examine different healing modalities. We invited practitioners of each modality to speak to us, to share readings with us, and to help us understand how they understood their discipline and their approach to diagnosing and treating patients. Not only were we educating ourselves, but we felt that in so doing, we would help hasten the day when all healing disciplines would collaborate together in the service of truth, health, and the best interests of our patients.

Chapter XIII

Senior Residency

He Taught Me to Listen to My Heart so I Might
Touch His with Healing

"Manuel Rosa is a..." The wide-eyed new intern checked his note card. "A 74-year-old Portuguese male who presented with four hours of substernal chest pain, positive enzymes, and EKG evidence of a large anterior MI. He progressed to respiratory failure last evening requiring intubation."

"Any arrhythmias, John?" I asked patiently after the end of his halting presentation of the patient he'd admitted his first night on call as an intern. I held a particularly warm spot for John because he'd just graduated from the same medical school I'd attended. John drew the ICU for his first month of internship and was clearly petrified.

"He's thrown a number of PVCs."

"Any of them concerning?"

"Concerning?" he looked back at me through his fatigue and uncertainty.

"Couplets? Runs of V Tach? R on T phenomenon?"

"I don't think he had V Tach, at least the nurses didn't say so." Then after a pause, "I'm not sure I know what the R on T phenomenon is."

"Look here," I said warmly, pointing to Mr. Rosa's EKG. You know which parts of the EKG are called R waves and T waves." John nodded. "Well, sometimes a PVC comes early enough that the R wave of the PVC shows up here, before the T wave of the previous beat is completed. In other words, a new depolarization of the heart happens before the muscle fibers have fully recovered from the previous one. That's a real danger sign because it's a setup for throwing him into V Fib. If you ever see a PVC close to an R on T, you jump on it with all the anti-arrhythmic medication fire power you can." Sam Wilson's ruddy face with that nervous smile briefly came to mind, but I pushed it away.

"What's his last blood gas show?" With this I led the ICU rounding team in a discussion of Mr. Rosa's ventilator settings and acid base status. I ended by saying, "John, Mr. Rosa looks like he might be going into ARDS[1] so he's going to need an A line.[2] After rounds I'll teach you to put one in."

---

[1] ARDS stands for adult respiratory distress syndrome. This is a condition with multiple causes all of which lead to a thickening of the tissue in the lung air sacs that makes getting oxygen into the bloodstream more difficult.

[2] A line is short for arterial line. This is a catheter placed in an artery, usually one of the arteries in the wrist. It serves both to provide a constant and exceptionally

Later that day John approached me. "Rick, you and I went to the same med school. So I know pretty much what you knew when you graduated." He paused. "When did you learn what you know now, just two years later?"

I looked warmly at John. "I completely understand. I didn't know most of this when I started my internship either. Trust me; you'll get to exactly where I am, too. You will be taught by junior and senior residents who really want to help you, as I do. They'll lead you by the hand through much of this. You'll also have the attending physicians. Some will be better than others, but you will learn from all of them. The whole residency system is set up to guide you to exactly where I've arrived. But you'll learn the most from your patients. Let them teach you well."

                    *   *   *

"Dr. Sheff," Maria said, exhaustion making it hard to get the words out, "I'm worried about Thomas, Jr. He cries all the time. Last night he wouldn't sleep at all. I don't know what's wrong with him."

I knew, but that didn't help Maria, and at this moment it didn't help me either. Through her prenatal care we'd talked about the new relationship with Tom, her kids' struggles with accepting Tom and the new baby, and her own ambivalence about having Tom's baby without marrying him. Not only had I cared for her two girls, but her mother had asked to see me as a patient as well. Then delivering Maria's baby had cemented a closeness between us that helped me understand why obstetricians and some family physicians put up with the pressure, stress, and sleep deprivation that come with delivering babies. And just a few weeks ago I'd been asked to take over care of Maria's grandmother in a nursing home. Together they became the first four-generation family in my practice, and I cherished the frequent and varied opportunities this presented for getting to know all of them and helping with their interrelated medical and family needs.

So it was with both fondness and frustration that I listened to Maria's outpouring. One week ago I would have given her my increasingly well rehearsed explanation of colic. I had examined her two-month-old baby numerous times and had found nothing wrong. Neither had my preceptors in the Family Care Center. They and I had agreed he had colic, a condition in which babies cried unusually often, hard and long, for no apparent reason. Colic usually came on in the first month of life, lasted up to four months, and then disappeared on its own. Medicine had little effective treatment for colic, though I'd been taught to prescribe several different formula changes and medications, none of which worked particularly well. Parents of a

---

accurate measure of a patient's blood pressure as well as a means of easily drawing out samples of arterial blood for analysis as often as necessary.

colicky baby walked the floors day and night with their screaming baby in their arms trying to keep their inconsolable, cherubic little monster from destroying their lives.

My previous speech to Maria would have included the true statement that we did not know what caused colic, but would have reassured her that nothing was seriously wrong with her baby. His colic would disappear soon (but never soon enough), and he would then be completely normal. She just had to be able to get through these next few months. That would mean sometimes having to put him down and let him scream. Otherwise she would exhaust herself and be of no use to her baby, her daughters, Tom, or herself. Then I would prescribe another formula or another medication to see if it would help, though it never really seemed to. Silently I would conclude, without much evidence, that if she had taken my advice to nurse rather than use formula her baby wouldn't have developed colic, but I'd keep that thought to myself as this was no time to add guilt to her unhappiness and frustration.

That speech, medically correct[3] and attempting to be empathic, now had to change. It had to change because I had visited Elaine, the wife of a fellow family medicine resident who had just had a baby. She herself had finished a pediatric residency several months ago at the same university medical center at which I'd spent several months of my residency training. Whenever I had done rotations there, she had been one of my favorite residents with which to work. Smart, funny, energetic, always sure of herself, she had taught me a great deal. That's why I was shocked to find her on the edge of tears and overwhelmed with her first three weeks as a new mother. When her baby cried, and didn't stop with her maternal offerings of the breast, changing a diaper, sweet songs, and loving snuggles, she degenerated into a frustrated, defeated, desperate mother.

This was just the condition of mothers I'd received nighttime calls from when I'd provided call coverage for the Family Care Center. During my second year, the junior resident covering pediatrics or obstetrics was responsible for covering telephone calls from Family Care Center patients on nights and weekends. This proved to be great training in handling the common calls family physicians receive in practice. I'd honed my responses to children with colds and fever, patients with intolerable back pain, and countless other medical conditions, including stressed out new moms. When in doubt, I could call on an experienced preceptor for back up, and we always went over the problematic calls with one of our teachers the

---

[3] In recent years, more sophisticated gastrointestinal testing has identified several potentially treatable causes of the symptoms of colic in some patients, such as acid reflux into the esophagus, but this was not known in the early 1980's.

following morning. So I had entered my senior year of residency quite sure I could handle whatever came in over the phone.

Until now. Despite all her training, Elaine sat before me in tears, her breasts painfully engorged because her son had gotten off his nursing schedule, pouring out the overwhelming challenges she faced every day and every night without let up. Her tirade ended with new found wisdom. "I can't believe I had the audacity as a pediatric resident to give advice to new mothers. Doctors shouldn't give advice to moms, grandmothers should!"

Chastened by Elaine's new found humility, I sat across from Maria wondering what I really had to offer her. True, it was a great relief for her as a mother to be reassured her child's apparent torment was "nothing serious" and would pass with time. But what of her own torment at not being able to help her infant child she loved so very much? It dawned on me that every parent, indeed anyone who loved another, had a critical need to help them in the face of pain and suffering. Telling Maria there was nothing she could do but wait for Thomas, Jr., to outgrow colic was not enough. Holding little hope they would help, I prescribed the formula changes and medications as I'd been taught. I didn't tell Maria this. I allowed her to hope that by putting a few drops of simethicone[4] in Thomas Jr.'s contorted, screaming mouth she was doing something important to relieve his distress. At least she was doing something to relieve her own.

As a physician nearing the end of my training, I was becoming increasingly aware of the importance of giving my patients and their families something to do in the face of a loved one's pain and suffering. We need to nurture those we love, and the medications and treatments I prescribed became the conduit for that nurturing.

This need for patients or their loved ones to *do* something in the face of illness and suffering contributes to doctors prescribing medications that are at best marginally indicated and at worst cause more suffering. Prescribing chemotherapy for terminal cancer patients is one of the most dramatic examples. But every day, in offices of pediatricians, internists, and family physicians, in emergency departments and walk-in centers across the country, physicians prescribe antibiotics for colds and marginal ear infections even though their own research literature tells them they won't help. In fact, over-prescribing of antibiotics is leading directly to increasing antibiotic resistance in bacteria of every type. The result is a deadly race between our ability to develop new antibiotics and the ability of bacteria to develop new resistance to our drugs. So far, the bacteria are winning.

Physicians are not alone to blame for this. Patients seek, even demand, treatment, and are not satisfied until they receive it. For a mother to take

---

[4] Simethicone was a medication that was thought to break gas in the intestine into smaller bubbles that would cause less pain.

time off work to bring her child to the doctor, wait endlessly in the waiting room, finally see the doctor, be told her child has a viral syndrome which will resolve without treatment, and leave the office empty handed, leads to a frustrated and disappointed mother who will soon seek another physician. So my preceptors taught me to be sure that that mother leaves with something in her hand she can use to treat her child, even if it wasn't an antibiotic prescription. An additional challenge was that it took far longer to explain to a mother why you were not prescribing an antibiotic than it took to write the prescription. I preferred to take the time, but I was still a resident in training. I could readily see how busy physicians could slide into the practice of prescribing unnecessary antibiotics. Patients would leave their office faster and happier. My best teachers explained to me how over time they had educated their patients about the dangers of unnecessary antibiotics and had trained them not to accept an antibiotic unless it was clearly indicated. They also made clear they'd lost patients with this approach. Those patients had shopped around until they had found a doctor who would give them what they wanted, even if it wasn't what they needed.

But there was more to this dilemma. Patients, and even physicians, need to hope. By prescribing something for Thomas, Jr.,--anything--I was allowing Maria to hope it might help, to hope her suffering baby would improve. Part of the power placed in a physician's hands is this power to give hope or to take it away. The more time I spent with patients and their families, the more powerful hope, or the lack of it, was proving to be. Yet there was a time to tell the truth when no hope existed. Holding out false hope was dishonest. It deprived patients and their families of the time needed to express their anger and resentment, feel their sadness, and hopefully come to resolution with what simply was. Elizabeth Kubler-Ross and others were teaching us this truth about the dying. Dorothy had taught me as well. Now I was learning the lesson applied to every instance of human suffering.

I thought I'd learned this lesson so clearly during my junior residency rotation in the ICU. I'd become increasingly comfortable with judging when to offer hope and when to help a patient or family prepare for the inevitable. But something happened recently that gave me pause. Julia, an 84-year-old woman who'd been one of my Family Care Center patients for two years, became very ill. She presented to our emergency room in congestive heart failure complicated by a severe infection spreading through her bloodstream. She was intubated and admitted to the ICU, where I rounded on her each day as her attending physician. The team of residents took excellent care of her, but we all realized she was not going to make it. She had slipped into a coma, her blood pressure hovering dangerously low despite pressor medications, and she showed no signs of recovering.

Remembering what Dr. Patel had said about Dorothy, I met with the ICU team one evening and suggested we wean her pressor medications and let her go. They all agreed there was no sense prolonging the inevitable. I discussed all this with her husband who sadly accepted my recommendation, and I went home that night expecting a call from the nurses at any time telling me Julia had died.

I was surprised to awake to my alarm and find I hadn't yet been called. Upon arriving at the hospital I went straight to the ICU. There I found Julia, her blood pressure still dangerously low, but her heart continuing to beat. We decided to continue only her antibiotics, stopping all other medications. Again, no call came that day, nor that night. The next day her blood pressure had somewhat improved. Three days later, she awoke from her coma, and we extubated her. Much to my complete consternation and her husband's joy, she recovered enough to go home. The only way this made medical sense was to postulate that the infection had kept her in shock, lowering her blood pressure so much that her failing heart didn't have as much resistance against which to pump. So blood continued to perfuse her vital organs while the antibiotics helped her immune system defeat the infection. As the infection waned, the strength of her heart simultaneously recovered. We, her physicians, could never purposely have orchestrated such a delicately timed sequence of events. Only nature could do such a thing. Even that felt like a stretch. For the first time in my medical career, I openly wondered if some higher power might have been at work here. Watching Julia roll out of the hospital in a wheelchair was enough to cause this devout atheist to begin questioning some of my bedrock assumptions. For some unknowable reason, it was just not her time.

Two months later, Julia was readmitted to the hospital and quickly died. It was her time.

In the face of so much uncertainty, who am I as a physician to deprive a patient or a family of hope? This line of thinking has guided medical wisdom for generations. Yet it has also led physicians to treat when it is well past the time to stop. The hospice movement was beginning to open physicians' eyes to the importance of giving up treatment for prolonging life and instead making the dying process more humane and loving. Unfortunately, the average duration of hospice services was only a matter of days.[5] In other words, physicians weren't ready to give up hope and refer a

---

[5] Some progress has been made in the frequency and duration of hospice services since my training in the early 1980's. Medicare introduced a hospice benefit in 1983. The average duration of hospice treatment for Medicare patients today is around 65 days. However, this figure is misleading because half of all Medicare patients use hospice for less than 24 days, even though Medicare covers hospice services for up to six months. The situation is even worse in the commercial population (individuals generally under age 65). In a study commissioned by the

patient to hospice until the patient was only a few days from death. In the overwhelming majority of cases, this end could have been seen sooner, and much denial and suffering avoided.

Looking at an exhausted Maria and a screaming Thomas, Jr., I heard the question I knew was coming. "When will Thomas, Jr. stop crying so much?"

"The vast majority of babies outgrow colic by three or four months of age," I answered, trying to hide my uncertainty.

Yet this uncertainty must exist whenever somebody asks the question on every patient's mind: "What will happen to me?" If I were truthful, I would answer, "I don't know." At best, I can offer an answer couched in my past clinical experience, which was still quite limited compared to a physician who'd been practicing a lifetime. Yet I found the most senior physicians even more hesitant than the residents to make prognostic pronouncements for patients. They'd had too many cases like Julia's to be so sanguine. Medical research can also never answer the question, "What will happen to me?" At best, research can tell us that of the next 100 patients in this age and disease category, what percentage will respond one way or another; but it can never tell us what will happen to the patient in front of me.

"So there's a chance he won't outgrow it?" Maria asked, anxious at my lack of certainty.

"Yes, but only a small chance." Maria still looked worried. "I see no reason why he wouldn't outgrow colic by then. He's perfectly normal, so that is what we can expect. Besides," I added, "he's growing beautifully. That's the best sign he'll do just fine." Now Maria was beaming. With this slight shift in language, her fears abated. I knew his colic might extend longer, but there was no need to worry Maria with this small possibility right now. If he was in this minority, we would address it when the time came.

As I neared the end of my training, I was becoming more comfortable with the uncertainty present in every patient encounter. This uncertainty, now no longer due primarily to the limits of my personal knowledge and training, was instead a part of the daily practice of medicine for all physicians. The very limits of medical science itself constrained me.

U.S. Department of Health and Human Services (HHS), Office of Disability, Aging and Long-Term Care Policy (DALTCP) and the Urban Institute which was published in 2000, over half the commercial patients receiving hospice services received only *one* day of care. In other words, they were not referred for hospice services until the absolute end was not only inevitable, but less than one day away. (The URL for this report is http://aspe.hhs.gov/daltcp/reports/empmktes.htm). Hospice services prescribed in this manner are grossly underutilized.

Helping my patients live with this uncertainty and still providing skillful treatment, guidance, and hope, was the essence of the healing art I was learning. As I handed Maria another prescription, offered reassurance that Thomas would be OK, and shared empathy for her trials, I felt the new-found humility Elaine had taught me. In the face of that humility, I, too, needed hope—hope that doing my best for Maria would be good enough. On this day, as on so many others, I was not sure that it was.

*   *   *

In the midst of sewing up a nasty laceration, I looked wearily at the clock. Almost 6 PM, I registered with relief. Then with a sinking feeling I realized I couldn't go home yet. Though I'd been up most of the night delivering one of my Family Care Center patients, and had now completed an 11-hour shift in the emergency department, my weekly house call to Jim still had to be done. Some weeks I looked forward to seeing Jim. Some weeks, like this one, that house call felt like nothing but an unwanted burden. Why had I agreed to see Jim at home weekly until he died? I wondered while placing the now familiar large black bag and portable scale into my car. I was so sure at the time he would be dead in a matter of weeks. A tired smile reflected my bemusement at yet another humbling lesson in the vicissitudes of medical prognosticating.

Eight months later, Jim still hung on. With daily dressing changes, continuous antibiotics by mouth, my weekly removal of dead and infected tissue, and his stubborn streak, he had fought off what I had assumed would be a certain death. Eventually his infected toe had turned to a dried, black stump--autoamputating, as the term was called. His kidneys were slowly failing, which made it harder to keep him out of congestive heart failure. Virtually every week I had to adjust one or more of his medications. Sometimes he would accumulate more fluid, so I'd increase his diuretic dose. The next week his potassium would have dropped, so I would increase his potassium dose. The very next week he would have become so dried out from the diuretic that his kidneys would not be able to filter his blood effectively, leading to a dangerous rise in his potassium and the need for yet another change to his medications. This became a never ending cycle, a delicate balancing act that managed to keep Jim alive.

Ellie answered the door, and there were my cookies on that same table, though it had been moved a decent distance from the couch ever since my first visit.

"I'm worried about Jim's foot," Ellie blurted out.

"What's changed," I asked, knowing Jim was so fragile he didn't have the reserve to weather even a small setback.

"It's gone black."

"The whole foot?" I asked, truly alarmed that Jim might have clotted off one of his remaining, barely functioning arteries. If this had happened, Jim had made it abundantly clear he wasn't going back to the hospital, even if it meant his death. Whatever I found, I'd have to handle it from his bedroom.

"No. No. He's got a black spot on the bottom of his foot where he can't see it."

I skipped the obligatory bite of Ellie's cookies, moving quickly to the bedroom. "Hi, Jim," I greeted him warmly, containing my alarm.

"Hey, Doc, it's so good of you to come see this ornery old man. No. No. It wasn't ornery you called me it was…ob…obstreperous. That's right, isn't it?" His eyes twinkled with this repetition of our weekly ritual.

"Ellie tells me there's a spot on the bottom of your foot I should look at."

"She's always fussin' over nothing. That's why I don' pay her no mind," came his usual barb, though these days it was softened by awareness of his vulnerability and the patience and acceptance Ellie had shown in the face of it.

"Still, I'd better have a look." Bending down at the edge of his bed, I found a 3 by 4 centimeter darkened area of skin on the ball of his foot with cloudy fluid oozing from its center, a bad sign Jim had suffered more skin breakdown and infection. With his diabetes and poor circulation, I worried this might be his final undoing. As much as seeing Jim every week had been a burden, I was struck with a pang of sadness that he might actually be dying. I quickly opened up the black bag and pulled out a scissors and forceps. Cleaning off the area with a sterile solution, I began to remove the dead tissue. I struggled awkwardly to get the lamp in that corner of the room to provide enough light to the area. Visibility was still difficult, so I got down on my knees for a more direct view of the wound.

"Aw, Dr. Sheff, you don't have to get on your knees like that. You shouldn't be gettin' on your knees."

"I don't mind. It gives me a better view of what I'm doing."

"Doc, I can't believe you're down on your knees to help me. Nobody's ever done that. I don't deserve it." He began to cry. "I've treated everybody so bad, even my own son. I just don't deserve such kindness, and from a doctor too. You've got better things to do with your time than get on your knees to help this poor, lame plumber."

"Jim, it's what you need, and I'll do whatever you need," I responded, touched by his outpouring. "You never mentioned your son before."

"That's 'cause I don't see him no more. Haven't for years. Lives in Florida now. He calls Ellie every now and then, but he won't talk to me." I could feel the heaviness with which he said these last words.

"What happened between you," I asked, continuing to carefully pull off small strips of soggy, dead skin.

"Don't know...I just don't know..." his voice trailed off.

"Sounds like you'd like to see him again."

"It won't happen," he said with finality, hardness coming back into his voice. His pride was back in charge.

"That should do it," I said, packing a dressing into his fresh wound and changing the subject. "Ellie," I called her into the bedroom, "this dressing needs to be soaked with quarter-strength Betadine and changed daily, just like his toe dressing." She nodded. "Here's another prescription to get Jim back on an antibiotic. Between the antibiotic and dressing changes, I hope we'll see a real improvement by next week." But as I went out the door, I was afraid of what I'd find when I returned.

A week later, the wound edges had broken down, and I had to remove more dead tissue, though not as much as I'd feared. As I worked on his foot, Jim couldn't stop talking. "Doc, I can't get over that you got down on your knees for me. Ellie, Dr. Sheff actually got on his knees. Nobody's ever done that. For the rest of my life, I'll always remember what you did for me."

"Jim, I'm glad you appreciate it, but..."

"No buts, Doc. Say whatever you want, I'll never forget it."

"Well, this dressing should hold you until tomorrow when Ellie changes it again. Now let's check your weight and vital signs and get a listen to your heart and lungs." I reached for his wrist to take a pulse and realized I hadn't been able to find my watch that morning.

"Here, Doc, use mine," Jim volunteered, handing me the watch off his wrist.

"Thanks...You're pulse is steady and strong." Jim beamed as I handed back the watch. "Lungs are unchanged, too." His lungs now always had some rales in the bases from low level, chronic congestive heart failure, but I'd come to expect this. "Even with that dressing, it's important we get you onto the scale...Weight's holding steady. Jim, I don't know how you do it, but you keep on going."

"Takes more than the sugar to kill this hard assed plumber," he bellowed.

The following week, I again had to borrow his watch since mine had not turned up. As I held the watch out to return it he said, "Doc, keep it. I want you to have my watch."

"Jim, I can't accept that. Mine will turn up, or I'll just get another one."

"No. No. Dr. Sheff, I insist. You should know better than to fight with this plumber. If you don't take that watch, I'm gonna have to get my lead pipe and force you to." He was smiling that toothless grin. "It's the least I can do for the doctor who got on his knees for me."

"Jim, thank you." I reached out and clasped his gaunt hand which responded with that remarkably strong plumber's grip. "I accept your gift." Tears were brimming in his eyes.

Six weeks later, Jim's wound had stabilized and even appeared to be starting to heal slowly. I had completed my rotation in the emergency department and was spending this month in an orthopedist's office, learning more about caring for the aches, pains, and strains I'd soon be seeing as a family doctor in practice. On this Wednesday afternoon, I'd accompanied the orthopedist into the OR for an emergency procedure, so I was delayed by an hour and a half in getting back to pick up the black bag and scale. As Ellie opened the door, I could tell something was wrong. Her eyes wouldn't meet mine, and she simply said, "Jim's been anxious about you coming today."

From the bedroom Jim's voice called, "Doc? Hey, Doc, is that you?"

"Yes, Jim, it's me," I said jauntily. But as I walked into the bedroom, Jim was sitting up, leaning forward at an unusual angle in the bed. "What's wrong, Jim? Are you short of breath?" I asked, instantly alarmed.

"I don't know. Maybe I am."

"Any chest pain?" Tugging out my stethoscope and quickly listening to his lungs, I could hear rales a quarter of the way up the lungs on both sides, a little higher than his baseline.

"No, I don't have no damn chest pain."

His heart rate was up. His blood pressure was low at 100 over 70, but this was around his baseline. Before I had finished checking these, I noticed his chest was heaving slightly and his respiratory rate had climbed. I listened again to his lungs, and in just the few minutes since my arrival the rales had risen to half way up his lungs. He was going into "flash" pulmonary edema again. I thought for a moment of calling an ambulance, but knew Jim wouldn't agree to go back to the hospital. I also knew from experience we didn't have time for an ambulance to get here before he would go into cardiac arrest.

"Ellie," I called out, "get me Jim's nitroglycerine pills and his nitro paste.[6] Quickly!" Ellie ran in with his nitro pills, and I put one under Jim's tongue. Next I smeared an inch of nitroglycerine paste on his chest. I tore open the black bag, looking to see what drugs I could find to use right now.

---

[6] As noted earlier, nitroglycerine relaxes the veins that bring blood back to the heart. In congestive heart failure, part of the problem is the heart has become over distended, and it can't contract effectively in this state. By relaxing the veins, the amount of blood returning to the heart decreases, which allows the heart to shrink back to a more normal size and to contract more effectively. Nitroglycerine can be delivered to the body as a pill that dissolves under the tongue or in a paste form that is applied to the skin. Nitroglycerine paste allows for slower, more sustained absorption than nitroglycerine delivered under the tongue. To treat Jim in this emergency, the nitroglycerine under the tongue delivered an immediate dose, and the nitroglycerine paste on the skin delivered a longer acting dose.

I grabbed a vial of furosemide,[7] drew up 80 milligrams. Moving as quickly as I could, I started an IV with a needle in the vein of his right arm and injected the furosemide. I listened again to his chest. The rales had now risen to fill both lungs completely. I looked around, realizing we had no oxygen to administer. Diving back into the black bag, I found a vial of morphine. This could be lifesaving, but if his blood pressure was too low, it could also kill him.[8] I rechecked his blood pressure, which had dropped to 90 over 60, the lowest he could tolerate before his narrow arteries would not allow enough blood to his heart and brain. His chest was heaving, his skin cool and clammy. I didn't have a choice. He was dying in front of me. I drew up the morphine and injected 2 milligrams into the IV in his arm.

"I gotta pee," Jim said through his gasps.

"That's a great sign. It means the diuretic is working," I responded, handing him the urine jug hanging at the bedside. The sound of large amounts of urine spurting into a plastic container was the first sign his body was responding. His respiratory rate slowed slightly. I rechecked his blood pressure, which held at 90 over 60. Cautiously, I administered another 2 milligrams of morphine. Between the sedation effect of the morphine and the combined impact of the diuretic, nitroglycerine, and morphine on his heart, Jim's breathing began to become less labored. I listened again to his chest and the rales now were only half way up his lungs. For the first time, I thought he might not die tonight.

While the medicines continued to work, I sought out Ellie in the kitchen. "What happened?"

"Dr. Sheff, when you didn't come at your usual time, he got real worried you wouldn't come at all. He started saying things, mean things about you and about me. Said you didn't think he was worth coming to see again. That's when he went for his beer."

"He drank beer tonight?"

"Yes, and the pretzels, like he used to."

---

[7] Furosemide, also known by its brand name of Lasix, is a potent diuretic that causes the kidneys to excrete urine at a high rate. When delivered intravenously, it can begin to act within minutes. A diuretic helps congestive heart failure by getting the kidneys to draw more water out of the blood, which reduces the volume of blood returning to the heart, which in turn allows the heart to return to a more normal size and pump more effectively.

[8] Morphine acts similarly to nitroglycerine in congestive heart failure, relaxing the veins and decreasing blood flow back to the heart. Both the nitroglycerine and the morphine lower blood pressure, so too much of these medications, especially in combination, could lower the blood pressure excessively, resulting in a stroke from not enough blood getting to the brain or a heart attack from not enough blood getting to the heart.

I stormed back into Jim's bedroom. "Did you drink beer and eat pretzels tonight?" I glowered at him.

"Dr. Sheff," he began to cry through his labored breathing, "I thought you weren't coming. I thought you wouldn't ever come again."

"Jim, what in hell gave you that idea?"

"Because you didn't come like you always did. I thought you finally decided old Jim wasn't worth it."

I sat down on the edge of his bed. "Jim," I took his hand, "I would never do that. I told you I'd come every week, and I'll keep my word." He was squeezing my hand now.

"Dr. Sheff, nobody's ever done for me what you've done. I feel like in you I got a son like I always wanted. And when you didn't come..." He broke off, unable to say any more.

I continued to sit at his bedside, holding his hand as his breathing eased. The knot in my throat began to ease as well. Jim wasn't going to die tonight. It wasn't his time.

*   *   *

In the final two months of this, my last year of residency, it was time to say goodbye to my patients in the Family Care Center and to prepare them to switch their care to the next crop of residents. I found this surprisingly hard, as did my patients. As patient after patient poured out how much I'd touched them, how much I'd helped them, and how much they didn't want me to leave, I was reminded of Smitty's outpouring at the end of my psychiatry rotation, his railing against the forced ending of a relationship that neither of us wanted to end. Saying goodbye to Maria, her precious children, her mother, and even her grandmother, left a hole in me at least as much as it did in them.

But none was so hard to leave as Jim. Miraculously, Jim had survived to the end of my residency. Though he'd gone home to die more than 14 months earlier, his stubborn heart had refused to give up. His kidneys were now functioning with less than a quarter of their capacity. He'd weathered several more dangerous foot ulcers. Weekly I was forced to adjust one medication or another. Yet every week, I came to see him. And every week he hailed me, "Hey, Doc. I'm still here." He was proud to be a survivor, but he also knew his time was limited. And when he used those words, "Hey, Doc," they held a special meaning for the two of us. Jim had fought not to trust me, but in the end had lost that fight, one of the few he'd ever lost. He had allowed another man into his wounded heart. I had done the same with Jim.

I recalled that first Family Care Center visit with Jim. Recalled his outburst, "Hey, Doc! Are you listening to my heart?" I realized that over the past three years I had listened to his heart more deeply than I had ever thought possible. And now another realization dawned. Jim had taught me to listen to my own heart. He taught me to let myself care, to bring an open and vulnerable heart to every patient encounter. Often a patient would only need my technical expertise. But I had to be ready to recognize and engage the moment when my patient, or their family, needed more from me. Though I had learned this lesson from countless patients, Jim had been my most important teacher. He'd shown me that if I wanted to truly heal others, I had to open to my own heart. From that source I could in turn touch the hearts of the patients who entrusted their care to me. And that, I now knew, was the essence of healing. Listening to my own heart, combined with the skillful application of technical medical knowledge, allowed me to become the physician I'd always hoped to be.

Picking a younger resident to assume Jim's care had been a challenge. I wanted someone who would care as much for Jim as I had. When I settled on Isaac, I wondered if he would see the preciousness I'd found in Jim. The first time I'd brought Isaac to Jim and Ellie's house, my choice was vindicated. Isaac opened to Jim's rough charm. He cooed over Ellie's cookies. And he instantly grasped Jim's fragile and complex medical condition. I knew I was leaving Jim in capable, caring hands.

But leaving Jim was still painfully difficult. He wanted me to say I would return, to say I would visit him again. I remembered my never fulfilled intention to return to the Institute. I remembered how wounded Jim had been when he thought fearfully and inaccurately that I wasn't coming back to see him. I was moving to another state. I didn't want to leave him wondering if and when I might return. It was time to bring an end to our tempestuous and profound relationship.

One last time I sat on his bedside. One last time I took his hand. In that moment, I realized how much Jim had helped make me into the physician I'd become. He had taught me much about the many clinical conditions that besieged his body. He'd taught me to appreciate the importance of home and family as the powerful contexts that mold how those clinical conditions present to doctors and the healthcare system and how they can best be treated. He'd taught me to trust in my newfound medical knowledge, skills and judgment, yet be humble about my ability to predict any patient's future. He'd taught me to accept the anger my patients expressed without taking it personally. He taught me to see into the heart beyond that anger so that I might touch it with healing. He taught me the importance of allowing a doctor-patient relationship to grow and deepen over time. Now he was teaching me the harsh truth that every human relationship must eventually come to an end. Perhaps most importantly, he taught me to open my heart to the love that undergirds every physician-patient interaction, if we would only let it.

As I released Jim's hand for the last time, I made my way to the door of that bedroom in which I'd spent so many Wednesday evenings. Jim looked up. "Hey, Doc. Thank you for everything you've done for me."

"Jim, thank you for everything you've done for me."

In that moment, I knew that for the rest of my life I'd never doubt that love has everything to do with medicine.

I walked out past the table with cookies on it, past the cracked wall, past the small yard. Driving away from Jim and Ellie's home for the last time, I felt sad beyond words. At the same time, having reached the end of my training, I felt a profound sense of completion. It was time for me to move on. A career in medicine awaited me. I was ready.

Epilogue

Isaac gave Jim exactly the care I'd hoped he would. Weekly visits, detailed attention to his fragile medical conditions, and warm attachment to the gruff but engaging heart Jim had now opened. Almost four months after I left residency, the call came. Jim had died at home. "Died with Harvard numbers," Isaac boasted, meaning all his test results were as close to normal as possible at the time of his death. In other words, Isaac had continued to give him the best technical care possible. But I also knew Jim had died feeling loved by his wife and cared for by his physicians. It was all I could have hoped for Jim, except perhaps a final phone call from his estranged son.

Marsha and I moved to the Boston area where I began my medical practice. To my great satisfaction, when patients came to see me I found I had the breadth and depth of knowledge and skills to provide the care they needed as their family physician. When I told my patients the type of medicine I practiced—seeing children, doing obstetrics and gynecology, treating adults and geriatric patients, being interested in their family and home lives—they uniformly exclaimed they thought physicians like that were a thing of the past. Immediately they wanted to bring their spouse, their children, their grandparents, and I welcomed them. Within a year, my practice was filled.

Even though my training had prepared me well to start in practice, for the first few years I continued to fill in gaps in my fund of knowledge about diseases and treatments to which I hadn't been exposed in medical school and residency. Then, as now, I would not hesitate to admit what I did not know. I was always willing to ask for help from my colleagues in other specialties whenever my patients needed it. I developed excellent relationships with the consultants with whom I worked closely.

Unfortunately, the same could not be said for me and Marsha. With great joy, just a few months after my residency ended she announced she was pregnant, and we celebrated the news together. I imagined the doting husband and loving father I'd be. Yet within weeks of learning of Marsha's pregnancy, I became irritable. During the first trimester Marsha was nauseous and fatigued, and I found myself feeling little patience or empathy for her. Instead of doting, I stayed later at the office. Irritability turned to anger, and anger turned to rage. I did not know what was happening to me, so I sought the help of a therapist.

In the therapist's office I sat in stony, raging silence. When not at work, I existed in only two states: anger and sleep. The need I now felt for sleep exceeded anything I'd experienced during residency, taking on an almost narcoleptic quality. At first I thought it was the cumulative effect of years of sleep deprivation. But as the months passed, I came to understand it was the result of years of anger and resentment I had never known was there. All

those times I'd used my hypertrophied will to override my true needs had
finally caught up with me. Years of pent up hostility now poured out at
every turn.[1] Somehow I managed to remain an empathic, caring physician. I
didn't know at the time I was proving the truth that impaired physicians will
do everything they can to preserve their work performance, even while their
home and personal lives unravel to extraordinary depths.

This is not meant to be the story of my relationship with Marsha. That is
its own rich and complex saga, with Marsha certainly contributing at least
her share to our difficulties. What is relevant here is that I thought I had
escaped medical school and residency training relatively unscathed. I had
sought to be constantly aware of the abuses and scars of my training, to
select a medical school with a flexible and more humanistic curriculum, to
choose a residency with a more lenient call schedule. Though all of this was
true, when called upon to open my heart to my own newly developing
family, I found that heart resentful and unavailable after years of building
walls around it, at least the parts that were not about caring for patients.

The truth was even more complex than this. My heart had been open to
Marsha, to friends, and to other family members over the years. Yet every
time I had reached inside, drawing on that prodigious will to get me through
yet another challenge—an endless list of Latin terms to memorize, a night
on call without sleep, a critically ill patient whose very life lay in my hands,
my own error that took the life of another—one more scar had formed. I
carried this chain of scars about me like the links in Jacob Marley's lifelong
chain of evil deeds,[2] without ever realizing it was so. Eventually I came to
understand I had begun forging these links long before medical training.
That extraordinary will and the capacity to override my very human needs,
in fact, had made me an ideal candidate to be a physician, at least as we train
them today.

This is the underlying tragedy playing out in the lives of so many
physicians. The very qualities that make us technically good physicians
simultaneously undermine our capacity to heal patients beyond being mere
technicians. The irony is even worse. Physicians are first called to a career
in medicine out of a desire to relieve human suffering, yet so often we make
others suffer needlessly through our own unrecognized scars. The ongoing
pressures of practicing medicine—the high stakes of every decision and

---

[1] Many years later I found a precise description of my experience during this time in
Jean Shinoda Bolen's book, *Gods in Every Man*. Reading Bolen's Jungian
description of the Neptune archetype, with its sudden feeling of being consumed
with storm and fury, I felt as though I was reading a personal account of this
troubled period in my life.

[2] Jacob Marley is the fictional partner of Ebenezer Scrooge in Charles Dickens' *A
Christmas Carol*. Marley's ghost visits Scrooge, howling that every evil deed he
committed in his lifetime forged a link in a chain he must wear after his death.

procedure, the frustrations of treating patients in systems that obstruct and fail, the constant threat of malpractice suits from the very patients we sacrifice so much to help—these play into our scars perfectly. We then feel entitled to strike out at others with excessively harsh criticism. We sometimes treat patients and our fellow healthcare workers with shocking insensitivity. In spite of our intelligence, we cannot see the impact our behavior has on others. Without sufficient consciousness and mastery of our inner turmoil, we are at risk of abusing the power in the doctor-patient relationship and the high status granted to us with fellow workers. At the same time we heal, we doctors also wound.

The chain of events that eventually produces doctors who wound others when their intention is to heal begins with the selection of candidates for medical school. Students are selected for medical training because of their capacity to achieve in the face of obstacles, to set aside personal needs when a job must be done, to act with clear logic by suppressing emotions. These qualities are then honed, lauded, and rewarded during seven or more years of intense training and socialization. Any signs of softness have little place, or worse yet, are ridiculed by more senior residents and attending physicians. Nowhere in today's physician training are we taught that to be able to truly heal another one must have loved and lost love, that this is what truly opens our hearts most deeply. No, just the opposite. We are taught to keep clinical distance and to show no weakness. In other words, today's medical training systematically hardens young physicians to their own vulnerabilities, to the very source in their hearts that is the wellspring of their capacity to heal others.

I'm speaking of healing in a non-technical sense, healing that touches another person's heart, perhaps even their soul, healing into which I've attempted to provide a glimpse in these pages. But this type of healing is not recognized, honored, nor valued by most of today's training of physicians. That training, indeed our entire healthcare system, honors and rewards technical competence, not compassion. For the past two generations, physicians and patients alike have been seduced by the Siren song of medical technology and the technical competence required to apply this technology safely and effectively. We all want the latest, expensive technology, but do not recognize our love affair with medical technology is driving the cost of healthcare beyond affordability for so many. The rising costs of healthcare are pushed further by our rush to apply new technology long before it is scientifically proven to provide better care.

The government and insurance companies compound this difficulty by reimbursing physicians for the time spent performing technical procedures at a far higher rate than for the same amount of time spent evaluating, thinking, and communicating, or performing a house call with all its richness and value. As a result, cognitive-based specialties, such as pediatrics, family medicine, internal medicine and psychiatry, remain the lowest paid medical specialties. Orthopedists, plastic surgeons, neurosurgeons, interventional cardiologists, and cardiac surgeons readily earn three to five or more times as much as their peers in the cognitive specialties. At the same time physicians today face increasing financial pressures because the costs of running their practices are rising faster than reimbursement from insurers, producing an ever stronger incentive for physicians to look for the most highly compensated services. Sadly, physician behavior follows this financial incentive. As long as procedures are reimbursed more than sitting with patients, consumers will receive more procedures and less time sitting with their physicians.

Of course, each of us wants a technically competent physician. Yet in the face of uncertainty, fear and pain, each of us also wants a caring, concerned physician who will take the time to sit with us, talk with us and our family, and meet us in the place of fear and questioning to which illness and suffering naturally give rise. These are the very qualities patients seek in a doctor when they find themselves ill and afraid. Yes, patients want technical competence, but they want caring just as much. Physicians in training, and often in practice, feel forced to choose between caring and competence. But if we select and train physicians well, neither patients nor physicians would have to choose between caring and competence. Unfortunately, all too often that is precisely the choice facing so many who are ill and the physicians who care for them. This is one of the reasons patients turn in such large numbers to alternative healing practitioners. Those practitioners at least seem to offer the caring patients so desperately seek.

Patients also sue alternative practitioners far less often than conventional physicians. That should not surprise us. Research has shown that physicians who have better doctor-patient relationships are sued less often than their peers. Surprisingly, only one in 20 patients who suffers an injury at the hands of their physician and the healthcare system ever sues. Many of these patients accept that medicine is an inexact science, that any medical intervention carries chances of an adverse outcome. They recognize that their physicians and all the others who provided their treatment did so with dedication and caring, even if something technically went wrong. They do not seek culpability nor retribution from their caregivers. But some patients do. And when asked why, they express anger. They are angry at being injured, yet that is often not sufficient to cause them to sue. They are angry at physicians who didn't spend enough time with them, didn't communicate with them, and didn't seem to care. They are angry because when an adverse outcome occurs, in retrospect they feel their physician didn't provide enough information and dialogue to allow for true informed consent in choosing a course of treatment, including a realistic grasp of its attendant risks. Today the vast majority of physicians certainly fulfill the technical requirements of obtaining informed consent, including explaining the risks and benefits of the proposed treatment and other options for treating the patient's condition. But they do so in a rote, technical manner which is far from a rich, personal process of mutually informed decision making. Such a process takes time, and communicating a sincere sense that the physician cares for this unique patient. Good physicians leave their patients with the sense they will do everything they can for that patient, including giving them a few extra minutes of their time. This is the greatest protection against a malpractice suit.

Unfortunately, years into practice I had the opportunity to test this wisdom. Elaine, a sweet, woman in her early thirties came to me after being frustrated by her previous treating physicians. She suffered from multiple sclerosis, a progressive neurological illness that eats away at the nervous system, causing patients to lose more and more function over time. Though far from an expert in multiple sclerosis, also known as MS, I agreed to serve as her family physician and to coordinate her care with the appropriate specialists. After our first few visits, she asked if she could bring her two children to me for their well child care, to which I readily agreed. During one of her visits, she casually asked if MS could affect her periods. When I asked why, she mentioned she hadn't had a period for several months. Minutes later we had confirmed she was pregnant.

Elaine then came to me for her prenatal care. I worked closely with her neurologist, helping her maintain an uneventful pregnancy. In fact, pregnancy was protective for her, causing a temporary pause in the progression of her neurological condition. I then had the joy of delivering her beautiful, healthy baby girl. Unfortunately, after the birth her MS flared badly, something her neurologist had warned might happen. She suddenly lost the ability to walk. With fierce determination she struggled desperately to care for her newborn and two other children, dragging herself up and down the stairs in their home with what little strength remained in her arms, all the while clinging precariously to her baby.

At her six-week post partum visit, we agreed that another pregnancy and ensuing severe flare of MS would be devastating, so I prescribed the birth control pill for immediate contraception and referred her to a gynecologist for a tubal ligation. The surgery went well, and she continued to see me for her MS as well as for the routine care for her children. Four months later she had not yet gotten her period. A pregnancy test confirmed she was again pregnant. Elaine would not consider abortion, even though she well knew what lay ahead. During her pregnancy she experienced the expected pause in progression of her MS. After the birth of yet another daughter, Angela, the disease flared terribly again. Suddenly she found herself confined to a

wheel chair with further loss of strength in her arms, struggling even more
desperately to care for her now four children.

A year went by during which I saw her frequently for her MS and the
ongoing care of the children, all of whom always came to every visit. One
day I stepped into her exam room and found her alone. This was unusual.
She had a troubled look on her face. "What's going on?" I asked somewhat
alarmed.

"Dr. Sheff, I wanted to see you alone. You know how upset I was
because my tubal ligation failed. You know how hard it's been since the
flare after Angela's birth. I was so angry at that gynecologist for botching
my tubal ligation that I wanted to sue."

"That is certainly understandable," I responded, sad at the thought of
another physician about to face a lawsuit.

"So I took copies of my medical record to a lawyer. He had an expert
analyze them. That expert came back saying the gynecologist had done
everything right. He had ordered a pregnancy test at the time of the tubal
ligation, which was negative. He had performed a D and C as part of the
procedure just to make sure I wasn't pregnant. But at what turned out to be
so early along in Angela's pregnancy, the expert said there was a certain
percent of cases in which a D and C might miss a tiny pregnancy.
Eventually I had the dye study that showed my tubes were closed off. He
said the gynecologist didn't do anything wrong so I couldn't sue him." She
paused. "But he told me to sue you."

I suddenly blanched, a cold, sinking feeling striking the pit of my
stomach. "For what?" I managed to get out.

"For putting me on the pill without telling me to use another form of
birth control for the first month."

My mouth went dry. I frantically leafed through her chart to find that
post partum visit. Sure enough, there was my note that said to start her oral
contraceptive, but no notation that I'd recommended barrier contraception
for the first month. I routinely recommended such protection for the first
month I placed any woman on the pill. But that day, for some reason, I had
forgotten to do so for Elaine. I had made a mistake, a costly mistake. I
looked at her, now confined to a wheelchair, weakness in both arms making
it almost impossible for her to be the mother she wanted so much to be for
Angela. I felt overwhelmed with the sense of responsibility for her further
deterioration, for the chaos in her home and in her children's lives. I also felt
fear, fear that I would be publicly shamed for my mistake through the
punitive legal process. My heart raced. I had never been sued before. This
was going to be my first time. Hard as it was, I already knew I could live
with having made a mistake. But to be publicly humiliated and punished for
it felt devastating.

Then I heard Elaine say, "I told my lawyer I couldn't sue Dr. Sheff." I stared at her, confused. "You've helped me through so much. You are my children's doctor. You delivered two of them. Together we've all been through hell, but you've always been there. You're my family doctor. I couldn't sue you. I just wanted you to know about this so you wouldn't make the same mistake with anybody else."

My mouth gaped open for a moment as I whipsawed from fear, to relief, to gratitude, and finally to deep respect and appreciation for Elaine's candor and forbearance.

*   *   *

The rage into which I'd found myself propelled after residency persisted for two years. Eventually, with time and the work of therapy, that rage released its fierce grip on me. Unlike many of my peers, I managed to keep it in check at work through that time, so nobody in my office or the hospital knew it ever existed. I did bring an edge to all my work, an edge others would describe as commitment, focus, a demanding of excellence from myself and those around me. That edge produced good care for my patients. Yet if the people I worked with were to read my story as I am now able to tell it, they would nod with recognition of that edge as tinged with the mixture of scars and anger I can now admit still stirs within me.

I was able to participate fully in my marriage, be an involved and loving father, and provide the care to my patients I had envisioned during the long years of medical school and residency. The scars healed, though never fully. What I hadn't realized was that those same years of training had left scars deep within Marsha as well. Over time the poisonous mixture of old wounds and ongoing stresses of professional and family life proved too much, and our marriage dissolved 14 years after it had begun, the seeds of that dissolution having been sewn during medical school and residency.

*   *   *

I passionately believe medicine is one of the highest callings for any person who wishes to serve others, to give of themselves. Just as passionately, I believe we must radically change how we train physicians and how we organize healthcare to undo the tragedy that results in so many unhappy, angry physicians and patients. Since completing training almost 25 years ago, I have been gratified to witness the emergence of courageous, loving physicians who have pioneered new ways to envision healthcare. In the 1980's Dr. Bernie Siegel stunned the medical world when, as a seemingly hard-core surgeon, he wrote *Love, Medicine and Miracles* about "exceptional patients" who show us all what is possible when we help our patients tap into their capacity to heal themselves. In the 1990's Dr. Rachel Naomi Remen became an eloquent spokesperson for heart-based medicine, urging us to attend to the well-being of both patients and physicians. And in the first decade of this century, Dr. Don Berwick, president and CEO of the Institute for Healthcare Improvement, has led a world-wide movement to make healthcare truly patient-centered.

Recent changes in medical training have been aimed at addressing at least some of the shortcomings I've tried to bring to light here. Resident hours have finally been limited to an average of 80 hours per week, though this change came about only after a patient's death was publicly blamed directly on the fatigue of a sleep deprived resident. (Of course, I knew of such a case over 20 years earlier. We can only imagine how many more cases have never come to light.) Policy makers are considering reducing residency hours further, though this will either add to the years of training for most specialties or produce physicians without the same level of knowledge, skill, and judgment as previous generations. Competency in the softer skills of empathy and communication are receiving increasing attention in medical schools and residencies, though this comes only at the cost of fighting with the basic science and clinical faculty for time in the overloaded curriculum.[3] There is even talk of testing for competency in these skills as part of the board certification process. A number of residencies, primarily in family medicine, have introduced peer to peer support groups that include both residents and faculty called Balint groups, named after a British physician who pioneered this approach. Balint group meetings are a focused time for physicians to share their inner experiences of working with patients, to receive support for managing the stresses of balancing work and home, and to reflect more deeply on the meaning of their work. They are a time physicians are encouraged to listen to their own

---

[3] The Accreditation Council for Graduate Medical Education (ACGME), the primary professional organization that accredits residency training programs for allopathic physicians, introduced the ACGME Outcome Project in 1999. This program requires residency training programs to train residents and assess their competence in the following areas: patient care, medical and clinical knowledge, practice-based learning and improvement, interpersonal and communication skills, professionalism, and systems-based practice. The Joint Commission, the organization that accredits a large majority of hospitals in the United States, adopted this same framework for physician competency in 2007. As a result of these changes, physicians are now struggling to design new ways to set expectations and measure physician performance that go beyond the mere technical. Unfortunately, we are still a long way away from these efforts achieving widespread support, especially when it comes to holding physicians accountable for practicing the art of medicine at the highest level.

heart. These are all changes in the right direction, changes pioneered by other physicians who, like me, emerged from their training fiercely committed to a better integration of the best of the art and science of medicine.

But these recent changes and those under consideration are, at best, only a start. I hope my story points the direction to broader and deeper changes that should be made to our current approach to selecting and training physicians.

*   *   *

When Jim Schmidt called out to me for the first time, "Hey, Doc," in his own angry, fearful, and uneducated way, he was calling upon me for both the art and science of my healing craft. Over time, I stumbled upon the proper portions of each to provide Jim the medical care and healing he so desperately needed. He, as did so many of my patients, taught me to be a physician in the very best sense of this term: a skilled practitioner grounded equally in the science of the human body and the art of human relationships. For this I am more grateful than words can express. More than 20 years have passed, and I still keep Jim's watch safe, though it has long ago ceased to run. This watch stands as a constant reminder of all I seek to offer as a physician to those who have entrusted their care to me.

Today Jim's voice echoes over the years since those precious visits to his home, calling not only to me but also to others. To patients and their families, he calls upon you to see your physician as the wounded healer he or she is in the hope that this awareness helps you and your physician to find each other in the healing relationship you both seek. To physicians in training and those considering a career in medicine, he calls upon you to draw wisdom from your patients and teachers that will help you survive your training with fewer scars than you otherwise would have and to deepen your capacity to heal yourselves and others. To those physicians in practice and the families who strive to love them, he calls upon you to heal the scars that shape your every day, whether you are aware of them or not. May hearing his voice help you to heal as telling of it has helped me. To my fellow healthcare workers who have also chosen to serve the ill and the needy, who daily experience the finest physicians have to offer as well as the destructive impact of their scars, he calls upon you to have a more compassionate understanding of the physicians with whom you work, while still holding them accountable to the highest standards of their chosen profession. Finally, to those leaders in healthcare and government with the authority and influence over how we train physicians and organize and pay for healthcare, Jim's plaintive cry puts a personal face on the impact your decisions make every day on countless physicians, healthcare workers, patients, and their families. I hope my sharing Jim's story and the others that fill these pages helps you, our leaders, to make your powerful decisions with a heightened attention to the inner experience of physicians and patients, the experience that lies at the core of all medical care given and received.

Each of us at some time will be a patient, trusting in the care of a physician. My hope for you, a hope Jim would affirm, is that at that moment of vulnerability you receive the care you seek and the healing you deserve.

ABOUT THE AUTHOR

Richard Sheff, MD, known to most of his patients as Dr. Rick, is a family physician with over 25 years of experience in medicine. He chose the specialty of family medicine because he wanted to see and treat patients as whole people whose illness and wellness are a result of the complex interplay of their biological, psychological, social, and cultural circumstances. The years have taught him that to this must be added recognition of each patient's spiritual circumstances if they are truly to be seen as a whole person, including understanding their illness and wellness.

Dr. Rick practiced family medicine in Massachusetts for 12 years, seeing adults, children, the elderly and, for the first part of his practice, delivering babies. Over the years he was asked to assume greater leadership responsibilities, including serving as medical director of his group practice, vice president for medical affairs of his hospital, president of a corporation that owned and operated physician practices, and vice president for medical affairs of an integrated delivery system. He left there to launch a new company, CommonWell, to help our healthcare system integrate the best of complementary and alternative medicine with the best of conventional medicine. At the same time, he began to consult with hospitals and physician organizations throughout the United States, and more recently internationally. He now serves as chairman and executive director of The Greeley Company, a highly respected healthcare consulting and education company dedicated to helping physicians and hospitals provide outstanding care to the communities they serve. Dr. Rick has consulted, authored and lectured on a wide-range of healthcare management and leadership issues, including quality, patient safety, physician performance and accountability, and conflict resolution. "I went into medicine to heal and teach," he says, "and today I find myself continuing this work, but with a national and even international ministry. My goal is to heal healthcare for those who provide care and for all the patients and families who entrust their vulnerability to physicians and our healthcare system."

Dr. Rick is a graduate of the University of Pennsylvania School of Medicine and the Brown University residency program in family medicine. He was an undergraduate at Cornell University and recipient of the Keasbey Scholarship for the study of politics and philosophy at Oxford University.

VISIT OUR WEBSITE...

If you enjoyed *Hey, Doc! Are you listening to YOUR heart?* and would like to join a community of people who seek to transform healthcare, visit our website at:

www.listentoyourheartmedicine.com